MILITARY
PSYCHOLOGY

MILITARY PSYCHOLOGY

Concepts, Trends and Interventions

EDITED BY
Nidhi Maheshwari
Vineeth V. Kumar

$SAGE www.sagepublications.com
Los Angeles • London • New Delhi • Singapore • Washington DC

First published in 2016 by

 SAGE Publications India Pvt Ltd
B1/I-1 Mohan Cooperative Industrial Area
Mathura Road, New Delhi 110 044, India
www.sagepub.in

SAGE Publications Inc
2455 Teller Road
Thousand Oaks, California 91320, USA

SAGE Publications Ltd
1 Oliver's Yard, 55 City Road
London EC1Y 1SP, United Kingdom

SAGE Publications Asia-Pacific Pte Ltd
3 Church Street
#10-04 Samsung Hub
Singapore 049483

Published by Vivek Mehra for SAGE Publications India Pvt Ltd, typeset in 10.5/12.5pt Minion by Zaza Eunice, Hosur, India and printed at Saurabh Printers Pvt Ltd, Greater Noida.

Library of Congress Cataloging-in-Publication Data Available

ISBN: 978-93-515-0630-0 (HB)

The SAGE Team: Shambhu Sahu, Saima Ghaffar and Ritu Chopra

To
The Soldiers of India

Bulk Sales

SAGE India offers special discounts
for purchase of books in bulk.
We also make available special imprints
and excerpts from our books on demand.

For orders and enquiries, write to us at

Marketing Department
SAGE Publications India Pvt Ltd
B1/I-1, Mohan Cooperative Industrial Area
Mathura Road, Post Bag 7
New Delhi 110044, India

E-mail us at **marketing@sagepub.in**

Get to know more about SAGE

Be invited to SAGE events, get on our mailing list.
Write today to **marketing@sagepub.in**

This book is also available as an e-book.

Contents

List of Illustrations

Tables

Case Studies

Figure

List of Abbreviations

ADI	Adaptation Index
AFP	Armed Forces of Philippines
AFP-HRO	Armed Forces of the Philippines-Human Rights Office
AFMC	Armed Forces Medical College
AFQT	Armed Forces Qualification Test
AFSPA	Armed Forces Special Powers Act
AGCT	Army General Classification Test
AICHR	ASEAN Intergovernmental Commission on Human Rights
AMC	Army Medical Corps
AMC-TAB	Army Medical Corps Trade Allocation Battery
APA	American Psychological Association
APFT	Army Physical Fitness Test
AR	Augmented Reality
ASAP	Armed Services Adaptability Profile
ASCI	Administrative Staff College of India
ASVAB	Armed Services Vocational Aptitude Battery
AVF	All Volunteer Force
AWL	Absence without Leave
BCI	Brain–Computer Interface
CARHRIHL	Comprehensive Agreement on the Respect of Human Rights and International Humanitarian Law
CBCA	Comprehensive Battery of Cognitive Abilities
CBT	Combat Brigade Teams
CF	Compassion Fatigue
CHR	Commission on Human Rights
CIA	Central Intelligence Agency
CLAWS	Centre for Land Warfare Studies
CMWG	Complaints Monitoring Working Group
CPP	Communist Party of the Philippines
CPSS	Computerized Pilot Selection System

CRMs	Comprehensive Resilience Modules
CSF	Comprehensive Soldier Fitness
CWC	Council for the Welfare of Children
DAH	Disordered Action of the Heart
DFR	Disaster First Responder
D-I	Developmental-Integrative
DIA	Defence Intelligence Agency
DIILS	Defense Institute of International Legal Studies
DIPR	Defence Institute of Psychological Research
DoD	Department of Defence
DOLE	Department of Labor and Employment
DPR	Directorate of Psychological Research
DRDO	Defence Research and Development Organisation
DRS	Dispositional Resilience Scale
DSM	Diagnostic and Statistical Manual of Mental Disorders
DSS	Department of State Security
DW	Disaster Worker
EBIS	Educational and Biographical Information Survey
EBO	Effect-Based Operation
ELF	Extra Low-frequency
EMP	Electromagnetic Pulse
EW	Electronic Warfare
FGN	Federal Government of Nigeria
FIND	Families of Victims of Involuntary Disappearance
FOIA	Freedom of Information Act
FY	Fiscal Year
GAT	Global Assessment Tool
GCRV	Grave Child Rights Violation
GPA	Grade Point Average
GPHMC	Government of the Philippines Monitoring Committee
GWOT	Global War on Terror
HO	Humanitarian Operations
HOI	History Opinion Inventory
HRO	Human Rights Office
HRV	Human Rights Violations
IAC	Inter-Agency Committee

IB	Infantry Battalion
ICRC	International Committee of the Red Cross
ICU	Islamic Courts Union
IDF	Israel Defense Forces
IEC	Information, Education and Communication
IHL	International Humanitarian Law
IIHMR	Indian Institute of Health Management and Research
ILO	International Labor Organization
IPAI	Industrial Psychiatry Association of India
IPSP	Internal Peace and Security Plan
ISCA	Indian Science Congress Association
ISIS	Islamic State of Iraq and Syria
IT	Information Technology
JTF	Joint Task Force
KDF	Kenyan Defence Forces
LAT	Language Aptitude Test
LAWS	Lethal Autonomous Weapons System
LDRSHIP	Loyalty, Duty, Respect, Selfless service, Honor, Integrity, Personal courage
LIC	Low-intensity Conflict
LOM	Lohamat Modi'in
LWE	Left-wing Extremism
MALAT	Mercaz L'Mivtzaei Toda'a
MAP	Military Applicant Profile
MAST	Military Adaptability Screening Test
MCRD-West	Marine Corps Recruiting Depot-West
MILF	Moro Islamic Liberation Front
MISO	Military Information and Support Operations
MMMs	Modern Military Missions
MNJTF	Multinational Joint Task Force
MOOTW	Military Operations Other Than War
MOS	Military Occupational Specialties
MRRS	Monitoring, Reporting and Response System
MRT	Master Resilience Trainer
MSI	Military Service Inventory
NAoP	National Academy of Psychology
NATO	North Atlantic Treaty Organization
NCO	Non-Commissioned Officer

NCS	Nigerian Customs Service
NCW	Network Centric Warfare
NDRF	National Disaster Response Force
NIS	Nigeria Immigration Service
NMM	National Monitoring Mechanism
NPA	New People's Army
NPF	Nigeria Police Force
NSG	National Security Guard
NYDN	Not Yet Diagnosed Nervous
OODA	Observation-Orientation-Decision-Action
ORTAS	Other Rank Trade Allocation System
OSS	Office of Strategic Services
PA	Palestinian Authority
PANAS	Positive and Negative Affect Scale
PDA	Prediction of Drug Use Admission
PET	Prediction of Emotional Instability
PISCES	Personal Identification Secure Comparison and Evaluation System
PNP	Philippines National Police
PoW	Prisoner of War
PRP	Penn Resilience Program
PRW	Psychological Research Wing
PsyOps	Psychological Operations
PTG	Post-traumatic Growth
PTSD	Post-traumatic Stress Disorder
PWML	People with Mental Illnesses
RBQ	Recruit Background Questionnaire
RSA	Resilience Scale for Adults
SB 5	Stanford Binet Test 5
SCREEN	Success Chances for Recruits Entering the Navy
SIWs	Self-Inflicted Wounds
SJT	Situation Judgment Test
SMS	Short Message Service
SMT	Sentence Making Test
SOPs	Standard Operating Procedures
SRT	Situation Reaction Test
SSB	Services Selection Board
ST	Stereotype Threat

TFDP	Task Force Detainees of the Philippines
TO	Table of Organization
UAV	Unmanned Aerial Vehicle
USMC	United States Marine Corps
VIA-IS	Values-in-Action Inventory of Strengths
VMT	Visual Memory Test
VR	Virtual Reality
VT	Vicarious Trauma
WAIS	Wechsler Adult Intelligence Scale
WMD	Weapons of Mass Destruction
WWI	First World War
WWII	Second World War

Foreword

With the ongoing global engagements of the soldier, military psychology is being acknowledged widely by the research fraternity who are involved in analysing the dynamics, determinants and development of soldier's well-being. Though military psychology has gained its identity as an area of applied psychology, one cannot deny its evolution through the psycho-analytic, behaviouristic and humanistic schools of thought. More so, with IT and computer science becoming an integral part of military missions and increasing involvement of cognitive sciences to military, psychologists' capabilities are put to test. Keeping this consideration, any research attempt, publication, monograph, etc., is not only highly appreciable but an equally challenging task.

The present volume titled *Military Psychology: Concepts, Trends and Interventions* has been a significant attempt to take on this challenge. The editors with their modest background in military psychology have figured a judicious attempt to justify their association with the subject. It is heartening to see that this volume on military psychology has tried to bring the able researchers of military psychology to a common platform for sharing their expertise on various pertinent topics of soldierly concern. Such an attempt will deserve a humble gratitude by those budding researchers who look forward to search any discussion or publication in this area of research.

The contributions from all over the world reflect India's competitive perspective in this area. I am sure such a publication will help all those students, researchers and professionals who wish to contribute their might towards military psychology on one hand and towards welfare of the main actor of military psychology, that is, the soldier, on the other hand.

Manas K. Mandal
Director General (Life Sciences), DRDO
Former Professor, IIT Kharagpur

Preface

Wars often destruct societies but evolve sciences. Every society has employed the psychological sciences for countering military challenges before, during and after participating in any war. Knowledge, so evolved by one society, has benefitted other societies for maintaining their forces. India is no exception. The country has witnessed many significant wars, battles and revolts, both as a result of foreign invasion, colonial participation, neighbours aggression or internal unrest, such as, the famous Battles of Panipat, First War of Indian Independence (1857), participation in World Wars, Azad Hind Fauz (Indian National Army) struggle, Indo-Pak Wars (1947, 1965, 1971 and 1999) and Indo-China War (1962). Alongside, she has professed various psychological concepts of military relevance to the world as well as imbibed many relevant concepts and interventions offered by the world into its military systems. Accordingly, it is always beneficial to place on record the relevant psychological concepts, trends and interventions as championed by various countries across the globe in order to provide impetus to the ongoing learning and research processes in the interest of the soldier. This book is one such attempt from India.

Marking the forthcoming century celebrations of military psychology in 2017, this book is presented with the perspective of apprising the researchers, scholars, service personnel as well as policy makers alike about the theoretical underpinnings and advances of military psychology so as to draw important conclusions for optimizing contemporary and future military performance. Contributors to the book are senior international and national military psychologists, experienced service officers, scientists, academicians, heads of various strategic departments and researchers who have tried to address some of the pertinent issues which are being faced or likely to be faced by various Armed Forces of the world. Besides psychology, contributors also belong to the domain of peace and conflict studies, disaster management, political sciences, international law, medicine, social justice, military law and psychiatry. Thus, the inter-disciplinary information can enrich our understanding about Indian as well as other soldiers in a multi-faceted manner so as to help them better cope with variety of operational challenges and optimize their performance in diverse military setups. More so, the work may suffice to the needs of Indian scholars and those of the Asian subcontinent amongst others, at an inter-disciplinary level (Military Psychology, Defence and Strategic Studies, Political Science, Sociology, HRM/HRD, Organizational Behaviour, etc.) as per their respective courses. Equally, knowledge so evolved can be incorporated into the selection, training and management of societal human resource even in corporate systems dealing with high-risk jobs.

It has been an enriching experience to work with the authors from varied professional and cultural background worldwide, while understanding the subject of military psychology. We thank all the authors for sharing their ideas and piece of work. Equally, sincere gratitude is extended to Dr Manas K. Mandal, DG R&D (Life Sciences), DRDO; Dr K. Ramachandran, Director, DIPR; and Sh N. P. Singh, Head, Strategic Behaviour Division, DIPR for giving their consistent support to the endeavour. Moreover, God has been generous to grace us enough wisdom, patience, determination and such a family that has supported us through the rough times to reach the desired end. Hope this book will be of interest to the readers and will lead to fresh ideas, research works and newer perspectives to the understanding of human behaviour in challenging contexts.

Introduction

Nidhi Maheshwari and Vineeth V. Kumar

The First World War (WWI), precisely 1917, signifies the formal alliance of psychology to military with the implementation of Army Alpha Tests for selection of men to Armed Forces. Consequent to this confluence, for the first time the psychological principles of human behaviour were applied to military performance in a scientific manner. Thereby, the discipline of military psychology has evolved during past 100 years transgressing various barriers and alluring many other disciplines, while its journey through a century from WWI to the Modern Military Missions (MMMs) giving rise to newer concepts along with observance of certain trends and derivation of various interventions in service of the soldier. Present book on military psychology is an attempt to put forth some of the relevant psychological concepts, trends and interventions that are of operational and strategic interest to the Indian as well as other forces.

The book is based on the presumptions of inductive as well as deductive learning processes for military scholars across the globe, with special reference to India. Following the deductive processes, several trends and concepts of universal occurrence in the Armed Forces have been mentioned for drawing vital guidelines for Indian soldiers and soldiers of similar countries. At the same time, general principles and intervention modes are induced by considering independent military cases from different parts of the world. Section I titled 'Psychological Concepts and Trends in Military Context' deals with the significant concepts and trends of military importance that qualifies a soldier and his ideal performance. Evenly, Section II titled 'Psychological Interventions in Military Context' deals with the interventions that help the soldier maintain and dispense his ideal performance. Various chapters under two sections cover topics like intelligence and personality testing, stereotype threats (STs), combat stress, psychological operations (PsyOps), Comprehensive Soldier Fitness (CSF), resilience building, hardiness training, countering terrorism, familial well-being, ethical leadership, psycho-social health of disaster responders, international humanitarian law and winning hearts and minds.

Following are the broad areas relevant to the Armed Forces in general and Indian Forces in particular that have been addressed through various chapters in the book.

Selecting the Right Soldier for the Right Mission

Armed Forces are undergoing metamorphosis with the advancement of technology. Refined technology has set new parameters of cognitive and non-cognitive abilities for military jobs. Due to the upgradation of gun power, selecting the right manpower for the job has become a challenge. The selection mode is of extreme significance in face of anti-terrorism operations and contemporary guerilla warfare, wherein night operations, ambush, raids, etc., are no exception for soldiers. And both their cognitive and non-cognitive abilities are at test to their best.

With special reference to the tests developed in India, Chapter 2 by Gurpreet Kaur, Dinika Anand and Soumi Awasthy and Chapter 3 by Suresh A., Arunima Gupta and Sucheta Sarkar focus on the cognitive and personality assessment procedures that can be used for selecting the right soldier. Chapter 2 casts a complete trajectory of intelligence and aptitude testing in Indian Armed Forces with a paradigm shift in the testing, definition of construct, mode of assessment and technique of analysis all through the years. Also, it highlights the use of computer and technology for adequate selection and placement of personnel. Chapter 3 enquires upon the use of biographical inventory in Indian military selection. It favourably compares the instrument with other personality measures used for predicting soldier's performance. Verifiability, generalizability and validity of biodata are discussed alongside the process of construction of such an instrument. More so, both the chapters provide essential inputs while restructuring the selection system of India or of any other country.

Building and Sustaining Resilient Soldiers

Guarding the territory and integrity of a nation against external invasion or internal aggression has eternally been an issue of concern for any forces. Whether it may be the Global War on Terror (GWOT), providing humanitarian assistance in African continent, disparaging the activities of the Islamic State of Iraq and Syria (ISIS), resolving Syrian conflict at the global level; or, back home, manning the counterinsurgency operations in Jammu and Kashmir, anti Left-wing Extremism (LWE) operations in the red corridor, executing the Yemen evacuation, high-altitude deployment in Siachen, Kargil, Arunachal Pradesh, etc. against Pakistanis and Chinese aggression in Indian soil; the forces are continually plagued with the concern for defying deployment stress in soldiers and sustaining their morale and resilience. Military leaders, trainers, policy makers, psychiatrists, operational psychologists and soldiers themselves are equally concerned about the traumatic and post-traumatic stress disorders (PTSDs) prevalent in the Armed Forces. Instances from similar socio-geo-political contexts can give an insight into the phenomenon of combat stress and its associated management strategies that apply universally to every soldier.

Chapter 4 by Joseph Miller discusses the mental agony named combat stress in a soldier during and after deployment by elaborating the case of an English veteran. This chapter enables the reader to understand PTSD from the perspective of a decorated soldier. It is expressed that wars change the perception of landscapes for its soldiers, who in turn try to change the landscape around them because of this changed perception. Thus, the significance of combat stress, its management and debriefing is upheld by the chapter which has become an issue of concern for Indian Armed Forces also.

Likewise, Chapter 9 by Eyal Lewin focuses on enhancing optimism in soldiers by studying the goal-oriented approach championed by the Israeli forces during the Arab–Israeli 1973 War. Doubting the common knowledge of social cohesion, collective fear and national ideology as the basis of soldiers' motivation to fight, chapter lauds the relevance of optimism. Discussing the positive psychology theme, case studies of four heroes at the northern front of Israel and five heroes at the southern front during the Arab–Israeli 1973 War have been analyzed to interpret and establish the significance of optimism. This empirical demonstration of optimism offers many lessons to soldiers in India and abroad for maintaining the weapon of optimism in their arsenal of mental robustness. The construct shall be of special interest to the soldiers working in counter-insurgency operations and naxal-prone areas where they are unaware of the duration, intensity and progress of their operations.

Similarly, Chapter 10 by Michael D. Matthews and Chapter 11 by Paul T. Bartone, Jarle Eid and Sigurd W. Hystad emphasize upon the Comprehensive Soldier Fitness (CSF) and hardiness training in soldiers, respectively, for their adequate coping with military stresses. Chapter 10 expresses concern over the rising psychological cost to combat and offers resilience-building through the magnificent CSF Programme as an alternative to the problem. Other elements of CSF as Post-Traumatic Growth (PTG), and extension of CSF to army families, children, civilians, other military and non-military organizations, have also been highlighted alongside the historical contribution of this programme to psychology. Chapter 11 draws the focus on hardiness training as a measure to build resilience in military. Psychological hardiness with commitment, control and challenge (3Cs) is presented as the primary resiliency factor which needs to be strengthened through training and leadership intervention. Also, self-help steps are provided as easy aids to build up the same. Indian Armed Forces is deliberating upon the concept of CSF and hardiness training for strengthening its forces; hence, knowledge of the same is an asset. Needless to say, such inputs are essential building blocks for designing the military training doctrine of any nation.

Promoting Mental Health of Soldiers Involved in MOOTW

Nowadays, for every challenging incident Armed Forces are brought in action. Whether it may be disaster evacuation in Nepal, Yemen, Uttarakhand or Kashmir, providing humanitarian aid in Africa or other South-east Asian nations, military comes as saviour to all. Consequent to such increasing involvement in Military Operations

Other Than War (MOOTW), Armed Forces across the globe including the Indian soldiers have to gear themselves up with newer roles and challenges thereupon. Chapter 12 by Sujata Satapathy illustrates the changing nature of Indian Armed and Para-military Forces in their stride to safeguard the nation and its citizen against various natural and man-made vagaries. It acknowledges the promotion of psycho-social health of soldiers involved in MOOTW often called as the Disaster First Responders (DFRs). The chapter argues that with the increasing deployment of personnel as DFRs during catastrophes and other humanitarian operations, much care must be taken to prevent and promote their psycho-social health. To begin with, several Indian disaster relief cases have been highlighted in the chapter followed by the work profile of the responders. Thereby, caregivers' trauma in terms of secondary traumatic stress, compassion fatigue, vicarious trauma (VT) and burnout is discussed. Specifically, preventive and promotive psycho-social care has been recommended for the responders through the design and dissemination of relevant Information, Education and Communication (IEC) materials which have proved their worth at earlier instances also in the country. The knowledge so shared can substantiate the significance and facilitation of psycho-social care of DFRs operating in any part of the world.

Guarding Forces Against Gender Biases

Indian Armed Forces and other militaries in various continents have opened or are reflecting on opening their gates for women in combat arms. The fair gender is now not only going to be a part of the medical corps but also that of engineers, artillery, air defence, submarine, etc. However, these warriors shall not be immune to the prevalent stereotypes which somehow might affect their military performance. Chapter 5 by Emerald M. Archer enquires upon the effect of stereotyping on the marksmanship performance of female Marines and highlights the consequences of the threat in terms of self-handicapping, reactance, distancing self from the stereotyped group and altered professional aspirations. Indian and other similar forces have important lessons to learn well in advance from this study on US Marines before such challenges become a reality for them.

Curbing Misconduct Behaviours

Extremes of behaviour have been dispensed by soldiers both within and beyond their territory during various peace-making and peace-keeping missions. For positive behaviours, the Armed Forces need to laud itself but for negative behaviours, the Forces must take preventive as well as corrective steps. It is universally acknowledged that preventive measures can be taken for misconduct behaviours before the situation worsens and slides beyond repair. Chapter 6 by Pankaj Kumar Sharma, Ashutosh Ratnam and T. Madhusudhan enunciates various misconduct behaviours as codified by the Indian Military Law. It casts a transparent picture about the forms of misconduct behaviours and their preventive and corrective measures that are being channelized in the Indian Military System. The chapter also explains why coping training is not working and what should be done to control such behaviours. Recommendations therein can have essential ramifications for armies worldwide.

Cultivating Ethical Leadership

Almost every military system is full of examples of ethical and value-based leadership. However, with the changing societal norms and aspirations, soldier of any culture might compromise his integrity and loyalty. This has become a primary concern for every military as it is considered a sacred institution. In Indian context also, we have come across cases such as the Bofors scam and the Tehelkha scam[1], that have raised eyebrows over the sanctity of the Services. Chapter 13 by Vidushi Pathak, Anju Rani and Sneha Goswami refurbishes the significance of value-based leadership in Armed Forces. Citing *slokas* (Sanskrit verses) from the sacred Bhagwat Gita and taking other instances from Indian and world military history, the chapter summarizes the core military values that need to be cherished with an aim of cultivating a shared value system and prospering authentic leadership in the national interest of any civilized military.

---

[1] http://www.ibnlive.com/news/india/a-brief-history-of-defence-scams-in-india-590770.html

Maintaining Healthy Military Families

Collectivist or individualistic societies both embrace the idea of unison of the soldier with his family, irrespective of its size. Effectiveness of a soldier is believed to be closely associated with the well-being of his family. Based on the respective culture of the nation, consistent efforts are made by the Forces to maintain the well-being of a military family, especially when the soldier is away. Like other countries, Indian Armed Forces are also ultra-sensitive to the familial aspect of a soldier and leave no stone unturned to keep it healthy. Thus, Chapter 14 by Archana and Updesh Kumar looks at the procurement of familial well-being and quality of life as pathways to achieving soldier effectiveness. Recurrent relocation, separation, frequent deployments, parenting stress, abrupt reunion, etc., are presented as some of the familial concerns in the military life. Thereby, measures are suggested to create resilient families and raise the quality of life in soldiers through building social support, sensitizing leadership, increasing positivity, enhancing resilience and striking healthy work–family balance. Similar measures can be practiced and preached in various military setups.

Countering Terrorism, Winning Hearts and Building Peace

Like other parts of the world, forces in the Asian subcontinent also are now battling against terrorism, building regional peace and winning hearts of the local populace in their area of operation. The task becomes exigent for Indian soldiers as they are often shunned by the local populace in places such as Jammu and Kashmir, Assam, Mizoram, Nagaland, etc., for allegedly violating human rights, fanning fake encounters and custodial deaths, and resorting to looting and other nefarious activities. There has been continued agitation by various social agencies against the forces for the illegal use of Armed Forces Special Powers Act (AFSPA) and violent protests have become the order of the day in the region. Conversely, the forces are intrigued with the question of whom to catch and how to catch in a region where the rival is rooted in strong social circles.

With the specific illustration of Boko Haram threat in Nigeria and deployment of Multinational Joint Task Force (MNJTF) against it, Chapter 15 by James Okolie-Osemene provides essential answers to questions of defining terrorism and ways of containing it. Boko Haram, like Al-Qaeda and ISIS, is often described by world leaders as a global threat. The chapter surfaces the antecedents and consequents to such anti-terrorism activities and the measures taken by the Forces while operating in such locations and fighting an unknown enemy. Observance of human rights issues and adoption of community-based approach are identified as workable strategies to restoring people's confidence in the forces and thereby strengthening the forces to crack the back bone of terrorism. Indian security agencies that are marred by similar activities of the sleeper agents, insurgents, terrorists and other non-state actors in Jammu and Kashmir, North-east, red corridor and other high-power locations can get useful insights for developing workable strategies for restoring people's confidence in the forces and operating conveniently in the terrorism and naxal-hit areas.

Likewise, Chapter 16 by Amparo Pamela H. Fabe endorses the importance of upholding International Humanitarian Law by the Armed Forces for winning hearts and building peace in the conflict-ridden areas. Discussing the specific case of Philippines which is stained by three-pronged insurgency war, the chapter enumerates the psycho-physical constraints of Filipino soldiers who have to "fight with their hands behind the back". Equally, it lauds the efforts made by the Forces in up-keeping humanitarian law by opening Human Rights Office (HRO), partnering with civil society organizations and training their soldiers on human rights. Also, interventions for protecting human rights have been discussed keeping in sight their positive outcomes for the Forces to conserve. Soldiers in the Asian subcontinent, especially India, have a lot to learn from such instances as they are also subjected to the "death by thousand cuts" warfare tactics and are "fighting with their hands behind the back" to such deadly tactics in their area of operations.

Preparing Soldier for Future Warfare

Globally, Armed Forces have realized that now the war is going to be more mental than physical. For India, facing the Chinese aggression in Kashmir (like Aksai Chin) or Arunachal Pradesh (like Tawang), containing threats of Pakistan-backed terrorists in Kashmir, stimulating 'hot pursuit' of militants in Myanmar (against infiltration in Manipur and Nagaland), eliminating naxals and de-radicalizing the disillusioned religious fundamentalists to join the mainstream are now more a matter of mental than physical might. There have been consistent efforts in Jammu and Kashmir, North-east, Naxalism-affected areas and other regions of the country to curb the rivals and win over our people through PsyOps such as social engineering, propaganda and brainwashing. Although PsyOps have always been a tricky issue for any Armed Forces to master, yet it is the strongest non-lethal weapon to train upon. Use of PsyOps requires immense skills and expertise to win over the situation in favour of the State and its success depends upon a number of factors for which Indian Armed Forces have to delve upon. Also, effort is required to upgrade and update the Forces with the enhanced use of technology and sophisticated weapon systems in future warfare scenario, while moving for global missions in the African, Middle-east, European and the South-east Asian regions.

Chapter 7 by Ron Schleifer throws light on the phenomenology of PsyOps through the analysis of Operation Cast Lead (27 December 2008 to 18 January 2009) executed by the Israel Defense Forces (IDF) against the PsyOps of Hamas, the Iran's client regime in Gaza. Various components of the two warring PsyOps agencies are analyzed on lines of their objectives, target, message content, transmission channels, media and thereby the extent of success achieved. Also, the direction of future PsyOps has been extrapolated through the analysis which is pertinent in current Indian context. Functionality of various factors such as the objective, the message, the channel and the media along with the target needs to be studied by Indian Forces also before designing any PsyOps strategy for its people within or behind enemy lines. Also, the input can act as an incubator while defining initiatives and formulating policies against naxalism or any other psychosocial threat worldwide.

Alike, Chapter 8 by Swati Johar and Updesh Kumar projects the disposition of future weapons for Indian scenario keeping the vision 2050 in mind. It forecasts the possibility of bio-genetic engineered systems, cyber weapons, remote neural monitoring and cognitive hacking through use of lasers, extra-low frequencies and mesh worms along with the need for enhancing unmanned technology with human intelligence interface, mind-enabled tools and psychological warfare strategies which are highly relevant and going to be a reality in future combat. The projections can also be helpful in globally facing the likely future challenges in the psychological testing, training and operating arenas.

Revisiting Centurion Concepts, Trends and Interventions

Future builds on the past. Without knowing the past labour in the arena of military psychology, progression and growth of the subject is meaningless. Chapter 1 by Nidhi Maheshwari, Vineeth V. Kumar and N. P. Singh manifests the centurion concepts, trends and interventions in military psychology across the globe. Objective is to aid a deeper understanding about the concepts associated with the psyche of a soldier fighting an unconventional war and prepare him better for future wars based on the past knowledge. Emergent concepts, trends and interventions during the centurion deployment of psychology to military are highlighted. Of special interest is the section on the rise and growth of military psychology in India which can provide an input to better understand our soldiers and enhance military performance. Thereafter, Chapter 17 by the editors provides future challenges and errands for military psychology. Past militaries of the world have together evolved such a grand knowledge base which shall be used by the future militaries of the world in soldierly interest.

Thus, it can be said that topics so covered in the book can aid the Indian as well as other Forces of the world to strengthen their structure and capability for optimizing performance during contemporary and future warfare.

SECTION I

Psychological Concepts and Trends in Military Context

1

Military Psychology: The Centurion Phenomenon

Nidhi Maheshwari, Vineeth V. Kumar and N. P. Singh

Unlike in any other science or art, in war the object reacts.

—Carl von Clausewitz

Military psychology owes its allegiance to both the science and an art. Science implies in knowing or deciphering the principles of military behaviour whereas art divulges in applying those principles to soldier's performance and associated systems. This science and art of military has evolved consistently with the evolution of society and wars. Kennedy, Hughes and McNeil (2012) argue that the growth of military psychology has occurred in spurts, each related to the demands, psychological as well as military, of the conflicts of different nations. More so, it has evolved from limited participation in wars of the past to today's war, where it has been an indispensable asset in combat readiness and policy development. According to Dandekar (2000), many societal changes resulted in a different view on the relevant competencies with regard to military. Downsizing of the military, the civilianization of military way of life, the end of the conscription in most of the West-European countries and recent terrorist attacks have influenced the way in which the military recruit, select, train and deploy in operations. Seligman and Fowler (2011) also express that First World War (WWI) and Second World War (WWII) drove explosive developments in aptitude testing, selection and classification.

Conversely, the society and warfare have also evolved with the implementation of psychological principles to military performance. Scales (2009) emphasizes that psychology and its closely associated disciplines are critical to success in contemporary war, and that this need will generate paradigm-shifting changes in the science and profession of psychology. He reveals that WWI saw dramatic developments in chemistry, WWII in physics (notably radar and nuclear technology) and the Cold War in information technology. Indeed, beginning with the major wars of the 20th century and continuing to the current conflicts of the 21st century, war pushed paradigm shifts in a variety of scientific disciplines (Laurence & Matthews, 2012). Further, Matthews (2014) argue that psychology is the science that will determine who wins and who loses the wars of the 21st century, just as physics ultimately led the United States to victory in WWII. Clearly, increasing necessity of the society and emerging challenges lead the discipline to expand itself to every nook and corner of military organization and behaviour. On certain occasions, the discipline even revolutionized clinical, industrial, social, health, vocational and organizational behaviour. More so, Mukherjee, Kumar & Mandal (2009) express that although largely guided by the users' requirements, military psychology does not restrict its research domain to applied fields; it provides researchers with ample opportunity to investigate basic issues of science and to test their application on ground. Thereby, Matthews and Laurence (2012) are justified in saying that military psychology, by necessity, is a heterogeneous field of inquiry. On one hand, it draws on all sub-disciplines of psychology to understand the variables that affect soldier performance, whereas, on the other hand, lessons learned from military psychology are of vital importance to all areas of psychology.

Definition

Military psychology is a discipline which is concerned with recruiting, training, socializing, assigning, employing, deploying, motivating, rewarding, managing, integrating, retaining, transitioning, supporting, counseling and healing military members (Laurence & Matthews, 2012). It is an area of study and application of psychological principles and methods to military environment (Gal & Mangelsdorff, 1991). It is also defined as the application of research techniques and principles of psychology to the resolution of problems to either optimize the behavioural capabilities of one's own military forces or minimize the enemies' behavioural capabilities to conduct war (Walters, 1968). Cronin (1998) goes on to define military psychology as the application of psychological principles to the military environment regardless of who is involved or where the work is conducted. In common parlance, it is the research, design and application of psychological theories and empirical data towards understanding, predicting and countering behaviours in either friendly or enemy forces or civilian population that may be undesirable, threatening or potentially dangerous to the conduct of military operations (Military psychology, n.d.). Moreover, Cronin (2003) goes on to equate it with a microcosm which embraces psychological disciplines and which affects almost all aspects of military settings. Further, Kreuger (2003) highlights the following nature of psychological research in military settings:

1. Recruitment, selection, placement, training and retention of military personnel.
2. Prediction and enhancement of combatant performance in harsh environments.
3. Human engineering design of complex weapon systems for effective use by soldiers, sailors, airmen and marines.
4. Training procedures to mould well-honed military teams by maximizing specialized differential skills to permit success on tough missions.
5. Soldier coping mechanisms for deployment to foreign lands, or to carry out extended hours of work, or to work under austere conditions.
6. Soldiers' abilities to adjust to countless intricacies of military lifestyle.
7. Collection and interpretation of large amounts of psychological data to assist military leaders and civilian authorities in making smart decisions and informed policies that affect millions of military members and their families.
8. Providing advice on integrating people of diverse ethnic and social backgrounds into the workplace.

Hence, it can be said that military psychology is both the science and art of understanding and employing the principles of cognition, affect and behaviour in military contexts so as to optimize the selection, training, adaptation and performance of soldiers in diverse roles of building peace and maintaining security.

Century of Military Psychology: Concepts, Trends and Interventions

Although psychological practices were deep seated in the military functioning since ages as is evident in the *slokas* (Sanskrit verses) of Srimad Bhagwat Gita, Kautilya's Arthashastra, Sun Tzu's *Art of War* and the Clausewitz's *On War*, etc. However, WWI can be considered as the defining moment for the official inception of psychology to military with the introduction of first intelligence test for selection of men in 1917. The episode will complete its 100 years in 2017. Since 1917, the discipline has travelled a long distance while experiencing many crests and troughs through its journey. There has been a war-wise spurt in the arena of military psychology largely depending upon the need of the warring nations, deployment of soldiers to various operations and military settings with changing geo-political and strategic milieu. Windle and Vallance (1964) opine that military psychology might be considered an offshoot of industrial psychology, but one which has come through luxurious growth to demand independent recognition. Consequently, newer concepts, trends and interventions were witnessed in its ambit. On the basis of the recorded history of soldiers' deployment to operations/wars and associated deployment of psychology against emerging challenges, centurion trajectory of military psychology can be studied under three phases which are described in the succeeding paragraphs.

Centurion Trajectory of Military Psychology

Looking at the nature of warfare and deployment of psychology along warfare, the trajectory of military psychology during the century can be divided into three phases: (a) Initial Phase (during WWI and WWII), (b) Middle Phase (during Korean, Vietnam and the Gulf War) and (c) Contemporary Phase [during the Modern Military Missions (MMMs)].

Initial Phase of Military Psychology (during WWI and WWII)

It all kick started with WWI when need was felt to psychologically screen the civilian youths who were drifted to participate in WWI by the consent of the American Congress. Yale's Biopsychology Professor and erstwhile American Psychological Association's (APA) President, Robert M. Yerkes, took the challenge and inspired the psychologists for providing their valuable contribution to the efforts of the nation. He expressed that, "It is obviously desirable that the psychologists of the country act together in interests of defence. Our knowledge and our methods are of importance to the military affairs of this country and it is our duty to work together to maximize the effective work aimed at increasing the efficiency of our army and navy" (Yerkes, 1918). Even before Yerkes, Sir Walter Dill Scott and his team had spent initial six months of war in designing tests that would adequately equate civilian skills and military tasks. Although Scott's programme to classify Army's personnel structure based on occupational skills received immediate approval from the Army. However, Scott and his team spent too much time in finalizing the classification cards and designing trade tests which adversely affected the interest of the Army in the proposition. Subsequently, Yerkes proposal for mass intelligence testing came to the fore. Obviously, this programme on mass intelligence testing received a lukewarm response from the Army (Keene, 1994) vis-à-vis the classification of personnel based on occupational skills. Nonetheless, pursuing the initiative and reflecting upon Binet's contribution in the arena of intelligence testing, Army Alpha (for literates) and Army Beta (for illiterates) tests were developed and implemented to screen the recruits for WWI in group setting under the guidance of Yerkes. This established the significance of 'intelligence' testing of recruits for the Forces which was otherwise an arduous task for the Army to embrace and convince upon. Later, Army Alpha proceeded to become the Wechsler-Bellevue Scale which led to the development of Wechsler Adult Intelligence Scale (WAIS). Besides intelligence testing, Woodworth Personality Data Sheet was also introduced at that time (Page, 1996) which was supposed to detect personalities that would crumble under fire. Although it did not work much, but the contribution was significant as it was the first demonstration of personality being quantified. Besides intelligence and personality assessment during that time, psychologists also ventured into developing instruments for selecting and training gun pointers, and developing techniques for predicting successful fighting aviators. Systematic methods for judging and rating qualifications of officer candidates and for training soldiers with differing ability to learn were also introduced (Uhlaner, 1978).

This was also the period of 'chemical warfare' and 'trench warfare' as it is called. Soldiers experienced the trauma of green gas, 'the mustard gas' in war close to their trenches, witnessing their brothers fuming in front of them, blown to pieces, hearing loud noises of round of fires moving over their trenches, etc. Consequently, they developed an unknown mental condition expressed physically by getting stuck during war, being unresponsive to sounds and feeling incapacitated. Originally, such soldiers were labelled as 'cowards' and were shunned for getting into a condition designated as 'shell shock'. But, gradually, terms such as 'gas hysteria', 'war neurosis', 'Not Yet Diagnosed Nervous (NYDN)' and 'Disordered Action of the Heart (DAH)' came to their rescue and these terms became the moniker of combat stress henceforth. With such a diagnosis, strides were made in the arena of neuropsychology, neurosurgery, cognitive rehabilitation (Boake, 1989; Franz, 1923) and cognitive restructuring for preventing and rehabilitating the soldiers to such vagaries. Gradually, forward psychiatry was implemented to advocate the PIE principle of proximity, immediacy and expectation of recovery for treating shell shock cases (Jones & Wessely, 2003). Thus, WWI enunciated significant psychological inputs to the art of war-fighting as well as handling war-ridden victims. Kennedy et al. (2012) go on to say that lessons learned in WWI continue to guide the mental-health professionals in addressing the response to fear of the current terrorist threats to employ chemical and biological warfare.

Also, problems related to morale, discipline and leadership came to the fore during this time. Besides the psychology division for intelligence testing, a morale division was also established under the aegis of Colonel E. L. Munson, Director of Medical Officers Training in the Office of Surgeon General, during the WWI period. The division intended to streamline the adaptation of drafted civilian youths to the army life. But, it could only exhibit staggered performances due to the reluctance of the officers at the top and their beliefs that they know better about their men than any other outsider to the system. However, during the post-WWI period it made significant strides by creating military's first internal soldier survey (Keene, 1994).

WWII again demanded psychologists to work towards selection of men (Melton, 1957) but with an enhanced focus on leadership development, psychological warfare techniques, motivation and its management (Gal & Mangelsdorff, 1991; Stouffer, Suchman, DeVinney, Star, & Williams, 1949) rather than purely on aptitude. Capshew (1986) expressed that while promoting themselves as experts concerning the 'human factor' in warfare, psychologists also found employment in military personnel work, propaganda analysis, survey research, equipment design and other areas during WWII. Screening for military service got transformed, and in 1940 the Army General Classification Test (AGCT) was introduced. It was used to measure the aptitude of recruits for selecting men for specialist courses (Drucker & Zeidner, 1988), and for officer training (Harrell, 1992). Standard test battery included assessments of general intelligence, arithmetic, verbal and non-verbal skills and instructions (comprehension). Nonetheless after WWII, with the introduction of Selective Service Act in 1948 for uniform aptitude testing procedures, the Armed Forces Qualification Test (AFQT) came into being in 1950 (Kennedy, Hughes, & McNeil, 2012).

Additionally, this phase (1944–1946) also saw the establishment of Division 19 for Military Psychology in APA which indicates the enhanced significance of the subject during that time. Essentially then, Office of Strategic Services [OSS, now Central Intelligence Agency (CIA)] came into being along with the first selection programme for OSS operatives dealing with espionage and propaganda (OSS Assessment Staff, 1948). This was also the time when importance was assigned to the psychological principles of military performance, morale, personnel selection, sexuality, leadership development and psychological warfare. The classic book on *Psychology of Men* was written by Boring (1945) highlighting such principles. Also, pigeons were trained by Skinner to be used as guided missiles (Gilgen, 1982) and possibility of bats was explored as possible messengers to drop miniature explosives over Japan (Drumm & Ovre, 2011). Although the efforts could not bear fruit due to logistical problems, soon the introduction of atomic bombs surfaced. Also, efforts were made to avoid malingering or gold bricking as it was called. Malingering was considered a felony and malingerers were supposed to have psychopathic personalities (Campbell, 1943).

On the other hand, the irresistible number of psychiatric casualties during WWII established that combat stress reactions were generally normal responses to the emotional trauma and stressors of war as opposed to a defect of character (Glass, 1969). This enhanced an earlier overlooked need for a dedicated psychologist exclusively for soldiers. Kennedy, Hughes, and McNeil (2012) express that during WWII the United States did not initially utilize lessons from WWI about combat stress reactions (i.e., need for timely intervention on the frontline) as little forward mental health was practiced, favouring reliance on psychological screening to avoid the negative psychological reactions to the war. To begin with, few physicians and psychiatrists were employed to meet the emotional needs of veterans, psychologists provided both individual and group therapy in Veteran Affairs facilities (Cranston, 1986). The first psychology internship programme was kick started in 1946 with initial enrollment of 200 interns in the VA system. Phares and Trull (1997) express that these efforts resulted in increased acceptance of psychologists, not just as researchers and experts in assessment but also as mental-health providers. Unlike WWI wherein psychologists were demobilized after war, post-WWII brought permanent active-duty status for psychologists in 1947 (Uhlaner, 1967). Importantly, with the emergence of newer symptoms, development of Diagnostic and Statistical Manual of Mental Disorders (DSM) (APA, 1952) came into being (Committee on Nomenclature and Statistics, 1952).

Certain endeavours made during WWI gained resurgence such as head injury rehabilitation, neuropsychology (Boake, 1989) and aviation psychology with the development of US Army Air Forces aviation programme in 1941 to assist the selection of aviation personnel (Driskell & Olmstead, 1989) during this war. Besides the selection of pilots, navigators and bombardiers, research was also conducted on the service member–equipment relationship (Koonce, 1984). Interestingly, the first modern simulator for WWII pilots, the 'Blue Canoe', was designed by Edwin A. Link during this period to train pilots to fly by instruments. The link trainer employed vacuum technology similar to that used in organs in the 1920s (DeAngelo, 2000). Job analysis, time and motion studies, studies of extreme climatic conditions in tropical and arctic conditions (Hughes, 2007) were also the subjects for psychologists during that time. In 1947, the US Air Force became a separate branch of the military within which industrial psychology got established (Hendrix, 2003).

Characteristic use of nuclear bombs and the surfacing of kamikaze pilots or Japanese suicide bombers made WWII distinct from any other war. Salter (2001) reports that survivors of nuclear attack developed both acute and chronic psychological reactions, including withdrawal, severe fear reactions, psychosomatic symptoms and post-traumatic stress disorder (PTSD).

Thus, by the end of the initial phase of military psychology, psychologists were not only employed for mental testing and assessment but also for enhancing morale, performing psychological operations (PsyOps), providing counseling and designing work stations and systems for the Forces.

Middle Phase (During Korean, Vietnam and Gulf Wars)

Middle phase of military psychology signified important strides with the challenges faced thereupon. Psychologists served overseas, in combat zones and on hospital ships (McGuire, 1990) during the Korean war. The war witnessed forced marches, severe malnutrition, inhumane treatment, continuous propaganda and re-education on communism (Ritchie, 2002) of soldiers along with significant use of torture as well as execution of US prisoners of war which gave rise to the concept of brainwashing (Ursano & Rundell, 1995). As a result, Survival, Evasion, Resistance, Escape (SERE) model was included extensively in the training programmes of soldiers who face the risk of being captured. The SERE model is still being followed to train soldiers who are employed in dangerous missions such as soldiers of special forces and aviation personnel. Additionally, lessons learnt from WWII regarding the need for mental-health providers in the combat zone were not forgotten. Also, after the first year in Korea, rotation policy of nine months was implemented that helped significantly reduce the number of psychiatric casualties (Glass, 1969). Psychological testing sustained importance during this time also. The Army and Air Force collaborated on a technical manual, outlining the role of military psychologists and proper use of psychological tests (US Departments of the Army and Air Force, 1951). Additionally, the goal of increasing the performance of military personnel given different equipment, various physical states (e.g., fatigue) and various environments gave rise to increased research in human factor engineering (Roscoe, 1997).

According to Defense Manpower Data Center (1999), the US implemented the Airmen Qualifying Examination in 1958 after the Korean War for administration to high-school students. Shortly thereafter, Army and Navy developed their own group ability tests, and ultimately in 1968 the Armed Services Vocational Aptitude Battery (ASVAB) was implemented to make a truly uniform aptitude tool. By this time, troops were witnessing another war—the Vietnam War. This war has been complex in terms of the nature of weapon, warfare technology and soldiering experiences. US troops found themselves engaged in Jungle warfare and often had terrific prisoner of war experiences. Implementing rotation policy, individual soldier was asked to move instead of the entire unit, which was detrimental to group cohesion and morale (Zeidner & Drucker, 1988). Moreover, soldiers also experienced hostile home front as the civilians were not in favour of such a war.

Although forward mental health care was available since the beginning of this war, there were barriers to care at individual as well as group level. Cases of combat neurosis were contained but the war saw an emergence of character disorders and substance abuse in the troops along with increased cases of PTSDs. Glaringly, some veterans are suffering from PTSD till today. Hence, need arose for a formal response to critical incidents, for example, death of a soldier during training, suicide, natural disaster, etc. In such times, Tryon (1963) emphasized that due to the increasing complexity of man–machine systems, social psychologists and personality psychologists have much to offer than the predominantly experimental psychologists drawn into military psychology.

Equally, warriors were challenged with unique stressors during the Gulf War. According to Martin, Sparacino and Belenky (1996), military personnel in Operation Desert Shield and Desert Storm were exposed to multiple combat stressors: greater number of enemy forces, possible use of chemical and biological weapons, environmental challenges, lethal animal life, inadequate hygiene opportunities and a culture that did not accept American values. Although stressors were multiple, yet there were lesser combat stress casualties vis-à-vis the PTSD cases. For the first time a psychologist was deployed on-board the Navy Aircraft carrier, USS John F. Kennedy. However, Gulf War Syndrome emerged as a major casualty due to the fear of vaccinations, exposure to toxic substances and psychological trauma (Kennedy, Hughes, & McNeil, 2012).

Contemporary Phase (During MMMs)

Contemporary phase of military psychology is fascinating as the Armed Forces engaged in MMMs bear immense diversity in their roles and objectives. Now, armies seem to reach out to other nations and populace in times of need, thus engaging into both combat and non-combat operations. Troops during such warfare are often

engaged defensively in counter-insurgency and anti-terrorism operations, whereas at other occasions, they might be offensively involved
in 'capture or kill' missions (like that for Saddam Husain and Osama
Bin Laden). On the other hand, they may be participating in operations related to disaster relief, humanitarian aid, Medevac (Medical
evacuation), non-combatant evacuation, strikes/raids, surveillance and
reconnaissance missions, hostage rescue, counter-drug/trafficking and
peace operations. Thus, MMMs may include Effect-Based Operations
(EBOs), Military Operations Other Than War (MOOTW) and Global
War on Terror (GWOT).

Sager, Van Iddekinge and Russel (2004) express that starting from
the experiences of the first Humanitarian Operations (HOs), Armed
Forces have realized that although war skills remain a central issue, yet
skills other than the traditional ones have become more important.
The evolving nature of peacekeeping duty in itself suggests that today
soldiers are faced with new psychological challenges (Litz, Orsillo,
Freidman, Ehlich, & Batres, 1997), and that it is no longer unusual for
contemporary peacekeeping missions to include exposure to traditional
war-zone experiences (Orsillo, Roemer, Litz, Ehlich, & Freidman,
1998). Under such conditions, much self-control, independent thinking and rapid decision making must be exercised. Additionally, soldiers
must be well-versed in dealing with local population and civilians
whom they are providing assistance. Thus, in the current context,
military psychology seems to incline towards understanding and
manipulating social processes and cultural systems besides focusing
on personal variables. For modern times, Vallance and Windle (1962)
were rationalized in predicting that just as the predominantly hardware
orientation of the military led to the confluence of psychologists and
engineers in 'human engineering' and 'man–machine systems analysis',
so the growing recognition of cultural variables may produce 'cultural
engineering' and 'culture-machine systems analysis'. While working
with Army's Human Terrain System, McFate, Damon and Holliday
(2012) report that what commanders want and need to know is remarkably consistent between theatres and over time: the prime categories
of knowledge are social structure, the political system (both formal
and informal), the economic system (both formal and informal) and
interests and grievances (pertaining to security, intra- and extra-group
conflict and the administration of justice).

Equally, role of positive psychology in enhancing soldier resilience
and preparedness is highlighted in these challenging times. Irrespective

of the gun power, the man behind the machine is strengthened, especially for missions in which each soldier functions as a system. Emphasis is placed on enhancing his resilience, morale, reliability, coping skills and stress tolerance capacities, operational stress control and readiness like never before. Soldiers are developed and trained to be tough—mentally, physically and emotionally (Christian, Stivers, & Sammons, 2009). Casey (2011) believes that disease-finding and treatment, though extremely important, is a flawed approach if the goal is increased military readiness and performance. Thus, instead of pathogenic, a salutogenic model is emphasized vehemently. In 2008, the Directorate of Comprehensive Soldier Fitness (CSF) was established (Cornum, Matthews, & Seligman, 2011) to institute a holistic fitness programme for soldiers, families and army civilians to enhance performance and build resilience. The programme is unique as it draws upon the principles of positive psychology and addresses not only soldiers but also families and civilians working in a military setup. It focuses on harnessing the signature strengths of soldiers to the benefit of the organization. CSF program is the largest application of psychological science in military history (Seligman & Fowler, 2011).

Importantly, this phase also saw an increased interest and concern for the military families, post-deployment health and reintegration issues. Familial well-being, psychological preparedness and quality of life became the core areas of research so as to streamline the soldier's performance by strengthening his support system. Many studies (Booth et al., 2007; Booth & Lederer, 2012; Castaneda et al., 2008; Stanley, Segal, & Laughton, 1990) on family members of military personnel emphasize the importance of familial well-being for the soldier. Similar results were obtained by Orsillo et al. (1998), who found that stressors such as being separated from family are predictive of psychiatric distress.

During this time, training in-sync with technology has gained bright attention of researchers. Some of the significant trends in this direction include serious games. They are training systems that utilize video game technology for training purposes. These games have been developed for training emergency medical procedures and combat casualty care, logistics, convoy operations, small unit tactics and many other topics. These games are designed to be engaging, challenging and motivating to military students (Belanich, Sibley, & Orvis, 2004). Singer et al. (2008) opine that improved game technology has dramatically lowered the cost of simulation. Such games while played over Internet, open the possibility of training with units that are already in conflict area, a so-called right-seat ride, or by units, now separated, that will be deploying together. Further, training sessions with coalition partners provide units with an opportunity to learn about differences in terminology and culture ahead of experiencing them in an area of operations.

Military Psychology in India

Indian martials have a rich history to observe psychology converging with warfare. Mahabharata, the Great Indian Epic, proclaims much of the concepts for military psychology to decipher. The *vyuh rachana* (fighting formations), leadership strategies, camouflage and deceit (PsyOps) were some of the notable emergences during the war. Likewise, Kautilya's teachings profess much about the implementation of psychology to war and leadership. The Great Indian Mutiny of 1857 was yet another exemplary outcome of an unsung PsyWar marked with a definite impact on the morale of the troops that were otherwise forced to serve the adversary. Although rise of Satyagrah and Indian National Movement too unfurled a host of avenues for military psychology to claim its existence in India, hitherto, officially Military Psychology had a subtle entry in India during the WWII by the establishment of experimental War Office Selection Board in Dehradun during 1943 for selection of personnel to Armed Forces.

It is documented that in 1949, the experimental board was rechristened as Psychological Research Wing (PRW) with the primary objective to evolve a scientific system for the selection of officers and update it through continued research programme. During this period, Gardner Murphy, a noted psychologist, visited India and held discussions with the then Prime Minister, Pt Jawaharlal Nehru, on the emerging operational challenges in conventional warfare and the scope of the institute's charter (Sainik Samachar, 2007). Thus, in 1962, PRW was redesigned as Directorate of Psychological Research to take on new areas of research related to morale, ideological conviction, group effectiveness, leadership behaviour, job satisfaction, high-altitude effect, motivation, attitude, anthropometrics, civil–military relations and other issues. The Directorate of Psychological Research grew into a full-fledged institute to be renamed as Defence Institute of Psychological Research (DIPR) in 1982. The Institute has traversed a long way toward its aim to achieve standards of excellence in the arena of military psychology.

Mukherjee, Kumar and Mandal (2009) report that the institute's research focus is on finding optimum solutions to problems pertaining to the selection of officers, placement and categorization of men, with a view to optimise the efficiency of the Armed Forces and to devise suitable standardized tests for personality, intelligence and aptitude assessment. The institute provides technical training to assessors who man the Services Selection Boards (SSBs) performing selection duties, and monitors and evaluates the selection system, vis-à-vis training and performance of the selected personnel during service career. It also conducts human factors research and applies it to selection of pilots, performance in extreme climates and under difficult conditions, designing simulation devices and assessing mental workload of personnel. Enhancing military leadership, building effective military teams, containing fragging and suicide, managing combat stress behaviours both during and post-deployment, understanding psycho-physiological correlates like that of biorhythm, night vision, camouflage detection, etc., are the contingent areas of research. Offering strategic services, the institute focuses on the personality profiling of significant persons in military sociocultural context, designing and countering PsyOps such as rumour and propaganda, identifying malicious intent in crowd, community profiling, conducting interrogation and underlining hostage negotiation tactics. Additionally, share of research focuses on enhancing civil–military relations, building effective small work teams and attracting youth to join the Armed Forces. More so, keeping in focus the demands and challenges of future conflicts, both extra-territorial and extra-terrestrial warfare, emphasis in research is being placed on PsyOps and cognitive neurosciences.

Some of the recent contributions of DIPR include the following: the development of Computerized Pilot Selection System (CPSS) (DIPR, 2005), Comprehensive Battery of Cognitive Abilities (CBCA) for the selection of officers in the Armed Forces (DIPR, 2007a), PBOR Selection and Trade Allocation Battery (DIPR, 2008), self-help guides on combat stress and its management (KamaRaju & Singh, 2006; Misra, Asnani, & Archana, 2006), Manual on Interrogation Techniques in

Counterinsurgency Operations (Singh, Tripathi, & Asnani, 2007), Deceit Detection and Interrogation Manual (Patnaik, Dhawalgi, & Mandal, 2007), Organizational strategies and self-help techniques to deal with the negative consequences of prolonged deployment in low-intensity conflict (LIC) areas and studying the suicides and fratricides in the Indian Armed Forces (DIPR, 2007b), etc. Alongside, research works are also being pursued in the area of combat stress management, special forces profiling, hostage negotiation, small-group leadership, PsyOps, rumor management, identification of malicious intent and behaviour in crowd, night vision for operational effectiveness, reliability testing of personnel, suicide terrorism and profiling of significant military heroes.

Equally researchers from other institutions such as Armed Forces Medical College (AFMC), IITs, NIMHANS, IDSA, USI and CLAWS have upheld the flag of military psychology. Research work includes the mental health care of paramilitary personnel (Verma, Mina, & Deshpande, 2013), stress in the Indian Armed Forces (Ryali, Bhat, & Srivastava, 2011), stressful life events of personnel (Raju, Srivastava, Chaudhury, & Saluja, 2001), impact of LIC operations on soldiers (Chaudhury, Chakraborty, Pande, John, Saini, & Rathee, 2005), PTSD in soldiers (Saldanha, Goel, Kapoor, Garg, & Kochhar, 1996), ecology of combat fatigue among troops (Puri, Sharma, Naik, & Banerjee, 1999), association of life events of serving personnel life with psycho-pathology in their children (Prabhu, Prakash, Bhat, & Gambhir, 2011), stress reduction of soldiers through meditation (Cheema & Grewal, 2013), terrorism, trauma and children (Harjai, Chandrashekhar, Raju, & Arora, 2005), changing socio-economic norms and its impact on India's Armed Forces (Gokhale, 2013), etc.

Conclusion

Military psychology beholds a rich history of almost a century and has a bright future beyond, for centuries. The legacy of the past paves the way and directions for applicability of psychology in future warfare scenario. Rather, it provides the launching pad for current research and applications to take off so as to serve the Armed Forces, expansively in the future. WWI introduced psychology to military in the arena of testing and selection; thereafter, it graduated to counselling and forward psychiatry, training, man–machine systems, motivation, leadership, team building, and now to cultural systems, cognitive engineering and operational and embedded psychology. Psychologists ventured into newer areas to take on newer challenges, and thus the boundary between disciplines and sub-disciplines got thinner. Clearly, host of arenas remain under the jurisdiction of military psychology to reign on and the century saga is just a premier to the epic.

References

Abbott, P. A. (1980). Social and behavioral sciences contributions to the realities of warfare. In J. Arima (Ed.), *Proceedings of Symposium on What Is Military Psychology?* (pp. 27–31). Monterey, CA: Naval Postgraduate School.

American Psychiatric Association (1952). *Diagnostic and statistical manual of mental disorders* (1st ed.). Washington, D.C.: American Psychiatric Association.

Belanich, J., Sibley, D., & Orvis, K. L. (2004). *Instructional characteristics and motivational features of a PC-based game* (ARI 1822). Arlington VA: US Army Research Institute of Behavioral and Social Sciences.

Boake, C. (1989). *A history of* cognitive *rehabilitation of head-injured patients, 1915 to 1980. Journal of Head Trauma Rehabilitation, 4,* 1–8.

Booth, B., et al. (2007). *What we know about army families: 2007 updates.* Prepared for the US Army Family and Morale, Welfare, and Recretion Command (F&MWR). Fairfax, VA: ICF International.

Booth, B., & Lederer, S. (2012). Military families in an era of persistent conflict. In J. H. Laurence & M. D. Matthews (Eds.), *The Oxford handbook of military psychology* (pp. 365–380). New York, NY: Oxford University Press.

Boring, E. G. (1945). *Psychology for the armed services.* Washington, D.C.: Washington Infantry Journal.

Campbell, M. M. (1943). Malingery in relation to psychopathy in military psychiatry. *Northwest Medicine, 42,* 349–354.

Capshew, J. A. (1986). *Psychology on the march: American psychologists and World War II (Professionalization, United States).* Retrieved from http://search.proquest.com/docview/303521707

Casey, G. (2011). Comprehensive soldier fitness: The vision of psychological resilience in the US Army. *American Psychologist, 66,* 1–3.

Castaneda, L., Harrell, M., Varda, D., Hall, K., Beckett, M., & Stern, S. (2008). *Deployment experience of guard and Reserve Families.* Santa Monica, CA: Rand.

Chaudhury, S., Chakraborty, P.K., Pande, V., John, T.R., Saini, R., & Rathee, S.P. (2005). Impact of low intensity conflict operations on service personnel. *Industrial Psychiatry Journal, 14,* 69–75.

Cheema, S. S., & Grewal, D. S. (2013). Meditation for stress reduction in Indian Army- An experimental study. *IOSR Journal of Business and Management, 10*, 27–37.

Christian, J., Stivers, J., & Sammons, M. (2009). Training to the warrior ethos: Implications for clinicians treating military members and their families. In S.M. Freeman, B.A. Moore, & A. Freeman (Eds.), *Living and surviving in Harm's Way: A Psychological Treatment Handbook for pre and Post-Deployment of Military Personnel* (pp. 27–47). New York, NY: Routledge.

Clausewitz, C. V. (1976). *On War* (Edited and translated by Michael Howard and Peter Paret). Princeton, NJ: Princeton University Press.

Cornum, R., Matthews, M., & Seligman, M. E. P. (2011). Comprehensive soldier fitness: building resilience in a challenging institutional context. *American Psychologist, 66*, 4–9.

Cranston, A. (1986). Psychology in the veterans administration. *American Psychologist, 41*, 990–995.

Cronin, C. (1998). *Military psychology: An introduction.* Washington, D.C.: Simon and Schuster.

———. (2003). *Military psychology: An introduction.* Boston, MA: Pearson Customs Pub.

Dandeker, C. (2000). The military and social change in the post cold war era: The need for a strategic approach to personnel issues in the armed forces. *Proceedings of the 42nd Annual Conference of the International Military Testing Association, Edinburgh (pp. 1–13).* Retrieved from http://www.internationalmta.org

DeAngelo, J. (2000). *The Link flight trainer: a historical mechanical engineering landmark (Roberson Museum of Science program).* New York, NY: AMSE International.

Defence Institute of Psychological Research.(2005). *Development of computerised pilot selection system.* (DIPR Technical Report). Delhi, India: DIPR.

———. (2007a). *Development of a comprehensive battery of cognitive abilities for selection of candidates for commissioned ranks in the Armed Forces.* (DIPR Technical Report). Delhi, India: DIPR.

———. (2007b). Suicide and fratricide among troops deployed in counter insurgency areas. (DIPR Technical Report). Delhi, India: DIPR.

———. (2008). *Development of new psychological test battery for the selection and trade allocation of other ranks in the Indian Army.* (DIPR Technical Report). Delhi, India: DIPR.

Defence Manpower Data Center. (1999). *Technical manual for ASVAB 18/19 Career Exploration Program* (rev. ed.). North Chicago: HQ USMEPCOM.

Driskell, J. E., & Olmstead, B. (1989). Psychology and the military: Research applications and trends. *American Psychologist, 44*, 43–54.

Drucker, A. J., & Zeidner, J. (1988). *Behavioral science in the army.* Alexandria, VA: US Army Research Institute for the Behavioral and Social Sciences.

Drumm, P., & Ovre, C. (2011). A batman to the rescue. *Monitor on Psychology, 42*(4), 24. Retrieved from https://www.apa.org/monitor/2011/04/batman.aspx

Dusek, E. R. (1980). *Personnel and training research in the military.* In J. Arima (Ed.), Proceedings of symposium on *what is military psychology?* (pp. 15–26). Monterey, CA: Naval Postgraduate School.

Franz, S. I. (1923). *Nervous and mental re-education.* New York, NY: The McMillan Company.

Gal, R., & Mangelsdorff, A. D. (1991). *Handbook of military psychology.* Chichester, UK: John Wiley and Sons.

Gilgen, A. R. (1982). *American psychology since World War II.* New York, NY: Greenwood Publishing Group, Incorporated.

Glass, A. J. (1969). Introduction. In P. G. Bourne (Ed.), *The psychology and physiology of stress* (pp. xiv–xxx). New York, NY: Academic Press.

Gokhale, N.A. (2013). Changing socio-economic norms and its impact on India's armed forces. *Journal of Defence Studies, 7,* 85–94.

Hughes, Hacker, J. G. H. (2007). *British naval psychology 1937–1947: Round pegs into square holes.* Unpublished master's thesis, University of London.

Harjai, M. M, Chandrashekhar, N., Raju, U., & Arora, P. (2005) Terrorism, trauma and children. *MJAFI, 61,* 330–332.

Harrell, T. W. (1992). Some history of the Army General Classification Test. *Journal of Applied Psychology, 77,* 875–878.

Hendrix, W. H. (2003). Psychological fly-by: A brief history of industrial psychology in the US Air Force. *American Psychological Society Observor, 16.* Retrieved from www.psychologicalscience.org/observer/getArticle.cfm?id=1451

Jones, E., & Wessely, S. (2003). Forward psychiatry in the military: Its origins and effectiveness. *Journal of Traumatic Stress, 16,* 411–419.

Keene, J. D. (1994). Intelligence and morale in the army of democracy: The genesis of military psychology during the First World War. *Military Psychology, 6,* 235–253.

Kennedy, C. H., Hughes, J. G. H., & McNeil, J. A. (2012). A history of military psychology. In C. H. Kennedy & E. A. Zillmer (Eds.), *Military psychology: Clinical and operational applications* (pp. 1–17). New York, NY: Guilford Press.

Kirkland, F.R., & Katz, P. (1989). Combat readiness and the army family. *Military Review, 69,* 63–74.

Koonce, J. M. (1984). A brief history of aviation psychology. *Human Factors, 26,* 499–508.

Krueger, G. P. (1998). Psychological research in military setting. In C. Cronin (Ed.), *Military psychology: An introduction* (pp. 15–33). Needham Heights, MA: Simon and Schuster Custom Publishing. Retrieved from http://issuu.com/joselewdw/docs/military_psychology_an_introduction_cronin

Laurence, J. H., & Matthews, M. D. (Eds.) (2012). The *Oxford handbook of military psychology.* New York, NY: Oxford University Press.

Litz, B. T., Orsillo, S. M., Friedman, M. J., Ehlich, P., & Batres, A. (1997). Post-traumatic stress disorder associated with peacekeeping duty in Somalia for US military personnel. *American Journal of Psychiatry, 154,* 178–184.

Martin, J. A., Sparacino, L. R., & Belenky, G. (1996). *The Gulf War and mental health.* Westport, CT: Praeger.

Matthews, M. D. (2012). Cognitive and non-cognitive factors in soldier performance. In J. H. Laurence & M. D. Matthews (Eds.), *The Oxford handbook of military psychology* (pp.197–217). New York, NY: Oxford University Press.

———. (2014). *Head strong: How psychology is revolutionizing war.* New York, NY: Oxford University Press.

Matthews, M. D., & Laurence, J. H. (Eds.) (2012). *Military psychology volume I: Selection, training, and performance.* London, UK: SAGE publications

McFate, M., Damon, B., & Holliday, R. (2012). What do commanders really want to know?: US Army Human Terrain System lessons learned from Iraq and Afghanistan. In J. H. Laurence & M. D. Matthews (Eds.), *The Oxford handbook of military psychology* (pp. 92–113). New York, NY: Oxford University Press.

McGuire, F. L. (1990). *Psychology aweigh: A history of clinical psychology in the United States Navy, 1900-1988.* Washington, DC: American Psychological Association.

Melton, A. W. (1957). Military psychology in the United States of America. *American Psychologist 12,* 740–746.

Military Psychology. (n.d). *Wikipedia.* Retrieved from http://en.wikipedia.org/wiki/Military_psychology

Misra, N., Asnani, V., & Archana. (2006). *Manual on stress and its management.* Delhi, India: DIPR.

Mukherjee, S., Kumar, U., & Mandal, M. K. (2009). Status of military psychology in India: A review. *Journal of the Indian Academy of Applied Psychology, 35,* 181–194.

Orsillo, S. M., Roemer, L., Litz, B. T., Ehlich, P., & Friedman, M. J. (1998). Psychiatric symptomatology associated with contemporary peacekeeping: An examination of post-mission functioning among peacekeepers in Somalia. *Journal of Traumatic Stress, 11,* 611–625.

OSS Assessment Staff. (1948). *Assessment of men.* New York, NY: Rinehart.

Page, G. D. (1996). Clinical psychology in the military: Developments and issues. *Clinical Psychology Review, 16,* 383–396.

Patnaik, P., Dhawalgi, S., & Mandal, M. K. (2007). *Training manual on deceit detection and interrogation.* Delhi, India: DIPR.

Phares, E. J., & Trull, T. J. (1997). *Clinical psychology: Concepts, methods, and profession (5th ed.).* Pacific Grove, CA: Brooks/Cole.

Prabhu, H.R.A., Prakash, J., Bhat, P.S. & Gambhir, J. (2011). Study of events in serving personnel and its association with psychopathology in their children: A multicentric study. *MJAFI, 67,* 225–229.

Puri, S. K., Sharma, P.C., Naik, C. R. K., & Banerjee, A. (1999). Ecology of combat fatigue among troops engaged in counterinsurgency operations. *MJAFI, 55,* 315–318.

Raju, M. S. V. K., Srivastava, K., Chaudhury, S., & Saluja, S.K. (2001). Quantification of stressful life events in service personnel. *Indian Journal of Psychiatry, 43,* 213–218.

Raju, M. S. V. K., & Singh, N. P. (2006). *Combat stress behaviours in LIC environment: principles and management.* Delhi, India: DIPR.

Ritchie, E. C. (2002). Psychiatry in the Korean War: Perils, PIES, and prisoners of war. *Military Medicine, 167,* 898–903.

Roscoe, S. N. (1997). The adolescence of engineering psychology. *Human Factors History Monograph Series, 1.* Retrieved from www.hfes.org/Publication Maintenance/FeaturedDocuments/27/adolescencehtml.html

Ryali, V.S.S.R., Bhat, P.S., & Srivastava, K. (2011). Stress in the Indian Armed Forces: How true and what to do? *MJAFI, 67,* 209–211.

Sager, C. E., Van Iddekinge, C.H., & Russel, T. L. (2004, October). *Select 21 Project Predictor Measures.* Paper presented at the 46th Annual Conference of the International Military Testing Association, Brussels, Belgium.

Sainik Samachar. (2007). *Defence Institute of Psychological Research: Selecting Mr. Right.* Retrieved from http://sainiksamachar.nic.in/englisharchives/2007/jul01-07/h15.htm

Saldanha, D., Goel, D.S., Kapoor, S., Garg, A., & Kochhar, H.K. (1996). Post-traumatic stress disorder in polytrauma cases. *MJAFI, 49,* 7–10.

Salter, C. A. (2001). Psychological effects of nuclear and radiological warfare. *Military Medicine, 166,* 17–18.

Scales, R. H. (2009). Clausewitz and World War IV. *Military Psychology, 21* (Suppl. 1), S23–S35.

Seligman, M. E. P., & Fowler, R. D. (2011). Comprehensive soldier fitness and the future of psychology. *American Psychologist, 66,* 82–86.

Shaffer, L. F. (1944). *The psychology of adjustment: An objective approach to mental hygiene.* Washington, DC: Houghton Mifflin, for the United States Armed Forces Institute.

Singer, M.J., Long, R., Stahl, J., & Kusumoto, L. (2008). *Formative evaluation of a massively multi-player persistent (MMP) environment for asymmetric warfare exercises* (ARI Technical Report 1227). Arlington, VA: US Army Research Institute for the Behavioral and Social Sciences.

Singh, N. P., Tripathi, D. N., & Asnani, V. (2007). *Interrogation techniques in counterinsurgency operations: Psychological perspectives and guidelines.* Delhi, India: DIPR.

Stanley, J., Segal, M., & Laughton, C. (1990). Grass root family action and military policy responses. *Marriage and Family Review, 15,* 207–223.

Stearns, A. W., & Schwab, R. S. (1943). Five hundred neuro-psychiatric casualties at a naval hospital. *Journal of the Maine Medical Association, 34,* 81–89.

Stouffer, S. A., Suchman, E. A., DeVinney, L. C., Star, S. A., & Williams, R. N. Jr. (1949). *The American soldier: Adjustment during army life.* Princeton, NJ: Princeton University Press.

Tryon, R. C. (1963). Psychology in flux: The academic-professional bipolarity. *American Psychologist, 18,* 134–143.

US Departments of the Army and the Air Force (1951). *Military clinical psychology, technical manual, TM 8-242, Air Force manual, 1600-45.* Washington, DC: Author.

Uhlaner, J. E. (1967, September). *Chronology of military psychology in the Army.* Paper presented at the Annual Convention of the American Psychological Association, Washington, D.C.

———. (1978). *The research psychologist in the army-1917 to 1973.* (Research Report 1155), Alexandria, VA: US Army Research Institute for the Behavioral and Social Sciences.

Ursano, R. J. & Rundell, J. R. (1995). The prisoner of war. In R. Zajtchuk & R. F. Bellamy (Eds.), *Textbook of military medicine: War psychiatry* (pp. 431–455). Washington, DC: Office of the Surgeon General, US Department of the Army.

Vallance, T. R., & Windle, C. (1962). Cultural engineering. *Military Review, 2,* 60–64.

Verma, R., Mina, S., & Deshpande, S. N. (2013). An analysis of paramilitary referrals to psychiatric services at a tertiary care center. *Industrial Psychiatry Journal, 22,* 54–59.

Walters, H. C. (1968). *Military psychology: Its use in modern war and indirect conflict.* Iowa: Brown Company Publishers.

Windle, C., & Vallance, T. R. (1964). The future of military psychology: Paramilitary psychology. *American Psychologist, 19,* 119–129.

Yerkes, R. M. (1918). Psychology in relation to the war. *Psychological Review, 25,* 85–113.

Zeidner, J. & Drucker, A. J. (1988). *Behavioral science in the Army: A corporate history of the Army Research Institute.* Alexandria, VA: US Army Research Institute.

2

Intelligence and Aptitude Testing

Gurpreet Kaur, Dinika Anand and Soumi Awasthy

Selection and placement of military personnel were among the earliest concerns of the Armed Forces. At the onset of First World War (WWI), effective selection and utilization of military personnel were key interest areas for those responsible for creating combat units. It has been in existence since times immemorial although the methods have varied from arbitrary and quick to detailed and lengthy. Selection of monks and priests through dreams, selection of a king by the method of elephant garlanding are examples of arbitrary selection. However, the evolving times and progress in domain of education and knowledge created a need for replacing arbitrary selection system. Appointment by heredity also lost its popularity and people craved for a system based on evaluation of suitability. This paved the way for evolvement of scientific selection system. Today, the importance and utility of psychological assessment in the selection of military personnel has increased manifold.

In the course of history, different criteria have been used for selection and placement of military personnel and intelligence testing has always been a focal point. Modern intelligence testing started with the work of Alfred Binet, who is hailed as the father of intellectual assessment. Binet in 1903 used the term 'intelligence' to refer to sum total of higher mental processes and in 1905, on the request of French Minister of Public Instruction undertook the task of developing a reliable diagnostic system to identify children with mental retardation. The first Binet–Simon scale was completed in 1905, followed by many revisions, with the latest one being Stanford Binet Test 5 (SB 5). SB 5 assesses individuals' fluid reasoning, knowledge, quantitative reasoning, visual-spatial processing and working memory with verbal and non-verbal tests (Roid, 2003). Binet–Simon testing propelled the study and measurement of intelligence into its current central position in the discipline of psychology.

Binet's work and its American adaptations by Robert Yerkes set the impetus for considerable growth prior to WWI in the field of intelligence testing. The combination of Army Alpha, a verbal test, and Army Beta, a performance test were widely used by the Armed Forces in the United States for intelligence testing. Thereafter, a number of tests were developed such as group intelligence scale (Otis, 1918a, 1918b, 1918c), Terman group tests of mental ability (Terman, 1916) and revision of army alpha test (Hendrickson, 1931).

Selection of Indian Military Personnel

In India, before Second World War (WWII), officers for the Armed Forces were selected based on a competitive examination and an interview conducted by the federal public service commission. The number of Indian officers annually recruited before the war was relatively small and no difficulty was experienced in filling the vacancies in this manner (PRW note no. 2, 1949). Recruitment to regular commissions was closed and there was a system of granting emergency commissions after the outbreak of the war. This lead to the increased demand for officers; thus, the recruitment of officers for the three services, Army, Navy and Air Force, was taken over by the respective service headquarters, and in 1942, recruitment of Armed Forces officers to the three services was integrated at directorate of recruiting in the Adjutant General's Branch at the General Headquarters (PRW note no. 2, 1949).

After this integration, the selection of candidates was conducted in two stages, that is, preliminary selection by provincial selection board and final selection by the central interview board. Although all officers for the Army were recruited through this integrated organization, the Royal Indian Navy continued to take in a number of direct entry candidates under their own arrangements and the Royal Indian Air Force appointed a few general duty recruiting officers who were authorized to recommend candidates directly to the central interview boards. But, gradually, it was noticed that a large proportion of the candidates selected by the method of personal interview failed to qualify at training schools. Subsequently, after discussions it was decided to establish a separate selection board at Dehradun in 1943 to apply scientific method of selection as an experimental measure. A small section to deal with the new technique of selection was also created in the recruiting directorate of the General Headquarters. The new system gained approval from the candidates who appeared before the two boards.

The Ghosh Committee was appointed in 1948 by the Government of India to evaluate the system of selection. The Committee noted the room for improvement in the prevailing protocol, stating that scientific studies might be conducted to bolster the existing processes. Thus, the Psychological Research Wing (PRW) came into being in 1949, and in August 1962 it was developed as Directorate of Psychological Research (DPR). Further, DPR came to its present status in January 1983 as Defence Institute of Psychological Research (DIPR). The only major modification introduced to date by the committee itself was the elimination of the interview by a psychiatrist. This system of selection has been able to cope with the requirements of the Indian Armed Forces.

The main advantage of psychological method of selection is that it does not merely differentiate the very good from the very bad, but it can also classify those in between in order of merit. A system has been adopted to assess these abilities and to relate these to the requirements of the job. The selection into the Armed Forces is based on various psychological parameters such as intelligence and personality. These two are believed to be contributing most to individual differences, which commensurate with any kind of selection. Thus, presently personnel selection includes assessment of intelligence, personality and aptitude testing for placement. The present chapter focuses on intelligence and aptitude testing for the Armed Forces.

Intelligence Testing for Selection

Intelligence is undoubtedly an immensely popular, complex and debated topic not only in the domain of psychology but also in daily discourse. Over time, the concept of intelligence has become more expansive and moved beyond the realms of IQ. Intelligence has been defined in different ways by scholars and researchers over the years. Burt (1957) opined that "intelligence is a quality that is intellectual and not emotional or moral: in measuring which the effect of the child's zeal, interest, industry, and the like are ruled out". In a similar vein, "intelligence is defined as composite of several functions that denote the combination of abilities required for survival and advancement within a particular culture" (Anastasi, 1992). Howard Gardner is credited with the theory of multiple intelligences and offers that "intelligence is the ability to solve problems, or to create products that are valued within one or more cultural settings" (Gardner, 1983). However, despite the numerous definitions of intelligence offered, the most popular definition is that of David Wechsler. Wechsler (1958) opines that, "intelligence is a global concept that involves an individual's ability to act purposefully, think rationally and deal effectively with the environment".

The selection system in the Armed Forces initially relied heavily on different types of intelligence tests. The selection protocol comprised of intelligence tests based on verbal, non-verbal and performance-based measures, as well as tests amenable to individual and group administration. Generally, only verbal and non-verbal measures of intelligence were administered to the candidates and performance-based measures were employed only if there was a doubt (PRW note no. 2, 1949). Candidates who applied for technical arms such as the Royal Indian Engineers and Indian Electrical and Mechanical Engineers had to take another test assessing basic knowledge of science subjects.

The focus of the verbal and nonverbal tests was primarily reasoning ability. It included analogies, classification, logical reasoning, abstract reasoning, coding/decoding, matrices, picture completion, speed and accuracy, numerical ability, etc. Some of these abilities are described in Table 2.1.

All the above-mentioned abilities have been assessed using various intelligence tests from time to time based on the requirements of the Armed Forces. A few of the intelligence tests used are explained as follows:

1. **Matrix 38:** This is one of the progressive matrices tests that provides a non-verbal series suitable for measuring intelligence. Each problem in the test consists of a design or matrix from which a small part has been removed and candidate has to tell the missing part (PRW note no. 2, 1949).
2. **S.P. Test 15 (V.I.T.):** This is a verbal intelligence test of the omnibus type in which questions based on reasoning analogy and number series, etc., have been included. Speed and accuracy of thought are required besides the reasoning capacity. Here,

Table 2.1
Definitions of abilities used in intelligence tests

Ability	Definition
Analogies	The skill of being able to transfer information from one concept to another.
Classification	The ability to organize, differentiate and understand given information.
Logical reasoning	The skill to deduce, infer and conclude based on the given information.
Abstract reasoning	The capability of an individual to use current information to make logical inferences and deductions about abstract concepts.
Fluid reasoning	The ability to think logically and solve problems in novel situations.
Quantitative Reasoning	Ability to understand numeric relationships and compute simple arithmetic functions.
Coding/decoding	An individual's skill to convert present stimuli into another form based on given set of rules and norms.
Matrices	An individual's ability to understand and integrate the information presented along with abstract reasoning and choosing the missing figure.
Speed and accuracy	The ability to perform on tasks with efficacy wherein the success is measured by a sum total of speed, accuracy and their interlinking.
Visual-spatial processing	Ability to organize and understand information into meaningful patterns and wholes based on the visual and other stimuli.

Source: Authors.

since intelligence is tested through the medium of language it is known as a verbal intelligence test or the V.I.T.

3. **I.S.P. Test 20:** It is a verbal intelligence test like S.P. Test 15 which distinguishes itself by presenting the items under separate headings, namely, analogy, opposites, mixed sentences, classifications, number series and coding and the like as opposed to an omnibus format.

4. **S.P. Test 2 (Bennett):** The test assesses knowledge of everyday science. The test booklet consists of 50 questions in the form of pictures along with an answer sheet in which questions are printed, each question corresponding to the picture of the same number in the booklet.

5. **S. P. Test 3-A:** It is a simple test of mathematics, combining speed with accuracy and quick thinking. It consists of two parts. Part I

comprises of questions on addition, subtraction, multiplication and division, whereas part II includes questions on arithmetic, algebra, geometry and trigonometry.

6. **Verbal Intelligence Test:** It includes problems related to verbal analogy, verbal comprehension, vocabulary, similarities letter/number series, number/letter classification and scrambled words.

7. **Non-verbal Intelligence Test:** The test is further divided into four subtests. They are figure analogy, figure series, figure classification and spatial relations.

Aptitude Testing for Placement

Aptitude can be defined as a condition or set of characteristics regarded as symptomatic of an individual's ability to acquire knowledge or skills with training. It is a pattern of traits, deemed to be indicative of one's potentialities and can be viewed as a readiness to acquire proficiency. In essence, aptitude is a construct which provides us with a forward reference based on an evaluation of current ability. An aptitude test samples certain abilities and characteristics of the individual as he is today. It helps to find out what he can do now and how well he can do it. Aptitude test responses offer an estimate of future possibility of accomplishment.

According to Freeman (1965), an aptitude is a combination of characteristics indicative of an individual's capacity to acquire (with training) some specific knowledge or skills such as the ability to speak a language, to become a musician or to do mechanical work. Moreover, aptitude may also mean aptness or quickness, because some people have practiced diligently for years under the best tutorship and still do not achieve what others appear to do so within less time and effort. The terms, special ability and talent, are frequently used synonymously with the term aptitude. Aptitudes are natural talents, special abilities for doing or learning to do certain kinds of things easily and quickly. Musical talent and artistic talent are examples of such aptitudes.

Aptitude tests are generally grouped into two categories, special aptitude tests that measure only one aptitude and multiple aptitude tests which consist of a set of tests that measure different aptitudes. An illustration of a special aptitude test would be a test that measures mechanical aptitude or electrical aptitude alone. The musical aptitude test (Seashore, Lewis, & Saetveit, 1939) is a special aptitude test. At the other end, multiple aptitude tests combine a set of separate tests together such that the individual tests measure relatively different independent abilities. This helps in maximum utilization of available manpower pool, wherein the same battery of tests, weighted in different combinations, provides predictive indexes for each applicant for several jobs. The Differential Aptitude Test, first published in 1947, is a popular and much used multiple aptitude test (Bennett, Seashore, & Wesam, 1974).

DIPR had developed various aptitude batteries during the last 50 years (PRW note no. 22, 1950; DPR note no. 167, 1957; DPR note no. 271, 1966; DPR note no. 321, 1973; DPR note no. 326, 1973; DIPR note no. 419, 1985; DIPR note no. 445, 1989; DIPR note no. 453, 1991; DIPR note no. 458, 1991; DIPR note no. 476, 1995; DIPR note no. 477, 1995; DIPR note no. 478, 1995; DIPR note no. 514, 1997; DIPR note no. 537, 1999), but they have essentially been paper-pencil and performance tests. DIPR has recently developed a fully computerized Other Rank Trade Allocation System (ORTAS) for allocation of 78 jobs of eight arms and services of Indian Army (DIPR note no. 640, 2010).

One of the aptitude battery developed by DIPR was for placement of leading radio operators into three branch specializations—telegraphy, special and tactical. The aptitude battery comprises five tests, namely, Verbint (for verbal intelligence), LAT (for language aptitude), VMT (for associate memory), SMT (for expressional fluency) and Coding (for associative learning) (DIPR note no. 386, 1982). Brief description of the same is as follows:

1. **Verbint:** This is a test of verbal intelligence. It consists of four types of questions, namely, verbal analogy, letter/number series, letter/number classification and scrambled words.
2. **Language Aptitude Test (LAT):** The test comprises of two sub-tests in which the first involves learning a new vocabulary based on their provided English equivalents. In each item pair, the first word is a new word and the second word is its equivalent in English. After giving time to memorize the English equivalents, the test taker is presented with a list of the same 20 new language words arranged in a different order and is asked to write the English equivalents. The second sub-test consists of presenting a new set of 20 words which the test taker is supposed

to memorize. The test given to the person is the same as in the subtest one but with a difference that in this case, for each new language word, five choices (English equivalents) are given to indicate the answer. This part of the test can be considered as a test of recognition. The test taker completes sub-tests one and two and then is provided with an answer sheet to test his learning on the language learned in sub-test one. After completion of which he is asked to answer sub-test two in the answer sheet.

3. **Visual Memory Test (VMT):** It measures associative memory or how quickly and correctly one can memorize codes. This test has been standardized on the sailors of communication branch of Indian Navy (DIPR note no. 386, 1982). A code is an arrangement of dots and dashes. Each code stands for an English alphabet. For example, five dots (. . . . .) represents A and the code five dashes (_ _ _ _ _) represents B. Letters from A to Y have been given specific codes in this test. The test taker is shown the memorization chart I on which appear the first 10 letters of the alphabet along with the respective codes. The chart is shown for two minutes during which the test taker tries to learn the association between the codes and the letters. Then, they are shown 20 cards one by one, each card having a code on it. Each card is shown for 10 seconds during which the test taker recalls the letter associated with the code and mentions it against the appropriate serial number in the answer sheet.

4. **Sentence Making Test (SMT):** It measures expressional fluency. In this test, each item consists of three to five letters of English alphabet in a particular form, for example

   B. . . . . .f. . . . . .n. . . . . .a. . . . . .a

   The test taker is told that the first letters of new words are given and that he should try to make connected words to form a meaningful sentence. Proper nouns should not be used and only the required number of words needs to be written.

5. **Coding Test:** This test is a measure of associative learning and quickness in coding. In this test, the test taker is given the codes and the letter of English alphabets associated with them, wherein the codes are different combinations of dots and dashes. For example, - - is E, 0-0 is F and so on. Twenty items are given with each item consisting of six letters. The task is to refer to

the codes given and write down the codes for the letters given separately in the answer sheet.

An aptitude battery for Artificer Apprentices was also developed at DIPR (DIPR note no. 477, 1995). Artificers are engaged in servicing, maintenance and rectification of all types of equipment used in ships, submarines and air-crafts. Following are the tests involved in the aptitude battery:

1. **Intelligence Test (Non-verbal):** This test measures the individual's ability to do abstract thinking, analysis and see relations among different objects. There are four types of items used in this test namely, classification, series, analogy and matrices.
2. **Space Relations Test (Non-verbal):** A test of space relations (paper–pencil type) measuring candidate's ability to judge shapes and sizes, the relations of objects in space and to manipulate them mentally and visualize the effect of putting them together or turn them around was developed. The objects may be seen in two as well as in three dimensions. Two types of items were used in this test, namely, form board (paper pencil type) for two-dimensional objects and space relations (paper pencil type) for three-dimensional objects.
3. **Mathematical Reasoning Test:** In this test, items consist of common mathematical problems. They require ability to reason with numbers, to manipulate numerical relations and to deal intelligently with quantitative materials. It also involves ability to solve problems of reasoning in general.
4. **Mechanical Information Test:** The test measures the individual's knowledge of principles of mechanical engineering.

Paradigm Shift in Testing

The 21st century is marked by the Armed Forces being involved with innovations in communication systems, weaponry and equipment cutting across the diverse missions. All these changes make it imperative to move away from the status quo in testing procedure (Rumsey, 2012). A dynamic, fluid system of selection and assessment which constantly adapts and advances in response to the demands of the Armed Forces is crucial. Non-conventional warfare and low-intensity conflict scenarios have replaced conventional warfare and wars today cannot be won in the physical domains, but can be conquered only through application of cognitive science, that is, how we perceive, feel, think and decide. Cognitive science has gained its due importance in recent years. The scale and diversity of tasks carried out by the military offer extensive opportunity for the application of cognitive science. With increased dependence on information technology by the military and a conscious paradigm shift to Network Centric Warfare (NCW), the quantum of information available to military commanders is increasing exponentially. The continuing growth of technology has made weapon system reach new heights of sophistication and today the high-tech operations are faster, more integrated and more complex, requiring much more than the conventional skills of a soldier. Assessment of thinking and decision-making speed, information processing speed and high level of communication skills have become mandatory. Various cognitive skills include a wide range of mental processes, for example, perception, memory, imagery, language, concept formation, problem solving, reasoning and decision making. Thus with this change in military operations and warfare scenario, there is a necessitated paradigm shift from intelligence testing to cognitive assessment.

Paradigm Shift in Construct Assessment

With respect to underlying construct being assessed, there is a broad distinction between intelligence and personality testing, but now there are tests developed to measure finer and specific abilities even within these domains. These changes mostly echo the dynamism of the current military operations and scenario. Moreover, the construct being measured by a test is defined in specific contexts for specific purposes. Thus, two tests that claim to measure the same construct may actually differ in a significant manner. For example, a test measuring courage for induction into military may measure a different underlying construct altogether as compared to a test of courage developed for a clinical population. The infinite variety of psychological test contents that are available today is overwhelming and adds to the complexity of the decisions involved in testing.

As mentioned earlier, in testing, construct being assessed earlier was intelligence and today the focus has shifted to cognitive testing. Cognition refers to mental activities, more specifically the process of knowing. It involves how one acquires, stores, retrieves and uses knowledge. Cognitive science investigates how information is processed (in faculties such as perception, language, memory, reasoning and emotion), represented and transformed in behaviour (human or other animals). It spans many levels of analysis from low-level learning and decision mechanisms to high-level logic and planning, and from neural circuitry to modular brain organization.

DIPR has been spearheading the movement for creating intelligence, aptitude tests and assessment batteries that are reflective of the paradigm shift in the Indian military context. The task of developing selection protocols which delve into cognitive abilities that comprise intelligence and incorporate contemporary understanding and knowledge has been accomplished by DIPR. Studies were taken up by DIPR to assess the general cognitive ability which includes comprehensive assessment of cognitive abilities for officer selection and also a computerized cognitive battery for officer selection. Comprehensive Battery of Cognitive Abilities (CBCAs) determines an individual's competence and levels of cognitive functioning and constitutes three levels, that is, registration, processing and higher order functioning (DIPR note no. 634, 2009). Thus, it can be adjudged as an efficient tool that meets the demands of the job and helps in assessing competent potential for the Indian Armed Forces. DIPR has also developed cognitive batteries for selection based upon three-stratum model of cognitive ability (Carroll, 1993). The cognitive battery assessing 'g' measures individual's fluid intelligence, crystallized intelligence, general memory ability, visual perception, retrieval ability, cognitive speediness and perceptual speed (Carroll, 1993).

Besides shift in assessment of selection tests, aptitude assessment has also undergone a major change. DIPR has now developed computerized aptitude batteries for various arms and services of Indian Army for allocation of trades. ORTAS, the psychological trade allocation battery is useful in allocating 78 trades to the recruits as per their aptitude, in eight arms and services of Indian Army, namely, Armored Corps, Regiment of Artillery, Mechanized Infantry, Corps of Engineers, Corps of Signals, Corps of EME, Army Service Corps and Army Air Defence. Eleven cognitive and two psychomotor tests were developed in it, namely, observation, spatial ability, form perception, perceptual speed, memory, English knowledge, visualization, reasoning, mechanical knowledge, visual discrimination, alertness and eye hand coordination (DIPR note no. 640, 2010).

Besides ORTAS, DIPR has also developed Army Medical Corps Trade Allocation Battery (AMC-TAB) for Army Medical Corps (AMC), which would also help in allocation of 13 trades of AMC (DIPR note no. 663, 2013).

Paradigm Shift in Mode of Assessment

There is also a change in the mode of assessment as contemporary tests move away from being limited to the paper–pencil format for administration or restricting to one kind of measure only vis-a-vis prevalence of multidimensional approach in testing. The tests developed today offer a comprehensive and multi-modal profile of the candidate encompassing performance on verbal, non-verbal and performance-based questions. Most of the tests developed in last five years are computerised. This shift is partly motivated by the need to assess the ability of soldiers to use the complex, technologically advanced weaponry that is available today. Computerized assessment facilitates in assessing one's reaction time in real time. Time fixation of tests has become more precise by computing reaction time of each candidate on each question. Tests have become more speed-oriented wherein one's decisionmaking is also assessed. Various computer accessories such as joy sticks, headphones and customised keyboards are being used at present to assess different sensory modalities of the candidate, which further facilitate appropriate assessment of candidates' cognition.

Paradigm Shift in Technique of Analysis

Item analysis in the early years was based on classical test theory which consists of analyzing the overall performance on the entire test instrument as a whole rather than delving in the nuances of item responses. These methods only allow for generating group-specific item statistics. However, with the advent of computerized testing the use of item response theory has become more popular. IRT refers to a set of mathematical models that describe, in probabilistic terms, the relationship between a person's response to a test item and his or her level of the 'latent variable' being measured by the scale. In IRT, besides difficulty and discrimination level of each item, guessing factor and ability score (person level statistics) is also assessed. IRT also facilitates computer-adaptive testing in which items are selected by the software based on the ability of the test taker.

Context-specific Selection

The validity of any research is determined by the measure being used, if the measure is questionable, the research carries no meaning. Job analysis-based test construction plays a vital role in the Indian military context because environment in which the Indian Army, Navy and Air Force operate is very different. It is possible that a gunner may require different abilities to perform in different services of Armed Forces besides the core abilities. As an illustration, the gunner posted on a naval ship uses cues that are distinct and unique to his station as opposed to those used by his counterpart on the ground. Job analysis provides answers to questions about the context in which the candidate shoots, types of guns or equipment he uses and the operations he performs and thus development of the test takes into consideration all this information. Job analysis includes need analysis and profile analysis to find out which are the abilities and skills required to deal with challenges in different contexts.

Different methodologies are required to construct a scientifically sound test for distinct and varied jobs. It is basically the interface between environment and person, which becomes the basis for job-based test construction. If the researcher intends to develop a test to select a candidate who can work effectively under water, then he should know almost everything about the underwater context. In-depth job analysis can aid in finding most effective abilities required for a candidate for underwater activities. On the contrary, to construct a test to choose candidate for high-altitude warfare, job analysis paves the way for a very different methodology as compared to the former because abilities and competencies required would be different. Thus, the tests developed in this manner are valued for their intensity and depth, but at the same time the researcher faces many challenges in developing such job-specific tests. Some of the challenges are as under:

OPERTIONALIZATION

Defining the construct and having an operational definition is fundamental for test construction. Operational definition should be very precise and ensures comprehensive knowledge of the terminology. It is important that everyone has the same understanding of the construct. Unless the researcher defines the meaning of the construct and its assessment in an operational manner, validity of the test cannot be established. Operationalization of the construct becomes all the more relevant when researcher has to assess some ambiguous and novel domain. This is a challenge that a social researcher confronts very frequently because they need to have agreement about the construct being assessed at the first place. Many tests do not get published because of unclear definition of the construct being assessed.

WRITING THE ITEMS

Comprehensive job analysis and operationalization of the construct paves the way for writing items that are capable of assessing the required skill or ability for a job. A critical aspect while writing the items is that no item measures a single ability, for example if a test item has been developed to assess memory then at the first step one requires reading ability to read the instructions and also requires some amount of psychomotor control to operate the mouse or the touch screen or the keyboard to click the right answer. Therefore, an item is likely to assess a major ability for which it is designed but at the same time, the same item may also measure other abilities, although not with that high magnitude. At this juncture, the researcher also needs to identify the format of the test, that is, paper–pencil or computerized format, individual test or group test, speed or power test and also verbal test or non-verbal test.

DESIGNING AND WRITING THE RESPONSES

A good item that measures the construct suitably constitutes not only well-constructed stems but also scientifically constructed response options. To have a good item, the researcher can combine open-ended and close-ended formats together. Initially, few stems assessing desired concept can be given to a said population. It can be in the form of situation reaction test (SRT) where candidates are asked to write the responses or reactions to the situations. Alternatively, the same stems are also given with response options. These are in the form of situation judgment test (SJT). SJT is in the form of multiple-choice questions wherein the candidate chooses the desired option. After taking responses from adequate sample on both formats that is SRT and SJT, responses are content-analyzed and the best responses are accepted. Best responses here can be tailored as per the requirement of the job for which test is being constructed.

Statistical Procedures for Job-based Test Construction

Computing reliability, validity and standardization are mandatory for any scientific test. Reliability is computed broadly for two reasons, either to establish internal consistency or for assessing stability over time. There are different methods to measure reliability, but selecting the right method and type of reliability depends upon the type and rationale of the test being constructed. The tests developed for the military user mostly fall under the domain of job-based tests as they are driven ideologically by the job descriptions and charter of duties that govern the various ranks and posts of the Armed Forces. Validity measures of job-based test are difficult due to the absence of established criterion against which comparisons may be drawn and thus, criterion validity is computed using external criterion like performance marks which are easily available. Standardizing the test, that is, setting the norms is an important step in the job-based test construction. Test construction is not complete without establishing the norms. Norms are quite flexible depending upon the requirement of the user or job. The cut-off for screening tests can be tailored as per the requisites of the job.

Way Ahead in Selection and Placement: Use of Computers

The advent of computers and technology has reshaped the way assessments are designed and delivered. The easy availability of computerized technology, its simplicity of use and applicability to global assessment practices have changed the landscape of selection and assessment. Computers have helped to reduce the extent of human interference and bias in testing at one end and modified the administrative processes tremendously at the other end (Kaur & Mukherjee, in press). The use of computers is also seen in scoring and analysis of the data gathered as evidenced in the plethora of data analytic tools available.

Technological advancement has most significantly altered presentation and appearances of test items and the response modes possible in testing and selection. Vast reservoirs of different types of items and response mode that greatly increase the validity of measurements are now available. Computers today can be used for creating models and procedures for scoring the responses of test-takers attempting different versions of the same test. They are used to assist in the selection and assessment process at different levels as discussed above (Bennett et al., 1997). At one level, the computer system is mainly used for recording the responses of the subjects electronically. Another situation in which computers are used is when part of the testing protocol is delivered through the machine. An illustration would be a situation in which the subject is made to view visual stimuli on the computer screen and then record his or her responses on an answer booklet provided. In some situations, the assessment protocols may include some computer-based questions along with other questions that are in the paper–pencil format.

Computer-aided tests provide recording of very detailed information about the test-taker's performance (Bartman, 2006). The physiological parameters in terms of GSR, non-verbal cues, eye movement and other such information can also be recorded today. Immersive testing experiences are capable of recording information about the process of arriving at a decision and the steps through which a person works during the sub-tasks while solving the assigned problem.

A special class of computer-based testing is *computer-adaptive testing* or *calibrated testing* in which the items are moderated according to the skill and responses of the respondent. The basic premise is the tailoring of testing protocol to each user by changing the difficulty level of questions. As an illustration, an incorrect response leads to the next item being easier, whereas a correct response increases the difficulty level of the next item (Wainer et al., 2010). Each test taker in this fashion obtains a true score which reflects his true ability on the attribute being measured (Rumsey, 2012). The format offers maximum item flexibility because the items constituting the test for each subject are always different and tailored eliminating the possibility of the subject being disheartened, discouraged or bored at any point. Also, the scores reflect the nuances of the ability level due to the item difficulty level matrix.

Computer-adaptive testing also involves a class of branched or multi-step design in which the responses on initial questions determine which of the available pathways will be followed during the remainder of the test. Drasgow, Olson-Buchanan and Moberg (1999) in a case study of a test opined that structured upon branches and the response of the test taker form the route that is followed through the branches. Each respondent starts the test by watching a video and answering questions based on it, but the responses decide the further course of the test in terms of which video is played next and so on.

The introduction of computers in testing also led to the development of *multimedia based tests*. Multimedia by definition stands for content in which the material presented is not limited to one particular sensory modality but involves a combination of different modalities. An illustration would be a test situation in which the examinee is supposed to answer a question based on a video clipping shown or a situation in which certain tasks are to be performed by integrating the information presented on the screen, as well as that being played through the headphones.

Media elements have the advantage of being highly dynamic, and congruent with situations the test taker may face outside of the assessment situation. Multimedia-based testing environments generally require the user to interact with the system more than the requirement of the multiple-choice format. The user may have to drag and drop certain objects or sometimes drag text or graphics to their appropriate locations on the screen. A key advantage of these tests is that they bridge the gap between the natural context and the laboratory setting offering a way to assess the perceptual skills of the test taker also.

Virtual Reality (VR) and Augmented Reality (AR)

These are two key areas that are increasingly being used in the assessment and testing environment in contemporary times. These facets of technology will form the benchmark of any assessment or testing protocol in the future due to their ability to create an extremely close and true approximation of the natural life experience of the test taker (Reynolds & Rupp, 2010). The use of computerized testing stands on the ground of reducing the gap between the artificial test environment and the real-world experience, and this gap is going to become negligible as researchers move further along the road of VR and AR experience.

Current trends in VR and AR include immersive video technologies which allow the possibility to 'navigate' within a video, exploring the scenario in all directions while the video is running. The VR technology may still be in nascent stages, but it is a promising area and thus, a future in which the current constraints pertaining to the quality, cost and availibility of VR are eliminated can be envisioned (Saggio & Ferrari, 2012).

VR, AR and simulation-based testing allow testing of actual behavioural response patterns that may be evoked in critical situations at work. This kind of testing is particularly useful for highly specialized, skill-intensive and stressful professional situations. Hanson et al. (1999) describe the development of a computer-based performance measure for air traffic controllers' selection. Similarly, Bartram and Dale (1983) describe the use of a simplified landing simulator for use in pilot selection. Vora et al. (2002) found that VR system was better and preferred over the PC-based training tool by the aircraft inspectors.

The widespread availability of the Internet and the progress in Internet technology in terms of hardware, bandwidth, speed and so on have also increased its use and applicability for the field of testing and assessment. Bartman (2005) classifies four different types of test administration over the Internet-based on the level of interaction between the test taker and the test administrator. While at one end in *open mode,* the test taker is unknown and there is no direct supervision, the *controlled mode* involves a test taker who has been identified taking the test in an unsupervised environment. The *supervised mode* involves a certain level of human supervision with the identity of the test taker and the test taking conditions being controlled and authenticated by the administrator/supervisor. Last, in *managed mode,* a high level of direct human supervision is assumed and there is also the need for control over the test-taking environment. These two conditions are achieved through the use of dedicated testing centres.

The popularity of Internet-based tests is also supported by a significant body of research that points towards the equivalence of results when tests are administered in the paper–pencil format and also through the Internet. In one such study, Preckel and Thiemann

(2003) compared the Internet and paper administration of a figure matrices test. They found that both test versions were comparable with regard to the contribution of item design features to task study. Salgado and Moscoso (2003) reported high levels of congruence between paper and Internet versions of a Big Five instrument, with coefficients ranging from 0.93 to 0.98 for the five scales. DIPR has developed Internet-based self-assessment tool for assessing one's military aptitude. This test assesses one's potential for Indian Armed Forces and provides instant feedback to the candidate about the same (DIPR note no. 626, 2007).

Advantages of Computer-Based Testing

1. It increases the efficiency of assessment delivery process and reduces the amount of time needed for administration of tests.
2. It saves resources utilized in printing and type-setting test material, as well as the potential errors that these procedures involve.
3. The inclusion of stimuli of different modalities such as auditory, visual and mixed stimuli becomes possible, thereby making the assessment experience more true to life (Basu, Cheng, Prasad, & Rao, 2007).
4. Computers facilitate in enhancing the precision of measurement because of the advanced administration techniques.
5. Item randomization, parallel form administration and other techniques to improve the security of the test and consequently the accuracy of the test are made accessible to the administrator.
6. Variables which are difficult to capture in paper–pencil tests such as eye movement or galvanic skin response can also be recorded and measured easily today with the use of computerized aides and software (Bennett et al., 1997).
7. There is flexibility in test management by creating room for tailored administration schedules.
8. The security of the test material is also improved as compared to physical testing situations in which tampering with answer booklets, broken seals, damage during transportation or printing and so on are always present.

9. Test scores and results arrive instantly or relatively sooner as compared to paper–pencil tests (Basu et al., 2007).
10. More engaging and interesting experience for the user thereby bears impact on candidate's motivation and interest.

Disadvantages of Computer-Based Testing

1. The initial cost of setting up a facility suitable for computer or technology delivered testing is significantly higher as opposed to that in the case of conventional paper–pencil tests. Also, the cost of maintenance and upkeep of computers, ancillary equipment such as printers, joysticks and so on also contributes to the financial burden of using computer-based testing.
2. Infrastructure requirements for paper–pencil tests are easily met in existing establishments such as schools, office spaces, university classrooms and so on, whereas a computer-based test needs a dedicated setup.
3. In comparison to conventional paper–pencil tests, computer-based tests take more time to develop and create because of the need to construct larger item pool and also because of the time spent creating the testing interface and software on the systems.
4. Last, it is difficult to adapt certain classes of tests such as projective tests for computerized testing and this limits its scope of application.

Conclusion

The domain of testing and assessment has undergone a significant change from focusing on measures of reasoning on one hand to testing more nuanced abilities within the fields of intelligence and aptitude. The process of testing today is not bound by definitive, rigorous boundaries which focus on evaluating every individual against a pre-defined and pre-determined set but is a more cohesive process that seeks to achieve a balance between the skills and abilities required by a person and characteristics of a particular job. Situational and contextual demands along with individual differences find expression in the construction of tests, as well as the overall process of selection and assessment. These factors may be collectively responsible for the evolution in the field and the influx of technology into the testing, selection and placement arena. Although there had been significant changes in the selection process, but these changes are not stable and will keep on changing with the dynamic requirements of Armed Forces. Thus, it can be said that the journey of intelligence and aptitude testing is never ending.

References

Anastasi, A. (1992). What counsellors should know about the use and interpretation of psychological tests. *Journal of Counselling and Development, 70*, 610–615.

Bartram, D., & Dale, H. C. A. (1983). *Micropat Version 3: A description of the fully automated personnel selection testing system being developed for the Army Air Corps* (Ministry of Defence Technical report ERG/Y6536/83/7). 14 pp. Hull, England: Ergonomics Research Group, University of Hull.

Bartman, D. (2005). Computer-based testing and the Internet. In A. Evers, N. Anderson, & O. Voskuijl (Eds.), *Handbook of personnel selection* (pp. 399–418). Boston, MA: Blackwell Publications.

————. (2006). Testing on the Internet: Issues, challenges and opportunities in the field of occupational assessment. In D. Bartram & R. K. Hambleton (Eds.), *Computer-based testing and the Internet: Issues and advances* (pp. 13–38). West Sussex, UK: John Wiley and Sons.

Basu, A., Cheng, I., Prasad, M., & Rao, G. (2007). *Multimedia adaptive computer based testing: An overview. Proceedings of International Conference on Multimedia and Expo (ICME) Special Session* (pp. 1850–1853), *Beijing.*

Bennett, G. K., Seashore, H. G., & Wesam, A. G. (1974). *Fifth edition manual for the differential aptitude tests, forms S and T.* New York, NY: Psychological Corporation.

Bennett, R., Goodman, M., Hessinger, J., Ligget, J., Marshal, G., Kahn, H., et al. (1997*). Using multimedia in large scale computer based testing programs.* New Jersey: Educational Testing Service.

Burt, C. (1957). *The causes and treatments of backwardness* (4th ed.). London, UK: University of London Press.

Carroll, J. B. (1993). *Human cognitive abilities: A survey of factor-analytical studies.* New York, NY: Cambridge University Press.

Drasgow, F., Olson-Buchanan, J. B., & Moberg, P. J. (1999). Development of an interactive video-assessment: Trials and tribulations. In F. Drasgow and J. B. Olson-Buchanan (Eds.), *Innovations in computerized assessment* (pp. 177–196). Mahwah, NJ: Erlbaum.

DIPR. (1982). *Construction of battery of psychological tests for classification of leading radio operators* (Note 386). Delhi, India: Defence Institute of Psychological Research.

————. (1985). *Development of a battery of tests for the selection of submariners* (Note 419). Delhi, India: Defence Institute of Psychological Research.

DIPR. (1989). *Development of a battery of specific aptitude test for allocation of other rank trade: Radio operator in the army* (Note 445). Delhi, India: Defence Institute of Psychological Research.

———. (1991). *The development of a new language aptitude test battery for defence personnel* (Note 453). Delhi, India: Defence Institute of Psychological Research.

———. (1991). *Construction of an aptitude test battery for recruitment of AEC (HAV) instructors* (Note 458). Delhi, India: Defence Institute of Psychological Research.

———. (1995). *Validation of the differential aptitude test battery for selection of airmen for technical trades in the Indian air force* (Note 476). Delhi, India: Defence Institute of Psychological Research.

———. (1995). *Development of a psychological test battery for selection of artificers for Indian navy* (Note 477). Delhi, India: Defence Institute of Psychological Research.

———. (1995). *Development of an aptitude test battery for the selection of infantry personnel to sniper cadre/courses* (Note 478). Delhi, India: Defence Institute of Psychological Research.

———. (1997). *Development of battery of psychological tests for trade allocation of soldiers (Tech) of EME corps* (Note 514). Delhi, India: Defence Institute of Psychological Research.

———. (1999). *Development of an aptitudinal criterion for MER sailors* (Note 537). Delhi, India: Defence Institute of Psychological Research.

———. (2007). *Development of military as a career: A self assessment tool* (Note 626). Delhi, India: Defence Institute of Psychological Research.

———. (2009). *Development of a comprehensive battery of cognitive abilities for selection of candidates to commissioned ranks in the armed forces* (Note 634). Delhi, India: Defence Institute of Psychological Research.

———. (2009). *Development of a computerized test of cognition for the recruitment of officers and a paper-pencil test for naviks (GD) in Indian coast guard* (Note 637). Delhi, India: Defence Institute of Psychological Research.

———. (2010). *Standardization of trade allocation battery for personnel below officer's rank in Indian army* (Note 640). Delhi, India: Defence Institute of Psychological Research.

———. (2013). *Development of aptitude battery for allocation of trades to soldier technical category of army medical corps* (Note 663). Delhi, India: Defence Institute of Psychological Research.

DPR. (1957). *Pilot Aptitude Battery Test Familiarity Factor* (Note 167). Delhi, India: Directorate of Psychological Research.

———. (1966). *A validation study of the navigators aptitude test battery* (Note 271). Delhi, India: Directorate of Psychological Research.

———. (1973). *Allocation of trades to airmen with the help of differential aptitude test battery* (Note no. 321). Delhi: Directorate of Psychological Research.

———. (1973). *Validation of Mt Drivers aptitude test battery no. II* (Note 326). Delhi, India: Directorate of Psychological Research.

Freeman, F. S. (1965). *The theory and practice of psychological testing* (3rd ed.). New Delhi, India: Oxford & I B H Publishers.

Gardner, H. (1983). *Frames of mind: The theory of multiple intelligences.* New York, NY: Basic Books.

Kaur, G., & Mukherjee, S. (in press). Computers in psychological testing. *Psybernews.*

Hanson, M. A., Borman, W. C., Mogilka, H. J., Manning, C., & Hedge, J. W. (1999). Computerized assessment of skill for a highly technical job. In F. Drasgow & J. B. Olson-Buchanan (Eds.), *Innovations in computerized assessment* (pp. 197–220). Mahwah, NJ: Erlbaum.

Hendrickson, G. (1931). An abbreviation of the Army Alpha. *School and Society, 33,* 467–468.

Otis, A. S. (1918a). An absolute point scale for the group measurement of intelligence: Part I. *Journal of Educational Psychology, 9,* 239–261.

———. (1918b). An absolute point scale for the group measurement of intelligence: Part II. *Journal of Educational Psychology, 9,* 323–348.

———. (1918c). *Otis Group Intelligence Scale: Advanced examination.* Yonkers, NY: World Book.

Preckel, F., & Thiemann, H. (2003). Online versus paper-pencil version of a high potential intelligence test. *Swiss Journal of Psychology, 62,* 131–138.

PRW (1949). *Work of the directorate of selection of personnel during the war* (Note no. 2). Delhi, India: Psychological Research Wing.

———. (1950). *The Pilot Aptitude Battery* (Note 22). Delhi, India: Psychological Research Wing.

Reynolds, D. & Rupp, D. (2010). Advances in technology facilitated assessment. In J. Scott & D. Reynolds (Eds.), *Handbook of workplace assessment: Evidence based practices for selecting & developing organizational talent* (pp. 609–641). San Francisco, CA: Jossey-Bass.

Roid, G. H. (2003). *Stanford - Binet Intelligence Scales* (5th ed.). Itasca, IL: Riverside.

Rumsey, M. G. (2012). Military selection and classification in US. In J. H. Laurence & and M. D. Matthews (Eds.), *The Oxford handbook of military psychology* (pp. 129–147). New York, NY: Oxford University Press.

Saggio, G., & Ferrari, M. (2012). New trends in virtual reality visualization of 3D scenarios, virtual reality - human computer interaction. In X. Tang (Ed.), *Virtual reality-human computer interaction.* Retrieved from: http://www.intechopen.com/download/get/type/pdfs/id/38742

Salgado, J. F., & Moscoso, S. (2003). Paper-and-pencil and Internet-based personality testing: Equivalence of measures. *International Journal of Selection and Assessment, 11,* 194–295.

Seashore, C. E., Lewis, D., & Saetveit, J. G. (1939). *Manual of institutions and interpretations for the Seashore measures of Musical Talents.* USA.

Terman, L. M. (1916). *The measurement of intelligence: An explanation and a complete guide for the use of the Stanford Revision and Extensions of the Binet–Simon scale.* Boston, MA: Houghton Mifflin.

Vora, J., Nair, S., Gramopadhye, A. K., Duchowski, A. T., Melloy, B. J., & Kanki, B. (2002). Using virtual reality technology for aircraft visual inspection training: Presence and comparison studies. *Applied Ergonomics, 33,* 559–570.

Wainer, H., Dorans, N. J., Eignor, D., Green, B. F., Flaugher, R., Mislevy, R. J., et al. (2010). *Computerized adaptive testing: A primer* (2nd ed.). New York, NY: Routledge.

Wechsler, D. (1958). *The measurement and appraisal of adult intelligence* (4th ed). Baltimore, MD: Williams & Wilkinds.

3

Biographical Inventory for Selection

Suresh A., Arunima Gupta and Sucheta Sarkar

Effective personnel recruitment and selection is one of the most essential steps in the enrichment of any organization and when it comes to military, the task becomes arduous, yet imperative. Many organizations spend ample time and effort to match the right person to the job as there are many costs associated with hiring personnel who subsequently are unsuccessful. These costs can be measured by their effect on the lives and careers of the people involved (Strauss & Sayles, 1972). There are a variety of selection techniques employed in the existing scenario. The interview method is one of the most popular (Harris, 1988) but has been criticized as being too subjective (Campion, Pursell, & Brown, 1988). The assessment centre approach asks the applicant to complete a battery of selection tests and work samples, but it is very time consuming and expensive (Adler, 1987). Biographical information from job applicants is the third technique that is assumed to be less subjective, inexpensive and effective (Mitchell, 1994). It is presumed to predict future job success. Researchers have argued that biodata signify a more valid predictor of job-related success than traditional personality measures (Mumford, Costanza, Connelly, & Johnson, 1996), as well as reducing aversive impact in comparison to cognitive ability tests (Stokes, Mumford, & Owens, 1994). This chapter inspects upon the feasibility of biographical inventory for personnel selection in Armed Forces.

'Biodata' (short for biographical data) refer to a pre-selection technology wherein applicants provide job-related information on their personal background and life experiences that tend to causally affect their personal growth. This assumption relies heavily on the 'consistency' principle, stating that the best predictor of future behaviour is past behaviour (Owens, 1976; Wernimont & Campbell, 1968). Fleishman (1988) noted that acquiring information about an individual's past experiences may result in a more efficient prediction of one's performance at work. The information acquired about the applicant may include (a) person's background and life history (e.g., recreational, educational or work experiences), (b) indirect indicators of achievements (e.g., awards, recognition, grades), (c) information about the situations surrounding these activities (e.g., family income, parents' education level) and (d) personal information (e.g., age, marital status, health, skills, interests and attitudes). Furthermore, biodata scales are used to assess numerous characteristics, such as temperament, assessment of work conditions, values, skills, aptitudes and abilities (Mount, Witt, & Barrick, 2000).

The items in biographical inventory are usually constructed in multiple-choice format and optimally weighted to predict criteria of interest (Mumford & Owens, 1987; Owens, 1976). They best describe the magnitude or frequency of an individual's past experiences. The job applicants respond to a more standardised version of paper and pencil tests focusing on individuals' past behaviour which typically includes the kinds of data obtained on weighted application blanks, life history data, personal history, individual achievement record, life experience inventories, biographical information blank and others.

In most Western countries, biodata are generally obtained through *application forms* (Dany & Torchy, 1994). However, there exist noticeable differences between application forms and biographical inventories. First, application forms involve a series of questions comprising basic information about the applicants' knowledge, skills, education or other job-related information, whereas biodata refer to a pre-selection method wherein the applicants provide exclusively job-related information concerning the desirable work criteria. Second, biographical information obtained in a statistically systematic way helps in preparing biographical profiles of job applicants for the purpose of classifying their potentialities for future work performance, whereas application forms do not provide any scope of profiling of job applicants. Finally, the information obtained through application forms is assessed in non-structured, informal and intuitive ways, whereas biodata encompass 'weighted scoring' in which the responses collected are empirically scored and considered individually as predictors of future job criteria or performance.

Why Biodata Predicts Performance?

The use of biodata—information about one's life experiences—as a means of predicting performance has become prevalent since the 1980s. Dean, Russell & Muchinsky (1999) attempted to extend the ecology model by focussing on how negative life events are associated with affective and cognitive reactions. They also examined the potential role of "resilience to negative life events" or moxie as a key moderator and/or mediator of negative life event–job performance relationships.

The ecology model has evolved from Owens' Developmental-Integrative (D-I) model (Owens, 1968, 1971, 1976), suggesting that biodata items capture prior behaviours and experiences affecting personal development on individual difference characteristics (e.g., knowledge, skills and abilities). According to the ecology model, individual differences resulting from a person's unique hereditary characteristics and exposure to situational circumstances determine one's propensity to react in a certain way and shape the preferences that individuals make.

The literature indicates that negative life events constitute key markers of human development. Taylor (1991) noted that negative events appear to elicit more physiological, affective, cognitive and behavioural activity and prompt more cognitive analysis than neutral or positive events. Peeters and Czapinski (1990) found negative events elicited more frequent and complex causal attributional activity than positive events, whereas others have shown that negative events are considered longer (e.g., Abele, 1985) and elicit more extreme attributions (e.g., Birnbaum, 1972). Models of affiliation (Schachter, 1959) and social support (House, 1981) also suggest negative or threatening events cause people to seek companionship, support and assistance from others. The above studies reveal that negative life events can either increase the commitment towards goal attainment by helping individuals overcome difficulties or can lower one's self-concept leading to failure at work.

Thus, an analysis of positive and negative life experiences plays a crucial role in the formation of individual identities which contributes towards cognitive and developmental understanding of behaviour indexed through biographical inventories.

Biodata and Personality Measures

Review of previous researches indicated that it is hard to distinguish between biodata scales and personality scales. Mumford and Stokes (1992) noted that biodata items are often used in self-report personality inventories. This observation implies that biodata items are often strong predictors of scores on personality scales (Rawls and Rawls, 1968). Also, Mumford and Owens (1987) found that biodata factors resembling the "Big Five" factors of personality (Digman, 1990) emerged. This is also supported by prior researches (Owens, 1976).

Nunnally (1959, p. 371) stated:

> The biographical inventory is probably the best measure of 'personality' presently available for personnel selection programs. Although it is not certain what kinds of personality attributes they measure, the inventories often add substantially to the predictive efficiency which can be obtained from tests of intellectual functions and special abilities.

Asher (1972) has outlined eight dimensions on which biodata items differ from personality inventories (Table 3.1). Other researchers (Goldberg, 1972; Hough, 1984; Mael, 1991; Mumford, Snell, & Reiter-Palmon, 1994; Mumford & Stokes, 1992) also attempted to propose few differences between biodata and personality scales. They have been compiled in Table 3.2.

Verifiability of Biodata

Since a biodata instrument is essentially a self-reported set of responses gathered for a specific purpose, there has been some concern about verifying the accuracy of these responses. Faking is an important issue in the use of biodata and all other non-cognitive measures. It has long been established that the items that are historical, objective, external, discrete, verifiable and firsthand serve to reduce the amount of response distortion that occurs in self-report measures. One of the major advantages that the biodata offer over other non-cognitive measures is that the empirical keying methods of scaling biodata help in identifying 'subtle' items for which the correct answer is not readily

Table 3.1
Asher's dimensions indicating differences between biodata and personality inventories

A Taxonomy of Biographical and Personality Items	
"A" (Hard Biodata)	*"B" (Personality)*
VERIFIABLE Did you graduate from high school?	UNVERIFIABLE How often do you feel like crying?
HISTORICAL How old were you when you started your first full-time job?	FUTURISTIC In your next job, what type of work would you like to do most?
ACTUAL BEHAVIOUR Did you ever build tree-houses as a child?	HYPOTHETICAL BEHAVIOUR If you were a professional athlete, what sport would you play?
MEMORY During your last two years in college, about how many hours a week did you spend on studies?	CONJECTURE If you were to buy a new car, which model were you to choose?
FACTUAL On the average, how many nights a week do you go out to eat?	INTERPRETIVE When you complete a task, what allows you to feel good about it?
SPECIFIC Which of your college courses was the easiest?	GENERAL Were summer trips a big part of your childhood?
RESPONSE Which of the following tasks have you performed in previous jobs?	RESPONSE TENDENCY When you need to solve a tough work problem, what do you usually do?
EXTERNAL EVENT As a child, how often did your siblings help you with the chores around the house?	INTERNAL EVENT Which best describes your feelings when you last made a speech in public?

Source: Kenneth S. Shultz (1996), p. 266.

known to the respondent. Conversely, researchers reviewed that the inclusion of more subjective and non-verifiable items in self-report measures increases the probability of faking (Becker & Colquitt, 1992). This issue raises concerns about the measurement accuracy of biodata studies.

Lautenschlager (1994) reviewed numerous studies and found that objective items are less susceptible to distortion than subjective items. In an examination of the relationship between item attributes and various measures of response distortion across five different samples,

Table 3.2

Differences between biodata and personality scales

	Biodata Scales	Personality Scales
➤	Biodata items emphasize on prior behaviour and experiences occurring in specific situations.	Self-report personality items generally solicit information regarding an individual's predisposition towards a specific situational state. The focus is therefore limited to personal identity and an individual's general behavioural tendency.
➤	Biodata items capture environmental, personal and the social factors that affect and are affected by the individual in terms of the frequency in a given time period.	Personality item responses are influenced only by dispositional factors.
➤	Biodata measures capture behavioural patterns that are explicitly tied to the decisions that individuals make when presented with a particular situational stimulus.	Personality measures are not tied to decision-making ability, but more to a preference.
➤	Biodata items often tap into content areas that are probably influenced more by individual knowledge or skills than by personality. Thus, biodata-type items are often used as a preferred vehicle for accessing job-relevant information.	Personality items do not capture the factors influencing an individuals' behaviour.
➤	Biodata items are externally constructed.	Personality measures are internally constructed.

Source: Authors.

Stanley et al. (2000) found that items that were internal, subjective and summative were more related to response distortion indices than were external, objective and discrete items. These findings were consistent with earlier researches (Becker & Colquitt, 1992; Mael, 1991).

Mael (1991) noted that it is essential to introspect the climate under which the biodata instrument will be administered. Lautenschlager (1994) found that when respondents were instructed to fake, they distorted their responses by either increasing or decreasing their scores as directed. Furthermore, Hough and Paullin (1994) reviewed research on both item subtlety and objectivity and concluded that item type was insignificantly related to faking. Certain considerations about the applicant population, as well as the employer, should be addressed. Under varying circumstances, it is reasonable to expect that some applicants may be highly motivated to fake or that some employees may mentor applicants. Thus, it can be concluded that in some situations, response distortion may not be controlled by item type.

Methods of Detecting Faking

Lautenschlager (1994) offered several ways to reduce response distortion:

1. The methods for reducing the likelihood of faking in respondents comprise inclusion of verifiable items and warnings (Pannone, 1984). Trent, Atwater and Abrahams (1986) reviewed that including a warning statement about the potential verification of responses and the consequence of detected faking, along with items that were verifiable, reduces intentional distortion amongst applicants. Doll (1971) found that warning that a follow-up interview will happen proved to be more effective in reducing response distortion than a warning that the instrument included a lie detection scale. However, Lautenschlager (1994) noted that caution should be taken to ensure that the warning is appropriate and does not lead to unintended consequences. Warning may become ineffective if the applicant is able to understand the notion behind its inclusion. Thus, it is important to note that warning applicants on the consequences of faking can mitigate the propensity to fake. Also, a biodata instrument must include verifiable items to make the warning effective, which further depends on the degree of job relevancy. Mael (1991) suggested that non-verifiable items should be combined with verifiable items, with many of the verifiable items loaded at the beginning of the instrument to make the biodata questionnaire more effective.
2. One way of attempting to detect faking is to include some type of response validity scale and the repetition of some items as a form of an accuracy check (Pannone, 1984). Mitchell and Stokes (1995) suggested the need to validate scales with the same care with which one validates one's selection instruments.

3. Snell, Sydell and Lueke (1999) discussed potential areas for reducing faking with item content, format and scoring. Some types of non-subtle items may be less fakable because the applicants may find it difficult to determine the scoring of items. For instance, Griffith, Frei, Snell, Hamill and Wheeler (1997) demonstrated that under fake good conditions, respondents used a "good thing versus a bad thing" strategy, which led to increases in scores on a conscientiousness measure. However, scores on a measure of openness to new experiences remained relatively stable as respondents could not identify the right response to fake for this construct with respect to the target job.

4. Snell et al. (1999) also observed that Likert-type scales largely determine fakability of non-cognitive measures. Once a respondent is aware of whether or not the trait measured by the item is positive or negative, the correct response is clear. There is some evidence stating that forced choice formats produce less score inflation than Likert-based scales when honest conditions are compared with fake good conditions. Owens (1976) recommended that the construction of biodata items must be placed on a continuum.

5. The keying method may also influence response distortion. Kluger, Reilly and Russell (1991) studied faking using item keying and option keying methods and found that option-keyed responses were less susceptible to inflation than item-keyed instruments in a sample of students applying for a fictitious job. This method serves to be efficient in reducing the applicant's ability to fake because the scored options are not known to the applicant and are at times counterintuitive. Hogan (1994) noted that option keying serves an advantage when item criterion relationships are significantly nonlinear.

The last method for detecting faking involves the use of a computer to measure the amount of time that a participant takes to respond to an item. On the contrary, many studies have produced inconsistent results (Dwight & Alliger, 1997; Gore, 2000).

To conclude, concerns over the degree to which an applicant indulges into faking behaviour and how this will affect the validity of the biodata instrument are the two major issues that must be addressed in the biodata form development process.

Validity of Biodata

A vast amount of research provides an insight into the biodata's high validity and the ability to predict a wide range of job criterion measures, as one of the most challenging aspects for personnel selection. Adoption of meta-analytic studies is viewed as an important tool in biodata research concerning the heterogeneity of the existing biodata studies and the generalization of validities across samples.

Criterion-related Validity

Meta-analyses comparing biodata against other technically validated predictor measures have been particularly useful in confirming biodata's potential validities. Van Rijn (1980) reported that over a long period of time, the nature of biodata questionnaires has highly supported the consistency principle, stating it to be one of the best predictors of future behaviour. Consistently, these reviews found that through the decade validities for biodata have varied significantly, for example, from the low-to-mid .20s in Hunter and Hunter (1984) and Schmitt, Gooding, Noe and Kirsch (1984) up to the .50s in Reilly and Chao (1982) studies. Although even the lower-bound validity estimates are higher than the validities reported for most personality scales, and Schmidt, Ones and Hunter's (1992) seminal meta-analysis of 85 years of validity studies estimated a validity of 0.35 for biodata. Thus, it is important to provide an accurate estimate of the validity of biodata, which requires identification of the factors that tend to moderate the relationship between biodata predictors and occupational criteria.

Generalizability of Biodata

Evidences for biodata's generalizability are more restricted and recent. Factors, such as changes in applicant groups, the nature of the target job, job success criterion, the labour market, workforce needs and personnel policies, appear to play a major role in contributing towards the generalizability and stability of biodata. Poorly constructed items, small sample sizes and the susceptibility of biodata's scoring procedure to chance factors have also been found to be responsible for lack of generalizability and stability of biodata (Anastasi, 1988; Hunter & Hunter, 1984). To establish generalizability of biodata, ample amount of evidence may be required that may provide the scope for administering the carefully developed biodata questionnaire amongst larger samples (Gandy, Outerbridge, Shraf, & Dye, 1989). Rothstein, Schmidt, Erwin, Owens, & Sparks (1990) showed that biodata generalizes across other variables that are frequently cited as moderators of biodata's validity. These include race, sex, education levels, years of company service, years of supervisory experience, and age.

Incremental Validity

The relatively high correlations of biodata with measures of job success and low correlations with most other predictor measures (Reilly & Warech, 1990) make biodata a particularly vital component for a composite predictor measure of personnel selection. Booth, McNally & Berry (1978) found that the inclusion of just three biodata items to a cognitive aptitude battery increased the multiple correlations with training performance from 0.35 to 0.48. Biodata's sensitivity to situational, extrinsic and other non-personal factors associated with past achievements supports and provides incremental strength to the construction of biodata questionnaire. While cognitive ability tests emphasize upon the potentialities of the applicants by ignoring the situational variations, biodata seem to reflect the situational factors held responsible for determining a person's behaviour. Hunter and Hunter (1984) suggested that biodata capture the futuristic aspect of a person's behaviour, whereas more traditional cognitive measures are intended to predict an applicant's ability to perform the job. By gaining an insight into a person's past experiences, greater prediction will be possible for what a person will typically do in the future. Incremental validity of biodata over cognitive ability tests has been demonstrated in samples of army recruits (Mael & Ashforth, 1995) and air traffic controllers (Dean, Russell, & Muchinsky, 1999).

With regard to personality, studies have shown that biodata scales predict performance outcomes incrementally in US cadets (McManus & Kelly, 1999). Moreover, Mount et al.'s (2000) study simultaneously controlled for the Big Five personality traits and general cognitive ability and found that biodata's contribution to the four occupational criteria, namely, problem-solving performance, quantity and quality of work, interpersonal relationships and retention probability varied from 2 per cent to 17 per cent.

Structure of Biodata

A narrow spectrum of research had been conducted on the structure underlying biodata (Schmidt, Ones, & Hunter, 1992) that deals with how large sets of personal data can be organized and reduced to a limited number of latent factors or meaningful clusters. Mumford et al.'s ecology model (Mumford, Stokes, & Owens, 1990) proposed that biodata can be structured in terms of core knowledge, expertise, aptitude, value and expectancy variables that describe the way people develop their characteristic patterns of behaviour at work and beyond. These constructs 'facilitate the attainment of desired outcomes while conditioning future situational choice by increasing the likelihood of reward in certain kinds of situation' (Mumford & Stokes, 1992: 81). Nickels (1990) posited that these constructs can be categorized into personality, social resources, intellectual resources, choice processes and filter processes. In a recent study (Dean & Russell, 2005), these constructs were replicated by administering 142 biodata items on a sample of 6,000 newly hired air traffic controllers. The results highlighted that the overall biodata significantly correlated with job performance and cognitive ability. Also, intellectual resources significantly predicted job performance, followed by choice processes, social and personality resources; filter processes were found to be least related to job performance. Dean and Russell's findings illustrate the usefulness of biodata inventories as an alternative self-report non-cognitive measure such as the Big Five (Sisco & Reilly, 2007). Since biodata scales stress on verifiable and objective items, they are less likely to be affected by respondents' faking and impression management as compared to existing personality measures.

Biographical Inventory in Military Selection

Prior research conducted in the field of biographical inventory revealed the role of factors such as pre-federal aviation administration air traffic control experience, high-school grades in mathematics and science, self-assessment of performance potential, previous military ATC experience and altruistic tendency as significant predictors of the air traffic controller success (Collins, Manning, & Taylor, 1984). They also examined the relationship between the biographical inventory and training performance with respect to two groups of applicants, mainly, minority versus non-minority groups. As a result, the determinants of air traffic controller success amongst minority group were ranked as age, high-school physical science grades and self-assessment of future ATCS performance, whereas, for minority group, the sequence varied placing high-school math grades first followed by age and self-assessment of future ATCS performance.

The Biographical Inventory has been used continuously since Second World War (WWII) till date to select naval aviators and other flight personnel (Frank & Baisden, 1993). The navy's aviation selection battery (Fiske, 1947) is composed of heterogeneous items empirically selected and keyed to predict the retention of students in aviation training. This form has been revised and developed several times. At present, the current form, used since 1992, assesses the retention of both student naval aviators and student naval flight officers with separate keys designed for each of the administrations (Frank & Baisden, 1993).

Although experimentations on the effectiveness of biographical measures in military services are being done, but there still remains a scope for improvement and development of more effective and efficient biographical instruments to enhance selection of successful military service providers. Below is a description of several biographical and attitudinal inventories.

1. **History Opinion Inventory (HOI):** The HOI comprises of 100 self-report, true/false items covering the broad aspects such as school adjustment, family stability, social orientation, emotional stability, physical complaints, motivation and expectations for achievement, and response towards authority. This inventory originated from the Military Adaptability Screening Test (MAST) and was developed

by Air Force. It has efficiently played a crucial role in overcoming the difficulties encountered by the MAST. MAST proved to be unsuccessful in predicting basic military training success. In a study, the History Opinion Survey was administered to 15,252 airmen during basic training, and its two scales—Prediction of Emotional Instability (PET) and Prediction of Drug Use Admission (PDA)—as well as the combined Adaptation Index (ADI) were found to be statistically significant in predicting attrition during basic training. This study concluded that the scales PDA and ADI did make a unique contribution in predicting attrition among Air Force recruits. In 1975, the scale was reduced to 50 items and was renamed as 'Military Service Inventory'. It began to be implemented as part of the Air Force Medical Evaluation Testing.

2. **Military Service Inventory (MSI):** The MSI is a 50-item subset of the HOI. In 1977, the MSI was administered on 53,000 military applicants over a four-month period and was found to be a valid predictor of attrition.

3. **Recruit Background Questionnaire (RBQ):** The RBQ is a 55-item self-report inventory developed by the Navy. This questionnaire has been designed to assess areas such as work and school experiences, hobbies, interests and family history. Approximately one-third of RBQ items are school-related. The RBQ consists of two alternate forms, namely, form 1 and form 2, which were administered on Navy applicants. This measure was found to be significantly correlated with attrition, particularly for male high-school graduates ($r = 0.28$ to 0.38). Correlations for male and female non-graduates ranged from 0.17 to 0.21 and 0.18 to 0.26, respectively. These results were found to be consistent with the cross validation samples with little variation. Multiple regression analyses indicated that the RBQ significantly predicted the attrition rate among the entry-level Navy recruits through a screening instrument known as Success Chances for Recruits Entering the Navy (SCREEN). This instrument attempts to predict the said criterion on the basis of background items such as age, education, number of dependents and aptitude score.

4. **Military Applicant Profile (MAP):** The MAP is a 60-item multiple-choice biographical questionnaire that deals with family, academic and work experiences, athletic/physical

competence, self-concept and social style/participation. There are four alternate forms of the MAP. This instrument was administered on a sample of Armed Forces applicants and the analyses indicated differential prediction for attrition among high-school diploma graduates and non-graduates. On the basis of further analyses by Eaton, Weltin and Wing (1982), the use of this instrument was extended to all non-high-school graduates.

5. **The Educational and Biographical Information Survey (EBIS):** The EBIS contains 34 structured response questions providing 120 items of information concerning education achievement, school attitudes, family relations, work history, status variables, arrest record and alcohol use. This instrument was developed by HumRRO and was also designed to tap the attrition rate amongst both the applicants and new recruits. The objective of such a measure was to improve the existing education and moral enlistment standards by predicting the attrition rate.

6. **Armed Services Adaptability Profile (ASAP):** The ASAP is a joint service biodata instrument currently under development. Each of the two ASAP forms consists of 130 multiple-choice items. These items were pooled from other biodata instruments already administered and validated on military applicants. Specifically, each form of the ASAP includes one complete form of the MAP and several questions from the RBQ intended to evaluate the adequacy of a common measure.

Construction of Biographical Instrument

Item Content

Mael (1991) categorizes the biographical items into mainly three groups. They are as follows:

1. *Historical:* The historical nature of biodata items typically concerns the past experiences of an individual rather than predicting how a person will react in certain situations.

2. ***Methodological:*** Certain methodological attributes such as externally focused, objective, first-hand and verifiable are thought to aid in obtaining accurate biodata responses (Asher, 1972; Mael, 1991). Externally focused items are presumed to capture an individual's reaction to events in circumstances in which he or she was actively involved. Questions such as 'How did your parents evaluate your academic achievement?' indicate obtaining second-hand information about how others would evaluate the respondent. Such items must be avoided and respondents should be asked for first-hand knowledge. Another attribute associated with item content is verifiability. Verifiable items are less prone to social desirability effect and encourage eliciting accurate and honest responses from the candidate (Cascio, 1975).

3. ***Controllability/Job Relevance:*** Controllability refers to the degree to which a person decides whether to engage or not to engage in an action (e.g., behaviours a person desires to perform, such as playing sports, versus circumstances that happen beyond a person's control, such as parental socioeconomic status). Mael (1991) posited that all life events (consciously chosen or not) have the ability to shape a person's future behaviour and should be included in a biodata instrument. There are differing views about whether to contain items such as parental behaviour and socioeconomic status due to applicants' lack of control over their early environment (Mael, 1991).

Biodata performance prediction generally does not involve literally predicting future performance by measuring identical past performance (Dean, Russell, & Muchinsky, 1999). Wernimont and Campbell's (1968) proposed "samples" versus "signs" distinction, in which 'samples' represent past behaviours that are used to predict future actions drawn from a single-performance domain. Behavioural signs are assumed to causally influence subsequent performance and are highly correlated with those causal influences. Biodata instruments generally use both signs and samples of past behaviour to predict future performance outcomes (Dean et al., 1999; Russell, 1996).

Item Development

Following are the different item generation techniques targeting criterion construct prediction. Mumford and Owens (1987) explained the following six sources of biodata items:

1. The literature on human development is assumed to provide theories, models, associated constructs and operationalizations that might help in selecting prior life experiences to target with biodata item content (*Human development literature*).
2. Interviews targeting prior life experiences of high and low performing incumbents are assumed to generate criterion valid biodata items (*Life history interviews with incumbents*).
3. Typical factor loadings of biodata items.
4. Life history correlates with the criterion domain.
5. Existing biodata items with known criterion validities.
6. Items generated from investigators' general psychological knowledge.

The latter four sources rely on investigators' subjective opinions or existing biodata inventories (Russell, 1994), which appears to play a vital role in ongoing validation research and subsequent item revision.

Theory-based item generation efforts (Dean et al., 1999; Mumford, Costanza, Connelly, & Johnson, 1996; Russell, 1994) serve as an alternative technique for generating biodata items. The main objective of this technique is to gather information about prior life experiences from job incumbents and subject matter experts, using several methods, such as interview, focus group discussion or writing assignment. The following are the traditional steps involved in harvesting such information:

1. Identification of key job requirements using standard job analysis procedures.
2. Communicating key job requirements to subject matter experts, that is, subject matter experts should have first-hand knowledge of job performance.
3. Description of prior life experiences in relation to key job dimensions by subject matter experts.

4. Extracting critical incidents from subject matter experts' prior life event descriptions. Items generated from these critical incidents typically include (a) behaviours engaged by the applicant in the incident, (b) key aspects of the situation circumstances (i.e., sources of assistance, obstacles, etc.), (c) attitudes associated with the event, (d) role responsibilities held by the candidate during the incident or (e) task outcomes accomplished.

Scoring of Biodata

A number of techniques have been used for scaling biodata items (Nickels, 1994). The methods can be broadly grouped into test-centred and person-centred. Methods that are test-centred include empirical keying, factorially derived keying and rationally derived keying. Person-centred methodology focuses on identifying particular recognizable groups of individuals that share similar background experiences and have common profiles. The method is most commonly known as subgrouping.

Empirical Keying Method

The empirical keying approach emphasizes on maximization of prediction through an external criterion and is most commonly employed in the scoring of biodata (Devlin, Abrahams, & Edwards, 1992). A variety of empirical scoring procedures exist for identifying and weighing items on the basis of their ability to differentiate between successful and unsuccessful employees (Aamodt & Pierce, 1987; Malone, 1977; Telenson, Alexander, & Barrett, 1983). These methods are primarily categorized as option keyed and correlational method. The methods also differ regarding how criterion performance is measured (e.g., a dichotomous criterion reflecting group membership), how item scores are related to criterion scores and how item weights are statistically derived (Hogan, 1994). The vertical percent method (England, 1971) is one of the traditional option-keyed methods to empirical keying. Here, each item or question is coded as 0 or 1 depending upon the nature of response sets. The items having response set 'Yes' are coded as 1 and those with response set 'No' are coded as 0. In the correlational method, weights are assigned to each item on the basis of their correlation with the job criterion (as derived from the previous samples). In both the methods, a composite score is finally computed for each candidate by summing up the scores obtained on all the items.

The major advantage of the empirical keying approach is that the items are directly related to an external measure of job success and it constitutes high incremental validity in the prediction of occupational success than existing cognitive and personality measures (Mount et al., 2000).

FACTORIAL KEYING APPROACH

The use of factor analysis has been a popular method for identifying the patterns of correlations between individual biodata items and job criteria (Schoenfeldt & Mendoza, 1994). Here, the biodata items are statistically grouped and reduced to a small number of factors, which are further used to identify the psychological structures and predict occupation success. The correlation matrix can be obtained in numerous ways. First, the items may be scored either rationally or empirically. Second, the items may be item-keyed (i.e., each item consists of multiple responses scored along a continuum) or option-keyed (i.e., each response option enters the correlation matrix as a scored binary item). Third, the matrix may be formed by combining both the techniques.

Although this method provides some insight into the dimensions underlying the structure of life experiences, but there are many limitations. First, this method typically ignores large amounts of variance (Hough & Paullin, 1994). In a review of several studies, the variance accounted for ranged from 19 per cent for a 10-factor solution (Schoenfeldt, 1989) to 49.6 per cent for 15 components (Lautenschlager & Shaffer, 1987). Second, it requires larger samples as that of empirical keying approach. Third, the scoring of the items used in the factor analyses may vary depending upon the job criterion and the nature of items used. Different approaches employed for keying the items, whether empirical or rational, could theoretically lead to a different factor structure. Fourth, the resultant factors depend on the items included in the analysis. Items keyed using empirical approach against a criterion would yield different underlying structures than analyses of the items keyed rationally. Fifth, the factorial keying approach makes biodata questionnaires equivalent to personality inventories, especially when subjective items are included. Researchers argued that in such cases experts would fail to distinguish between personality scales and factorial keyed biodata (Robertson & Smith, 2001). Finally, this method does not appear to be independent of situation-specific considerations.

Inspite of the above-mentioned drawbacks, the factor analyses approach has been considered a useful tool for identifying the psychological structures of both empirically and rationally scaled biodata questionnaires.

RATIONAL KEYING APPROACH

Rational approaches to scoring biodata are internally based methods in which biodata items are scaled based on the relationship amongst the items and not on the external criterion. Fine and Cronshaw (1994) proposed that a thorough job analysis informs the selection of biodata items (see also Stokes & Cooper, 2001). Hough and Paullin (1994) discussed two strategies: direct and indirect. In the direct approach, the individual difference variables are defined using expert opinions. In the indirect approach, the distinguishing job-related behaviours are related to job-related psychological constructs through factor analysis. This is followed by the construction of the biodata questionnaire aimed at measuring the underlying dimensions. Both the techniques emphasize on forming scales that are internally consistent and homogeneous. The primary difference between the two methods is that in the direct method, expert opinions are utilized based on theory and research to select and weigh items, whereas in the indirect method factor analysis is employed to select and weigh items. Researchers reviewed that although the indirect method has little correspondence to the measure of job success, the approach tends to have greater construct validity and greater psychological meaningfulness. Another limitation is that the items developed using this method may be objectionable to applicants as the association between the items and the job success measure is not clear. On the other hand, the direct measure yields greater face validity and content validity and is more acceptable. However, this measure is equally likely to be susceptible to falsification. Mitchell (1994) referred to using factor analysis in a broad, exploratory, post hoc approach to discovering the dimensionality of the biodata items as "more akin to rationalization than to rationality" (p. 486). Practitioners often use both the techniques of rational keying approach, with direct method used to develop scales followed by the indirect method to refine the scales identified. Mitchell (1994) noted that this approach has numerous advantages. First, biodata questions can be developed and validated as that of any paper–pencil cognitive tests. Second, Mitchell's biodata questionnaires should readily generalize to occupations with profiles of antecedent behaviours similar to those of the occupation used in the development of the key. Moreover, reduced need for larger samples in the development of biodata scale questionnaire opens up the application of biodata to jurisdiction. However, one of the major disadvantages associated with this approach is that making 'correct responses' becomes too obvious for respondents which may increase the likelihood of faking (Lautenschlager, 1994).

Subgrouping

The fourth and the final approach for the scoring of biodata is subgrouping (Mumford & Owens, 1987). This approach seeks to categorize individuals based on profiles created via autobiographical information. The key objective here is to accurately assign individuals to specific groups discovered based on biographical information provided. Evidence suggests that this approach has a general predictive system (Mumford & Stokes, 1992; Owens, 1976), using homogeneous subgroups as the unit of investigation. It was found that subgroup status was predictive of verbal abilities (Eberhard & Owens, 1975), drug use (Strimbu & Schoenfeldt, 1973), over and underachievement, Rorschach responses and vocational interests (Mumford & Stokes, 1992). In addition to ongoing research that supported the predictive ability of the technique from a longitudinal perspective (Davis, 1984), subgrouping also served as a basis for "maximal manpower utilization" (Brush & Owens, 1979), and served as an alternative to moderator group analysis (Feild, Lissitz, & Schoenfeldt, 1975).

Conclusion

Utility of biographical information in personnel selection is based on the notion that the best predictor of future performance is past experiences. Although biodata have been applied to a wide range of occupations such as clerical jobs, mechanical equipment distributors, hotel staff, civil servant and managers, its role in military services is also practical with the help of efficient biographical instruments designed to enhance selection of successful military service providers.

References

Aamodt, M. G., & Pierce, W. L., Jr. (1987). Comparison of the rare response and vertical percent methods for scoring the biographical information blank. *Educational and Psychological Measurement, 47*, 505–511.

Abele, A. (1985). Thinking about thinking: Causal, evaluative, and finalistic cognitions about social situations. *European Journal of Social Psychology, 15*, 315–322.

Adler, S. (1987). Toward the more efficient use of assessment centre technology in personnel selection. *Journal of Business and Psychology, 2*, 75–93.

Anastasi, A. (1988). *Psychological testing* (6th ed.). New York, NY: Macmillan.

Asher, J. J. (1972). The biographical item: Can it be improved? *Personnel Psychology, 25*, 251–269.

Becker, T. E., & Colquitt, A. L. (1992). Potential versus actual faking of a biodata form: An analysis along several dimensions of item type. *Personnel Psychology, 45*, 389–406.

Birnbaum, M. H. (1972). Morality judgments: Test of an averaging model with differential weights. *Journal of Experimental Psychology, 93*, 35–42.

Booth, R. E., McNally, M. S., & Berry, N. H. (1978). Predicting performance effectiveness in paramedical occupations. *Personnel Psychology, 31*, 581–593.

Brush, D. H., & Owens, W. A. (1979). Implementation and evaluation of an assessment classification model for manpower utilization. *Personnel Psychology, 32*, 369–383.

Campion, M., Pursell, E. D., & Brown, B. K. (1988). Structured Interviewing: Raising the psychometric properties of the employment interview. *Personnel Psychology, 41*, 25–42.

Cascio, W. F. (1975). Accuracy of verifiable biographical information blank responses. *Journal of Applied Psychology, 60*, 767–769.

Chamorro-Premuzic, T., & Furnham, A. (2010). *The psychology of personnel selection.* Cambridge, UK: Cambridge University Press.

Collins, W. E., Manning, C. A., & Taylor, D. K. (1984). *A comparison of prestrike and poststrike air traffic control specialist trainees: Biographic factors associated with academy training success. In Studies of poststrike air traffic control specialist trainees: I. Age, biographic factors, and selection test performance related to academy training success* (Report No. FAA-AM-84-6). Washington, D.C.: Federal Aviation Administration, Office of Aviation Medicine.

Dany, F., & Torchy, V. (1994). Recruitment and selection in Europe: policies, practices and methods. In C. Brewster & A. Hegewisch (Eds.), *Policy and practice in european human resource management: The price waterhouse cranfield survey* (pp. 68–85). London, UK: Routledge.

Davis, K. R. (1984). A longitudinal analysis of biographical subgroups using Owens' developmental-integrative model. *Personnel Psychology, 37*, 1–14.

Dean, M. A., & Russell, C. J. (2005). An examination of biodata theory-based constructs in a field context. *International Journal of Selection and Assessment, 13*, 139–149.

Dean, M.A., Russell, C.J., & Muchinsky, P.M. (1999). Life experiences and performance prediction: Toward a theory of biodata. *Research in Personnel and Human Resources Management, 17*, 245–281.

Devlin, S. E., Abrahams, N. M., & Edwards, J. E. (1992). Empirical keying of biographical data: Cross-validity as a function of scaling procedure and sample size. *Military Psychology, 4*, 119–136.

Digman, J. M. (1990). Personality structure: Emergence of the five-factor model. *Annual Review of Psychology, 41*, 417–440.

Doll, R. E. (1971). Item susceptibility to attempted faking as related to item characteristics and adopted faking set. *The Journal of Psychology, 77*, 9–16.

Dwight, S. A., & Alliger, G. M. (1997, April). *Using response latencies to identify overt integrity test dissimulation.* Paper presented at the 12th Annual Conference of the Society for Industrial and Organizational Psychology, St. Louis, MO.

Eaton, N.K., Weltin, M., & Wing, H. (1982). *Validity of the Military Applicant Profile (MAP) for predicting early attrition in different educational, age, and racial groups* (Technical report 567). Alexandria, VA: U.S. Army Research Institute for the Behavioral and Social Sciences. Retrieved from http://www.dtic.mil/dtic/tr/fulltext /u2 / a130939.pdf.

Eberhard, C., & Owens, W. A. (1975). Word association as a function of biodata subgrouping. *Developmental Psychology, 11*, 159–164.

England, G.W. (1971). *Development and use of weighed application blanks (Bulletin No. 55).* Minneapolis: Industrial Relations Center, University of Minnesota.

Feild, H. S., Lissitz, R. W., & Schoenfeldt, L. F. (1975). The utility of homogeneous subgroups and individual information in prediction. *Multivariate Behavioral Research, 10*, 449–461.

Fine, S.A., & Cronshaw, S. (1994). The role of job analysis in establishing the validity of biodata. In G. Stokes, M. Mumford, & W. Owens (Eds.), *Biodata handbook: Theory, research and use of biographical information in selection and performance prediction* (pp. 39–64). Palo Alto, CA: Consulting Psychologists Press.

Fiske, D.W. (1947). Validation of naval aviation cadet selection tests against training criteria. *Psychological Bulletin, 31*, 601–614.

Fleishman, E. A. (1988). Some new frontiers in personnel selection research. *Personnel Psychology, 41*, 679–701.

Frank, L. H., & Baisden, A. G. (1993).The 1992 navy and marine corps aviation selection test battery development. *Proceedings of the 35th Annual Conference of the Military Testing Association, 35*, 14–19.

Gandy, J. A., Outerbridge, A. N., Shraf, J. C., & Dye, D. A. (1989). *Development and initial validation of the individual achievement record.* Washington, D.C.: US Office of Personnel Management.

Goldberg, L. R. (1972). Parameters of personality inventory construction and utilization. A comparison of prediction strategies and tactics. *Multivariate Behavioral Research Monograph, 72*, 2.

Gore, B. A. (2000). *Reducing and detecting faking as a computer administered biodata questionnaire.* (Unpublished doctoral dissertation). University of Georgia, Athens, GA.

Griffith, R. L., Frei, R. L., Snell, A. F., Hamill, L. S., & Wheeler, J. K. (1997). *Warning versus no warnings: Differential effect of method bias.* Paper presented at the 12th Annual Meeting of the Society for Industrial and Organizational Psychologists, St. Louis, MO.

Harris, M. M., & Schaubroeck, J. (1988). A meta-analysis of self-supervisor, self-peer, and peer-supervisor ratings. *Personnel Psychology, 41*, 43–62.

Hogan, J.B. (1994). Emperical keying of background data measures. In G. Stokes, M. Mumford, & W. Owens (Eds.), *Biodata handbook: Theory, research, and use of biographical information in selection and performance prediction* (pp. 69–107). Palo Alto, CA: Consulting Psychologists Press.

Hough, L. M. (1984). Development and evaluation of the 'Accomplishment Record' method of selecting and promoting professionals. *Journal of Applied Psychology, 69*, 135–146.

Hough, L.M., & Paullin, C. (1994). Construct-oriented scale construction: The rational approach. In G. Stokes, M. Mumford, & W. Owens (Eds.), *Biodata handbook: Theory, research and use of biographical information in selection and performance prediction* (pp. 109–145). Palo Alto, CA: Consulting Psychologists Press.

House, J. A. (1981). *Work stress and social support*. Reading, MA: Addison-Wesley.

Hunter, J. E., & Hunter, R. F. (1984). Validity and utility of alternative predictors of job performance. *Psychological Bulletin, 96,* 72–98.

Kluger, A., Reilly, R. R., & Russell, C. J. (1991). Faking biodata tests: Are option keyed instruments more resistant? *Journal of Applied Psychology, 76,* 889–896.

Lautenschlager, G. J. (1994). Accuracy and faking of background data. In G. S. Stokes, M. D. Mumford, & W. A. Owens (Eds.), *Biodata Handbook: Theory, research and use of biographical information in selection and performance prediction* (pp. 391–419). Palo Alto, CA: Consulting Psychologists Press.

Lautenschlager, G. J., & Shaffer, G. S. (1987). Re-examining the component stability of Owens' Biographical Questionnaire. *Journal of Applied Psychology, 72,* 149–152.

Mael, F. A. (1991). A conceptual rationale for the domain and attributes of biodata items. *Personnel Psychology, 44,* 763–792.

Mael, F. A., & Ashforth, B. E. (1995). Loyal from day one: Biodata, organizational identification, and turnover among newcomers. *Personnel Psychology, 48,* 309–333.

Malone, M. P. (1977). *Predictive efficiency and discriminatory impact of verifiable biographical data as a function of data analysis procedure.* (Unpublished doctoral dissertation). Illinois Institute of Technology, Chicago, IL.

McManus, M. A., & Kelly, M. L. (1999). Personality measures and biodata: Evidence regarding their incremental predictive value in the life insurance industry. *Personnel Psychology, 52,* 137–148.

Mitchell, T. W. (1994). The utility of biodata. In G. S. Stokes, M. D. Mumford, & W. A. Owens (Eds.), *Biodata Handbook: Theory, research and use of biographical information in selection and performance prediction* (pp. 485–516). Palo Alto, CA: Consulting Psychologists Press.

Mitchell, T. W., & Stokes, G. S. (1995, April). *The nuts and bolts of biodata.* Workshop conducted at the Annual Meeting of the Society for Industrial and Organizational Psychology, Orlando, FL.

Mount, M. K., Witt, L. A., & Barrick, M. R. (2000). Incremental validity of empirically keyed Biodata scales over GMA and the five factor personality constructs. *Personnel Psychology, 53,* 299–323.

Mumford, M. D., & Owens W. A. (1987). Methodological review: Principles, procedures, and findings in the application of background data measures. *Applied Psychological Measurement, 11,* 1–31.

Mumford, M. D., & Stokes, G. S. (1992). Developmental determinants of individual action: Theory and practice in applying background measures. In M. D. Dunnette & L. M. Hough (Eds.), *Handbook of industrial and organizational psychology* (pp. 61–138). Palo Alto, CA: Consulting Psychologists Press.

Mumford, M. D., Costanza, D. P., Connelly, M. S., & Johnson, J. F. (1996) Item generation procedures and background data scales: Implications for construct and criterion related validity. *Personnel Psychology, 49,* 361–398.

Mumford, M. D., Snell, A. F., & Reiter-Palmon, R. (1994). Personality and background data: Life history and self-concepts in an ecological system. In G. S. Stokes, M. D. Mumford, & W. A. Owens (Eds.). *Biodata handbook: Theory, research and use of biographical information in selection and performance prediction* (pp. 583–625). Palo Alto, CA: Consulting Psychologists Press, Inc.

Mumford, M. D., Stokes, G. S., & Owens, W. A. (1990). *Patterns of life history: The ecology of human individuality.* Hillsdale, NJ: Lawrence Erlbaum Associates.

Nickels, B. J. (1990). The construction of background data measures: Developing procedures which optimized construct, content, and criterion-related validities. *Dissertation Abstracts International, 51*(4-B), 2099–2100.

————. (1994). The nature of biodata. In G. S. Stokes, M. D. Mumford, & W. A. Owens (Eds.). *Biodata handbook: Theory, research and use of biographical information in selection and performance prediction* (pp. 1–16). Palo Alto, CA: Consulting Psychologists Press.

Nunnally, J. C. (1959). *Tests and measurement: Assessment and prediction.* New York, NY: McGraw-Hill.

Owens, W. A. (1968). Toward one discipline of scientific psychology. *American Psychologist, 23,* 782–785.

————. (1971). A quasi-actuarial basis for individual assessment. *American Psychologist, 26,* 992–999.

————. (1976). Background data. In M. D. Dunnette (Ed.), *Handbook of industrial and organizational psychology* (pp. 609–644). Chicago, IL: Rand McNally.

Pannone, R. (1984). Predicting test performance: A content valid approach to screening applicants. *Personnel Psychology, 37,* 507–514.

Peeters, G., & Czapinski, J. (1990). Positive-negative asymmetry in evaluations: The distinction between affective and informational negativity effects. *European Review of Social Psychology, 1,* 33–60.

Rawls, D. J., & Rawls, J. R. (1968). Personality characteristics and personal history data of successful and less successful executives. *Psychological Reports, 23,* 1023–1034.

Reilly, R. R., & Chao, G. T. (1982). Validity and fairness of some alternative employee selection procedures. *Personnel Psychology, 35,* 1–63.

Reilly, R. R., & Warech, M. A. (1989). *The validity and fairness of alternative predictors of occupational performance.* Paper invited by the National Commission on Testing and Public Policy, Washington, D.C.

Robertson, I. T., & Smith, M. (2001). Personnel selection. *Journal of Occupational and Organizational Psychology, 74,* 441–472.

Rothstein, H. R., Schmidt, F. L., Erwin, F. W., Owens, W. A., & Sparks, C. P. (1990). Biographical data in employment selection: Can validities be made generalizable? *Journal of Applied Psychology, 75,* 175–184.

Russell, C. J. (1994). Generation procedures for biodata items: A point of departure. In G. S. Stokes, M. D. Mumford, & W. A. Owens (Eds.). *Biodata handbook: Theory, research, and use of biographical information in selection and performance prediction* (pp. 17–38). Palo Alto, CA: Consulting Psychologists Press.

————. (1996). Toward a model of life experience learning. In R. B. Stennett, A. G. Parisi, & G. S. Stokes (Eds.) *A compendium: Papers presented to the first biennial biodata conference* (pp. 17–31). Applied Psychology Student Association, University of Georgia: Athens, GA.

Schachter, S. (1959). *The physiology of affiliation.* Stanford, CA: Stanford University Press.

Schmidt, F. L., Ones, D. S., & Hunter, J. E. (1992). Personnel selection. *Annual Review of Psychology, 43,* 627–70.

Schmitt, N., Gooding, R. Z., Noe, R. A., & Kirsch, M. (1984). Meta analyses of validity studies published between 1964 and 1982 and the investigation of study characteristics. *Personnel Psychology, 37,* 407–422.

Schoenfeldt, L. F. (1989, August). *Biographical data as the new frontier in employee selection research.* Paper Presented at the Annual Meeting of the American Psychological Association, New Orleans, LA.

80   Suresh A., Arunima Gupta and Sucheta Sarkar

Schoenfeldt, L. F., & Mendoza, J. L. (1994), Developing and using factorially derived biographical scales. In G. Stokes, M. Mumford, & W. Owens (Eds.), *Biodata handbook: Theory, research and use of biographical information in selection and performance prediction* (pp. 147–169). Palo Alto, CA: Consulting Psychologists Press.

Shultz, K. S. (1996). Distinguishing personality and biodata items using confirmatory factor analysis of multitrait-multimethod matrices. *Journal of Business and Psychology, 10,* 263–288.

Sisco, H. & Reilly, R. R. (2007). Development and validation of a biodata inventory as an alternative method to measurement of the five-factor model of personality. *Social Science Journal, 44,* 383–9.

Snell, A. F., Sydell, E. J., & Lueke, S. B. (1999). Towards a theory of applicant faking: Integrating studies of deception. *Human Resource Management Review, 9,* 219–242.

Stanley, S. A., Hecht, J. E., Montagliani, A., Stokes, G. S., Barroso, C. R., & Hausa, O. R. (2000, April). *Biodata item attributes in multiple samples: Validity and response distortion.* Paper presented at the Annual Meeting of the Society for Industrial and Organizational Psychology, New Orleans, LA.

Stokes, G., Mumford, M., & Owens, W. (1994). *Biodata handbook: Theory, research and use of biographical information and performance prediction.* Palo Alto, CA: Consulting Psychologists Press.

Stokes, G. S., & Cooper, L. A. (2001). Content/ construct approaches in life history form development for selection. *International Journal of Selection and Assessment, 9,* 138–151.

Strauss, G., & Sayles, L. (1972). *Personnel: The human problems of management.* Englewood Cliffs, NJ: Prentice-Hall, Inc.

Strimbu, J. L., & Schoenfeldt, L. F. (1973). Life history subgroups in the prediction of drug usage patterns and attitudes. *JSAS Catalog of Selected Documents in Psychology, 3* (MS. NO. 412), 83.

Taylor, S. E. (1991). Asymmetrical effects of positive and negative events: The mobilization-minimization hypothesis. *Psychological Bulletin, 110,* 67–85.

Telenson, P. A., Alexander, R. A., & Barrett, G. V. (1983). Scoring the biographical information blank: A comparison of three weighting techniques. *Applied Psychological Measurement, 7,* 73–80.

Trent, T. T., Atwater, D. C., & Abrahams, N. M. (1986, April). Biographical screening of military applicants: Experimental assessment of item response distortion. In G. E. Lee (Ed.), *Proceedings of the Tenth Symposium on Psychology in the Department of Defense* (pp. 96–100). Colorado Springs, CO: US Air Force Academy, Department of Behavioral Sciences and Leadership.

Van Rijn, P. (1980). *Biographical questionnaires and scored application blanks in personnel selection.* Washington, D.C.: US Office of Personnel Management, Personnel research and development Center.

Wernimont, P. F., & Campbell, J. P. (1968). Signs, samples, and criteria. *Journal of Applied Psychology, 52,* 372–376.

4

Human Landscape of War

Joseph Miller

In America, Post-traumatic Stress Disorder (PTSD) has become a moniker of contemporary experience in Iraq and Afghanistan veterans and a catchall term for psychological reactions to warfare. From the very beginning of the psychological understandings of Shell shock, definitions have included the altering of ideas relating to the environment. Social psychologist and First World War (WWI) veteran Kurt Lewin's field theory was articulated in an essay titled *War Landscapes* (Lewin, 1917). His conceptualization of life space was premised on how war changed the perception of landscapes for its participants.

Shell shock and syndromes relating to stress have always been linked to the environment through triggering or the currently defined intrusive spectrum of symptoms. It also highlights how fevers offer a great opportunity to better understand war's effect prior to the 20th century when understood in relation with Hans Selye's "syndrome of just being sick" or general adaptation disorder (Selye, 1973). That is to say, under excessive and untreated stress some people get sick.[1] While describing research works, internal psychology and physiology of stress, this chapter examines the life of General, Governor and Custom Collector James Miller. A detailed case study of the life of territorial Governor James Miller illustrates that he defined an illness that was an assault on his constitution by nature. His fevered reactions after battles, storms and a return to the frontier were called ague, and correlate to what psychologist Hans Selye called general adaptation disorder: a disease of just being sick in relation

---

[1] Psychologists and medical professionals still largely respect Selye's early 20th century research and Young's research fundamentally contradicts it and basically misunderstands how the limbic system rather than the hippocampal, or conscious memory center, is what the body uses to manage stress. That is to say, a problem of a malfunctioning Hypothalamic Pituitary Adrenal axis largely defies Young's solely socially constructed model, especially in terms of survival responses that are much more driven by evolutionary push factors.

to stress. His absenteeism as the Governor of the Arkansas territory endures as a symbol of Arkansas on the periphery in its historical journal to this day, but his utter contempt for the landscape of a frontier has been overlooked as an expression of avoidance of a frontier that would have reminded him of his service in the War of 1812, and in skirmishes with the Shawnee confederation that preceded it. It can be argued that when interpreted as representation of the intrusive spectrum of PTSD symptoms, Miller's failure as a territorial Governor is better understood and the chapter concludes by placing his particularly public example in context with other failed territorial Governors (William Hull, William Henry Harrison and Meriwether Lewis), as well as how conflicts have continually reshaped the American landscape.

This also illustrates how military heroes who failed in civil life were a commonly misunderstood phenomenon and that the tendency to highlight exceptional individuals, such as General Washington and Andrew Jackson, is a problem of selective memory rather than a fair representation of the experience in the early national period of the United States (Resch, 1999).

Examining the Historical Origins of PTSD

In his famous work on the WWI, Paul Fussell (2000) described a radical reevaluation of place that came inside of the trenches. For British soldiers the shear proximity of the trenches to home created a surreal irony that forever shaped their notions of home. How could home be so close to the hell of the trenches? Shell shock became the first widely accepted work of mental illness relating to combat. However, if we look at the works of historians such as Dean (1997) and Sommerville's (1995, 2011) on combat and condition, then WWI's primacy for the origination of PTSD must be challenged. Psychiatrics Shay (2002) and Wessely (2006) have also challenged this recentness of today's war syndrome by finding evidence of PTSD during the Boer War and Trojan War. Similarly, other research works have also challenged the notion (Jones, Palmer, & Wessely, 2002; Shay, 2002).

Recent research illustrates how Shell shock was not the observation of a new phenomenon rather a change in the observation of an existing condition. Shell shock was not new, psychology was new and there was finally a scientifically grounded method to categorize and analyze a timeless syndrome.[2] Great work is being done to both understand war's effect on the psyche and help existing veterans and survivors of trauma reintegrate following extreme circumstances. An important, yet overlooked aspect of war's affect is how war reshapes the soldier's view of nature.

Despite the widespread acceptance of Shell shock as a war neurosis during WWI, the PTSD diagnosis is still in its infancy. This has only occurred after Vietnam veterans lobbied for years in order for there to be clinical disorder that was not the result of weakness or lack of fortitude. There is still controversy regarding a syndrome designed to highlight an acceptable syndrome for male (or females conforming to male ideals), so even with a concrete diagnosis criterion there is still widespread and at times contentious debate (Andreasen, 2010; Micale & Lerner, 2001; Schaller, 2012). Also, in the social undercurrents of contemporary debate the condition has become a catchall to describe any form of mental illness relating to war (Hoge, 2010). Within the flurry of debate that comes at the initial phase of research on a newly redefined disorder/injury, an important aspect that has endured from the outset of the understandings of war's effect on the psyche has been greatly overlooked: how survival shapes views of the environment.

Humans have a strong evolutionary drive to survive when they are exposed to life-threatening events, so it is natural that such events would forever shape the way that survivors view similar environments. Kurt Lewin's field theory is almost always associated with physics because the propensity of psychologists to associate themselves with the "harder sciences", but his initial trajectory as a scholar came from his experiences in WWI (Goodwin, 2011). Lewin (1917) argued how a specific landscape could be inspiringly beautiful for some, but for veterans solely a source of impending danger and a place for opponents to hide. His overall concept of field theory in social psychology would be that behaviour is a function of personality (experience included) and environment $B = f$ (P, E) (Goodwin, 2011). Although Lewin's social psychological research would never make great strides in war's effect, it began with his intuitive knowledge of how war shaped a human being's view of the environment. Understanding how Shell shock changed views of land and by viewing these changes in actors before hand, it is

---

[2] See Allan Young's *The Harmony of Illusions: Inventing Post-Traumatic Stress Disorder* (1997).

possible to broaden both the understandings of war and how humans interact in their environment.

Another key figure in early understandings of the stress research was greatly influenced by similar intuition, observation and personal experience. Hans Selye was medical doctor who spent his life researching the physiological aspects of stress. He was not a veteran of WWI, but his father was and he observed how difficult it was for him following his service as a trauma surgeon. One experience from his youth that would shape his views on stress was hearing the stories of bayonet charges and becoming physically ill afterwards. His autobiography was titled *The Stress of My Life* in order to demonstrate how his own life influenced his research and discovery of general adaptation disorder that was simply the somatization of illness in relation to continuous or extreme stress. The evolutionary drive to survive was so strong that it would make the body ill without the presence of any actual sickness and Selye's first exposure to this was from stories of war in his childhood (Selye, 1977).

In the early 20th century, malaria would be better understood and the fevers that were widespread during previous conflicts began to be isolated from Selye's new syndrome of just being sick because of stress. In Britain, Boer War vets who did not show the modern signs of the malaria bacteria were isolated from most of the cases of rheumatic fever (Jones et al., 2002; Jones & Wessely, 2004). Historians would be wise to recognize that before the 20th century, fevers were not so carefully defined and it is problematic to presuppose that all fevers of that age are the same as the fevers of our age, in the same way that it is dangerous to draw linear boundaries before nations formally existed. Greer (2010) has called the tendency to presuppose the national boundaries as 'hypernationalism' and in the same vein hypermalarial ideas are similarly problematic when giving modern definitions of previous illnesses of constitution when other mental maladies could just as easily explain the sickness, but better explains the failure of veterans in civil life. Before psychology, mental disorders went wholly untreated and would have more often resulted in the general adaptation disorder identified by Selye. Within today's understanding of PTSD as a master catch-all condition that explains all war-related mental illness, cases of fever in relation to combat trauma help illustrate what Selye initially observed about the body's physiological reactions to stress in an era long before the proliferation of psychology as an accepted scientific discipline.

In a careful examination of the contentious suicide of Meriwether Lewis (another failed Governorship after a harrowing adventure on the frontier), James Holmberg cautiously accepted the necessity of examining Lewis's life in terms of manic depressive/bi-polar disorder. He stated, "While we must be careful in assigning a modern diagnosis to Lewis's documented symptoms, it is also appropriate that we do so" (Guice & Buckley, 2007, p. 20). This author finds the act of diagnosis fruitless given the fact that a medical diagnosis' purpose is to establish a regimen of treatment for a man who has long been dead, while respecting Holmberg's careful analysis of an early modern suicide in relation to mania. However, by employing contemporary medical and psychological understanding of illness historians can better understand the intimate details of the life of early modern actors. The case of James Miller illustrates the intrusive spectrum of PTSD symptoms that fully reshaped his views about nature. This illustrates how the place of combat, violence, or any life-threatening event will often be entangled and negatively associated with those memories.

James Miller witnessed and participated in terrible acts of violence on the frontier and would forever avoid and think negatively about it. Joanne Bourke has observed that the language of trauma and PTSD did not exist for men until the late 19th century and women until the 1960s (Bourke, 1996, 2009; Micale & Learner, 2001). Prior to Charcot's isolation of emotionally traumatic experiences through hypnosis, and Freud's expansion of this process, hysteria was linked to the uterus and trauma was a word solely used to describe physical injuries. Examining PTSD in a time that lacked a language to describe war as an emotional rather than physical trauma is problematic at best, yet by examining Miller's completely oppositional temperament towards the environment, or more as he would say the climate, of the Arkansas Territory provides an early modern example of the avoidance spectrum that comes with PTSD even without the benefit of the language of emotional trauma. Furthermore, examining the writing of participants of the campaign fought by James Miller demonstrates a pattern, in contradiction to the medical notions that trauma was solely injury, of employing stoic descriptions of wounds, but detailed demonstrations of illness and violence. Prior to the professionalization of psychology, and science in general, the intuition of soldiers often provides better records of psychological trauma than medical writing.

James Miller: A Case Study

James Miller: As New England's Most Distinguished Soldier (1808–1819)

James Miller's performance during the War of 1812 was nothing short of spectacular. He was awarded four Congressional Gold medals for leading Bayonet Charges at Brownstown, Chippewa, Lundy's Lane and Fort Erie. Four 'awards of the medal' that preceded the Congressional Medal of Honor are not equal in the US Military history. In 1850, Hawthorne described him as New England's most distinguished soldier. His reputation was so great that he was named a Territorial Governor, he easily won a Connecticut Congressional seat, and his vote of approval and character statements about William Henry Harrison in 1840 increased his credibility as a presidential candidate. Despite his fanfare and popularity in the US during the early national period, Miller has been largely lost to larger memory: save for the US Army's Fifth Infantry Regiments motto of "I'll Try Sir" that commemorates Miller's charge at Lundy's Lane.

His war record may be impressive, but the most inspiring thing about James Miller was how frail and sick he was during his leadership, all the while constantly inspiring his men through some of the most difficult fighting on the frontier between the United States and what would later confederate into Canada. 'Ague' was Miller's term for fevered reactions that are often assumed as solely the result of malarial parasite, but a close analysis of Miller's paroxysms illustrates that they started after harrowing events in combat and that began with his first exposures to violence. Private Adam Walker's journal (1816) was published with the name of his popular commander in the title. Walker could not objectively resist the urge to commend James Miller, but his account gave incredibly specific details of Miller's bouts of sickness. His first case of ague came with many other soldiers of the Fourth Infantry Regiment after a cunning tribal warrior penetrated their perimeter under cover of darkness. Walker would also describe his commander's collapse after a battle at the village of Maguaga, which the Americans called Brownstown. This will be returned too, because Miller would later describe the sickness as a result of being exposed to a storm, but it is very much important to understand Walker's perception of ague.

Walker's description of the ague is surprising because of how much they differed from physical injuries. For Walker the ague was something terrible, he like most soldiers was an exemplar of stoic manhood, but this was different when it came to ague. Walker would list the quotations of wounded men. "I received a wound and was obliged to retire" is how Sergeant Orr responded after getting shot by a musket at Tippecanoe, whereas Walker would describe the ague in starker terminology (Walker, 1816, pp. 26–28). "We daily obliged to wade the river and haul the boats after us over the rapids, which occasioned of our men at the arrival of Vincennes, to be disordered with that painful disease, the fever and ague" (Walker, 1816, p. 43).[3] Walker would further describe how Miller performed the "duties of a common soldier" to safeguard the health of his soldiers (Walker, 1816, p. 43). It is important to note that for Walker, and we will see this later in Miller's descriptions, the environment caused the illness, but he lists high sickness rates after harrowing river crossings, after a tribal warrior breached a perimeter, and after battles (Walker, 1816, pp. 17–18). Contemporary actors believed they were afflicted by a sickness caused by the frontier itself. So in a time when the only trauma was physical injury and there was not yet a language for emotional trauma, a private was upset with this commonly held medical idea. He described a sickness that came in fevered fits directly after battles in harsher language than actual physical traumas. Walker's language illustrates how the ague was something far more difficult to suffer from than a simple fever: a harrowingly traumatic response to violence. Both Miller's words and in greater measure his life would challenge hypermalarial ideas about the sickness he suffered until his last days, especially given the way it changed his ideas about nature and the frontier.

For Miller, weather, climatological factors and summer in the frontier would trigger his illness, and although it is clear that being exposed to infected mosquitoes causes malaria, malaria will not explain the full spectrum of climatological and traumatic triggers that exacerbated Miller's sickness. Any shell-shocked veteran understands how frightening lightening can become after combat. In general, being exposed to a thunderstorm is intimidating, to say the least. The natural process of static electricity building and building into a sudden boom is palpable.

---

[3] River crossings persist as one of the most dangerous aspect of US Ranger training in both the Mountain and Swamp phases of the Ranger qualification course.

The silence just before a strike is equally intimidating, but it is never understood in terms of how similar it is to being exposed to combat in the early 19th century. The silent pause and the waiting for the static build-up to a load explosion is akin to the tension and sense of time moving slowly that occurred in the ranks when units stood face to face and fired volleys at each other. Worse the war on the northern frontier was characterized by ambushes to movements, so the exposure to combat would be similar to being caught in a storm. Miller's veterans of Tippecanoe would fall victim to ague again after the Battle of Maguaga.

At the trial of Miller's commander, General William Hull, it was stated that the sickness occurred because of a storm that occurred on the march, but it was the second time Miller fell sick after a battle. Miller confirmed the proximity of the battle to his illness, "The men were very much sick, and fatigued. I had a relapse of ague the day after the battle" (Forbes, 1814, p. 171). So, it is not only natural to assume that men would attach the sickness to storms rather than combat because of the similarities between such a storm and the combat that characterized their campaign. It is not coincidental that Miller fell ill after a tribal warrior attacked his perimeter and then following the leading of a bayonet charge, yet these traumatic experiences before the language of emotional trauma would be expressed in terms of an aversion to the climate, weather and the environment from the very outset of their campaign until their descriptions in later life. The New England regiment's culture merged with a painful disorder that they could not quite understand, but that was in ways worse than physical injuries, in a way that made the frontier itself seem dangerous. Miller would bring this aversion to his leadership in Arkansas, but only after one of the most impressive examples of courage, leadership and daring in the American military history.

What is most impressive about the interaction with William Hull during his court martial was Miller's recognition of his own infirmity and his determination to fight despite it. Trials in which accused defend themselves often have intense moments of confrontation and Miller's commander's trial did not disappoint. After admitting his illness and those of his men, Miller essentially called his commander a coward for not ordering them to fight anyways. When asked if his commander's actions to safeguard the health of Miller's regiment were justified and not cowardice, Miller stated, "Yes! Such an immediate surrender I think was indicative of a want of courage" (Forbes, 1814, p. 151). Miller would continue to fight and earn his greatest laurels after his infirmity.

James Miller: As the First Governor of Arkansas Territory (1819–1824)

Despite his universal appeal to his soldiers, Miller's service as the Governor of the Arkansas Territory was far from laudable and he very appropriately fostered contempt for his performance by the people of Arkansas. This is largely because he was absent from the state more often than he was present. During his tenure, including a tribal war, Miller was in New England more than in the territory. For him the climate was the source of his illness, and he would cite that precedence from the very beginning until the very end of his tenure as the Governor. He began his tenure in Arkansas by arriving late because as he stated in 1820, "I had spent the prime of life in the Service [Army]; my constitution was impaired by privation and exposure to the inclement seasons of the North during the late war". Again, Miller would cite his illness as the result of weather and climate rather than repeated exposures to combat. His tardy arrival would not be an isolated event. His absence due to illness and his words in criticism of the Arkansas territory still represent a strong historiographical tradition in Arkansas.The two biographies of Miller's tenure as the Governor highlight Miller's failure due to his aloof temperament and absenteeism. *The Arkansas Historical Quarterly* has published two articles relating to Miller's poor performance as the territory's first Governor. Ledbetter (1988) and White's (1960) articles both conclude that Miller's absence from the state for the majority of his term as the Governor led to his inevitable failure. Both biographies cited Miller's initial popularity and universal acclaim, but maintained that his fear of illness influenced his absence and decision to abandon the appointment. Further, the *Arkansas Historical Quarterly* has published two of Governor Miller's personal letters to highlight his dislike of the frontier climate (Miller, 1820). In this vein, Miller has become a symbol of the Arkansas Territory on the periphery; a place that's first federally appointed Governor despised. The focus employed in the *Arkansas Historical Quarterly* recognized Miller's failings and sickness, but made no effort to reconcile that with Miller's success in the military.

In fact, the most thorough description of Miller's ague is made in his letter to "his friend in Peterboro" found in the *Arkansas Historical Quarterly* (1820). He described the difficulties of understanding persistent illness in the early 19th century, and his weakness due to the ague that was "a slow bilious fever" (1820). His fever was persistent and he was often incapacitated by constant illness, yet he never succumbed to disease. Miller stated, "Very few deaths occur by disease, but people remain weak and fit for nothing for a long time" (1820). As a romantic writer, Nathaniel Hawthorne would use more creative language to state that Miller and other veterans were "by no means less liable than their fellow-men to age and infirmity, they had evidently some talisman or other that kept death at bay" (Hawthorne, 1850, p. 24). The talisman was simply the fact that their illnesses were somatic expressions of trauma, and were not as likely to result in death. Miller was struggling to reconcile his constant illness that incapacitated him without taking his life.

Miller's performance would continue to suffer and he remained at his New England home for most of the duration of his governorship. During Miller's absence in the summer of 1821, he neglected to name a "Sub-Agent either to the Quapaws or Cherokees", and when war broke out between the Cherokee and Osages the acting governor, Robert Crittenden, was left without any instructions. Miller was an increasingly aloof figure, but he understood the dangers of such a conflict. In a letter that Miller sent to Calhoun he concluded that without the intervention of the federal government to amend the Jackson-Hinds treaty that "if it passes in its present shape the Territory is destroyed forever (sic)" (Carter, 1953). During the winter and early spring, Miller's role as the Chief of Indian affairs was marked by conflicts, yet when he returned to New England, he left no successor (Miller, 1820). The Army Commander, Major William Bradford, spoke of Miller's absence without guidance on the impending war by stating, "War will exist this summer unless I take hand with them" (1820). Bradford was left without guidance on how to settle an impending war while Miller traveled home. A man, who earlier condemned his commander's cowardice for recalling Miller, due to illness, was now leaving his territory at the outset of tribal warfare. No persistent fever could explain this slow progression of behavior to avoid conflict better than the progression of PTSD-related avoidance symptoms that occur with continued misuse of the HPA axis (DSM-5). Again for Miller, it was a result of frontier climate and summer.

Miller and his contemporaries regularly cited climatological factors as the cause of his illness and reclusive behaviour. As the Governor in 1820, Miller suffered sickness when travelling on the "Arkansas" river. Climatological factors were cited as the cause of Miller's sickness at General Hull's court-martial, and in the writing of Adam Walker, as well as in Miller's speech to the Arkansas legislature, in all of his requests for leaves or absences from the Arkansas territory, and in his decision to step down as the Governor (Forbes, 1814; Walker, 1816).

James Miller: As Collector of Customs in Salem, Massachusetts (1824–1849)

Perhaps the best account of Miller's affliction following the War of 1812 was in the "Custom-House" introduction of Nathaniel Hawthorne's (1981) *The Scarlet Letter*. Notably, this was a nonfictional account, with obvious stylistic license and political bias, and Hawthorne utilized information about the men of the Custom-House gained from his entire life in the small town of Salem rather than solely from his tenure at the Custom-House. It is not the sort of essay that lends itself to charts and empirical methodology; however, its sentiment and expression of a novelist, who challenged the values of his own generation, are valuable. For Hawthorne, Miller was an old reclusive soul who became increasingly aloof, because of his bayonet charges and not his age. He used the metaphor of the old, deteriorated and overgrown Fort Ticonderoga. Miller still had all the essential aspects of the fort intact, yet with time and without necessity for such attributes had decayed. Long before science was a profession, Hawthorne would grasp that Miller's skills in battle were still immersed into the older man, yet without that danger it had become cumbersome. However, this would not even be his best metaphor describing PTSD.

For Hawthorne, Miller's service in combat was both admirable and burdensome, yet it was onerous in precisely the way PTSD is understood today. Hawthorne would describe Miller's performance positively by stating that the "benevolence, which, fiercely led the bayonets on at Chippewa or Fort Erie, I take to be of quite as genuine a stamp as what actuates any or all the polemic philanthropists of the age" (1850, p. 32). What is most revealing was that Miller's wartime service was arduous long after the fact. According to Hawthorne Miller's "earlier days; of integrity, that, like most of his other endowments lay in a somewhat heavy mass, and was just as unmalleable and unmanable as a ton of iron ore" (1850, 32). This precisely describes the prognosis of untreated PTSD in that the hippocampus, or the part of the brain that controls conscious/declarative memory, is damaged progressively by the overuse of adrenal chemicals (Karl et al., 2006). In his description of PTSD, Jonathan Shay used remarkable similar terms to describe lost declarative memory. Shay stated that "the veteran has lost authority over his own process of memory" (Shay, 1994, p. 38) and Hawthorne's recognition of Miller's burdens grasped this struggle long before there was a named medical condition (Shay, 1994). Hawthorne both understood and observed a phenomenon before scientific understanding codified it. He also understood how it shaped James Miller's sense of nature and to a lessor extent his masculinity.

In his description of Miller in terms of the metaphor of Fort Ticonderoga, Hawthorne recognized Miller's affinity for the cultivation and aroma of domesticated flowers. The idealized image of a man who "had slain men with his own hand", yet had a great "fondness for the sight of fragrance of flowers" (Hawthorne, 1850, p. 32). Hawthorne challenged the laurels of bloody battle and the image of "New England's most distinguished soldier" who also "seemed to have a young girl's appreciation of the floral tribe" (1850, pp. 23, 33). Hawthorne's metaphor of "Old Ticonderoga" captured the complexity of Miller's life in a way that understood the aftermath of war and the complexities of trauma in Miller's memory. Hawthorne claimed that "Nature" did not "adorn the human ruin with blossoms of new beauty, that have their roots in the chinks and crevices of decay, as she sows wall-flowers over the ruined fortress of Ticonderoga" (1850, p. 32). It is with this recognition that Miller's decline and longing for beauty stemmed from the decay to his health brought on by his leadership during the War of 1812 that Hawthorne recognized how his life was forever enriched by being "brought into the habits of companionship" of Miller.

Analysis of James Miller's Case

Miller suffered from somatization of PTSD into an extreme form of general adaptation disorder, when understood in terms of Jonathan Shay's research on PTSD. Shay argues that "simple combat PTSD is best understood as the persistence into civilian life of valid adaptations to combat" (Shay, 2002, p. 40). That is to say that in combat PTSD is useful: overuse of the process that manages the chemical cortisol, which is usually understood as adrenaline that becomes less useful when the same reactions are attached to other reminders of the traumatic experience (Ozzer & Weiss, 2004). The most common intrusive symptom being similar environment, or climatological factors, so Miller's and his regiments continued high performance, but hatred of the frontier are better understood as a result of traumatic emotional experiences.

The writings of Miller's contemporaries confirm this as well. Walker and the accounts he quoted used few words describing the injuries, but he would describe the emotion of the violence on the frontier like no second hand observer could.

> To attempt a full and detailed account of this action, or portray to the imagination of the reader the horrors attendant on this sanguinary conflict far exceeds my powers of description. The awful yell of the savages, seeming rather the shrieks of despair, than the shouts of triumph, the tremendous roar of musketry, the agonizing screams of the wounded and dying, added to shouts of the victors, mingling in tumultuous uproar, formed a scene that can be better imagined than described. (Walker, 1816, pp. 24-25)

So in a time that preceded the language of emotional trauma soldiers from New England were defying the period's definition of traumatic experience by favouring constitutional illness and emotionally draining experience to physical injuries in their writing. An anonymous account stated, "The wounded leaped at the idea that they would soon have an opportunity of avenging their wrongs, and besought the surgeons to report them fit for duty" (An Ohio Volunteer, 1812, p. 58). Yet, they still fought and were highly embittered by General Hull's surrender at the Battle of Detroit. Malaria does not make soldiers sick and incapable in battle, but PTSD should precisely do that. So, in this case, it is unwise to assume that all cases of fever were the fevers of the 20th century. The best evidence would come after the wars end, just as Shay maintains today, and this is where Miller's ideas about the environment are most visible.

This association of a fever caused by the environment was also present in a contemporary medical article by Dr Chambers. He described a case of persistent fever that was also caused by climatological factors. In patients who struggled with quinine overdose, ague symptoms recurred again in consequence of an exposure to a cold breeze, while on a boating excursion (Chambers, 1846). This also confirms the repeated research projects of Harvard cultural psychologist Devon Hinton who has seen malarial fits as masking PTSD as well as Cinchonism cause by Quinine-sulfate overdoses subjects with unidentified PTSD in regions that share malarial risk factors and a lack of awareness of PTSD (Hinton, Pham, Chau, & Tran, 2003; Hinton, Pich, Chhean, & Pollack, 2004). This illustrates how climate can come to personify the triggering symptoms common in PTSD as well as providing a rationale for a culturally accepted ailment caused by the forces of nature (DSM-5). It is not a matter of coincidence that Miller happened to suffer from the same fits of ague during a settlement of tribal warfare on the frontier, and that Miller subsequently selected the summer months to retire to his home in New England in order to avoid sickness. It is still common for veterans to suffer their worst symptoms of PTSD in the anniversaries of traumatic experiences (US Veterans Administration, 2013). The Arkansas frontier in the summer shared climatological similarities with Miller's battles with the Shawnee on the Ohio frontier. Miller's ague and his fears of illness during a similar period represent the intrusive triggering symptoms common in PTSD. His behaviour demonstrated, "Persistent avoidance of stimuli associated with the traumatic event(s)" present in cases of PTSD, and served as culturally accepted means for men to express symptoms of mental illness during the early nineteenth century (American Psychiatric Association, 2013; Micale, 2008).

The dissonance, between Miller's heroic military career, and his governorship, was present in the writing of Ledbetter (1988). He quotes the *Arkansas Gazette* that stated, "We regret to learn that the health of our gallant and worthy Governor is still impaired but we can hardly believe that it is owing to the unhealthy climate of Arkansas" (Ledbetter, 1988, p. 110). It was certain to individuals in Arkansas that fear of more fits of illness was causing Miller to avoid his responsibilities despite Walker's description of an officer who shared "the duties of the common soldier". Miller may have cited the climate of Arkansas as his reason to seek election as a Connecticut Congressman, only to reject the position in order to take over the role of an obscure custom collector in Salem Massachusetts (Ledbetter, 1988, p. 111). His anxiety about his health pushed him to fully withdraw from politics, yet it never prevented him from being a leader who inspired his soldiers in combat. For Miller, the frontier was a dangerous place, but only after the war ended did such fears of his health influence his performance. This illustrates how war's effects usually are not visible until the danger subsides, and manifest themselves as something entirely different. Freud would later call these precedence defense mechanisms to avoid the actual sources of emotional trauma (Goodwin, 2011). For Miller, it was the climate of the frontier in the summer months, largely because the northern frontier only offered combat during the summer months. Importantly, these apprehensions about the climate were only limiting after the danger subsided because they verify the research of Jonathan Shay (Shay, 2002). Miller's regiment and Walker's careful addition of accounts of those that suffered wounds in battle make it apparent that in this case PTSD's intrusive spectrum of symptoms reinforced cultural sensibilities of New Englanders. The frontier was an inherently dangerous place.

Miller's affinity to domesticated nature was oppositional to the wild frontier that caused his illness. This desire to cultivate and understand a nature that bore little resemblance to the source of violent acts witnessed by Miller should be viewed in terms of its own unique set of experiences [B = f (P, E)], as well as the more common story of soldiers struggling to redefine place following the experience of warfare. The nature of Miller's exposures to violence as well as the places, seasons and climates during the exposures were significant influences of his sense of place. Miller's primary observers and every one of his biographers have recognized this opposition of the frontier: a point of view certainly reinforced New England understandings of the wilderness, but was also greatly shaped by the experience of combat.

Understanding Miller's Contemporary Frontier Governors in Context

Within a case study it is a fair criticism to note how one life might not speak of a common story (though that seems less justified within a topic that explores something like human physiology in relation to traumatic emotional experience); so it is important to recognize other individuals with similar experiences. Through a brief look at some of Miller's peer frontier Governors, it becomes apparent that others were shaped in similar ways: accounting for the difference and idiosyncrasies of individuals. Briefly understanding of the failure of General and Michigan Territorial Governor William Hull, the suicide of Meriwether Lewis and the death of William Henry Harrison strengthen the dialogue about Miller's life and provide better understanding of how violence on the frontier affected individuals differently. It also casts a strong glimpse about ideology relating to the wild and masculinity.

The ill advised speech made by William Henry Harrison without the benefit of coat in his inauguration has been a comical, albeit actually very tragic, item of trivia pertaining to the American Presidency. However, it should be understood in terms of medical ideology about a person's constitution. Unlike Harrison's peer, and friend James Miller, he had never succumbed to any persistent constitutional illness. For Harrison, his life perfectly fit into the ideals of virile masculinity and medical ideology about his constitution (Chambers, 1846; Hindle, 1843; Hunter, 1841). War and exposure on the frontier had not weakened him but only made him stronger. During a time before germ theory, a man who had largely escaped most of the age's illness through isolation on the frontier weakened his immune system by giving a long speech in the cold without a jacket: a mistake that cost him his life. For Harrison, violence made him confident with unrealistic expectations of himself that he could be exposed to anything and come out stronger and in a less forgiving time period he lost his life. If he had survived in the wild how could he die in a city?

The still controversial suicide of Meriwether Lewis, pertinent to Lewis and Clark expedition, is perhaps the most dramatic example of a frontier governor's failure. After a political controversy and the loss of his papers regarding his famous exploration, Lewis was described by all first-hand accounts as psychotic and self-destructive. On his journey he brutally killed himself, at least according to everyone present. Scholars have highlighted this in terms of mania, and the frustrations of dealing with politics, yet they often overlook a key aspect of why such a journey would be difficult. It was another journey through the wilderness for a man that had expressed great anguish about rewriting his reports of his first expedition, and this is never interpreted as something that triggered memories of one of the most perilous journeys in the American history. The drunkenness and soul-loathing torrent that occurred prior to his death could just as easily be explained as self medicating through the avoidance spectrum of PTSD as it could through mania: mania itself has strong comorbidity with PTSD. Lewis was certainly armed like a man who dreaded such a journey, so it is important to note that he was expressing the same sort of anxiety about the wilderness expressed by Miller, but just through arming himself to the teeth and drinking himself into depression-laden stupors. Again, the diagnosis is less useful than recognizing how danger on the frontier was shaping the actions of these individuals, and the same kind of aversion to a dangerous region could be expressed in very different ways by different people. In all cases, we can see how violence shaped the understanding of place (Guice & Buckley, 2007).

Last, it is important to discuss Miller's commander, William Hull. One because of the way that Miller's criticism of Hull could have been just as easily leveled against his own governorship, but also because of the way that Hull expressed his own failure. William Hull presided over an embarrassing failure by surrendering 2,500 soldiers to a force of 1,200 (Forbes, 1814). This contentious surrender is still a hotly debated historiographical tradition with some scholars supporting Hull's surrender. The problem with this support is that it often supports Hull's racist logic for surrender, or at least fails to criticize it. For Hull, he had to surrender his force because he felt that Native Americans could not control their bloodlust and that the civilians sheltering inside the fortress would be slaughtered. In fact, he felt that tribes of North Amerindians were the most ferocious force imaginable. In his explanation for his surrender to the Secretary of War, he stated, "The bands of savages which had joined the British force were numerous and beyond any former example. Their numbers have since increased, and the history of the Barbarians of North Europe does not furnish examples of more greedy violence than these savages exhibited" (as cited in Hall, 1824). Again, an individual's understanding of the people of a hostile space, being inhumanly vicious and unstoppable force, better represents the tendency for veterans to view the world as an inherently and constantly dangerous place than the cool calculated decision of a General who had, in the Revolution, fought tribal opponents without the same ominous over representation of their ability. Although this author (Miller, 2012) has argued in the past that Hull's failure is better explained by PTSD, it is also important to note how the redefinition of oppositional space following his service in the American Revolution influenced his aversion to tribal opponents, as well as justifiably criticizing a school of thought that supports Hull's racist logic by highlighting the subsequent massacre at Raisin River without context and as if Hull was predisposed to such foreknowledge (Gilpin, 1958). It is more important to note how the intrusive spectrum of PTSD symptoms encouraged pre-existing cultural prejudice by created grossly poor estimations of the capability and savagery of tribal opponents.

Miller, like other frontier governors in the early national period whose life and leadership were influenced by violence, each expressed an aversion to the wild frontier differently. Miller consistently avoided it at first through somatic manifestation of an illness that originally occurred after battle, by retiring to an easier existence, and finally through a redefined sense of space found in a child-like affinity to flowers. Lewis became so overwhelmed on a journey that he drank himself into unhealthy stupors and ultimately killed himself. William Hull changed his views on the people of the frontier after the American Revolution and subsequently surrendered to a force much smaller than his own. Harrison perfectly expressed views about masculinity that believed that danger and wilderness made men stronger, only to tragically die in an outward expression of his vigorous nature. Each man exhibited changed views about the frontier because of violence they experienced on it.

The recognition of Miller's and his peer soldier governors' redefinitions of place and aversion to the frontier not only strengthened the understandings of PTSD prior to the language of trauma, but has also enabled better understanding how war has shaped the American landscape. Significant changes to the United States have occurred following major conflicts. For years, the cultivation of crops was thought to be an excellent place for soldiers to toil by turning their swords into plowshares. George Washington is perhaps the best arbiter of this trend, but its climax occurred in the land grant institutions that shared military training and agricultural education following the Civil War as well the use of farming in veteran homes in order to help struggling veterans return to civil life. Expats such as Ernest Hemingway's rejection of his nation almost entirely, and redefinition of values in eastern terms, is a well-understood reaction to the trenches of the WWI (Hemingway, 1929; Maugham, 1946). Many scholars have pondered why suburbs dominated the post Second World War (WWII) landscape because there was nothing that predated it (Rome, 2001). This new space of life could also be a generation's similar expression of aversions to the cities, towns and wilderness where they fought the largest conflict in history. Vietnam veterans, like others in their generation, have sought the open road through the freedom of powerful motorcycles and seek a mechanized nomadic existence (Veterans of Vietnam MC, n.d.). Iraq and Afghanistan war veterans have come to despise the roads that carried improvised explosive attacks. Charities committed to trail running camps are growing larger everyday (Team Red White and Blue, n.d.). With hope this generation's aversion for urban combat may help surge a new base of advocacy for conservation, and non-urban space.

The Human Landscape: Reshaping Mind, Negotiating Site

Hans Selye and Kurt Lewin were two of the many scholars who tried to understand human experience in relation to the theories of Charles Darwin. It is no mystery how such theories are built on adaptations to the environment. Still, Lewin's complicated life space includes cultural phenomenon into his psychological equation. By understanding a singular phenomenon at its core, such as the difficult reintegration of soldiers or PTSD in relation to both the environmental factors that caused the condition and the cultural background of the soldiers provides an excellent mechanism to understand culture and trauma. There is often conflict between ideals for soldiers and the realities of warfare's aftermath, but this dissonance breaks down the ideas of an age in a way that makes them more visible for scholars. Moreover, when narrowing the search even further to how these conflicts occur in relation to conceptualizations of nature it becomes apparent just how significant something like PTSD or normative physiological and psychological reactions to danger have made to the landscape of America. The human body is also seen as a landscape that is both affected by environmental factors as well an agent that impacts the environment. Specifically, soldiers as the common arbiter or archetypes of masculine ideals reshape ideas about landscape and the landscape itself as a result of the common renegotiation of home and place that occurs after sustained exposure to combat and danger. The wars themselves and the way they are fought bring the greatest influence to these changed beliefs about land in a Darwinist-adaptive drive to avoid images and places that remind soldiers of danger that they have survived. In every generation the veterans most emotionally affected by conflict reshape their space or favor residence in spaces that bare the least resemblance of the horrors of their burdensome memories of combat. The mind itself is reshaped by war in ways that often reshape the land around it.

Conclusion

Combat trauma, often exchanged with PTSD, is not the legacy of World Wars. Rather similar symptoms were reported even before World Wars but unfortunately, they were either not recorded or were undermined in the name of cowardice. Within today's understanding of PTSD as a master catch-all condition that explains all war-related mental illnesses, cases of fever in relation to combat trauma help illustrate what has initially been observed about the body's physiological reactions to stress in an era long before the proliferation of psychology as an accepted scientific discipline. Presence of such symptoms signifies the effect of war on its participants. War changes the perception of landscapes for its soldiers. Examining Miller's completely oppositional temperament towards the environment explains how and why many military heroes fail in civil life. War climate can subsequently trigger symptoms common to PTSD which in turn insist the soldier to change his perception about his landscape. Clearly, it signifies how mind itself is reshaped by war in ways that often reshape the land around it. Besides the ecological landscape, the human body can also be seen as a landscape that is affected by both environmental factors and acts like an agent that affects the environment. In a contemporary scenario, anticipating these perceptual changes in soldiers before or after the actual combat would strengthen their preparedness, as well as gear them up for healthy reintegration with the environment. Obviously, erstwhile significance of combat stress management stands in current conflicts also and will be a requisite for future as well because it is the soldier who pays a heavy price for all the conflicts he is engaged in.

References

American Psychiatric Association. (2013). *Diagnostic and statistical manual of mental disorders* (5th ed.). Washington, D.C.: Author.

An Ohio Volunteer. (1812). *The capitulation, or, A history of the expedition conducted by William Hull, brigadier-general of the Northwestern Army.* Chillicothe [Oh.]: James Barnes.

Andreasen, N. C. (2010). Posttraumatic stress disorder: A history and a critique. *Annals of the New York Academy of Sciences, 1208,* 67–71.

Bourke, J. (1996). *Dismembering the male: Men's bodies, Britain, and the Great War* (1st ed.). Chicago, IL: University of Chicago Press.

Bourke, J. (2009). *Rape: Sex, violence, history.* London, UK: Counterpoint.

Carter, C. (Ed.). (1953). *The territory of Arkansas 1819-1825* (XIX). Washington, D.C.: The National Archives.

Chambers, R. (1846). Observations on ague. *Provincial Medical and Surgical Journal, 10,* 129–131.

Clarence, C. (Ed.). (1953). *The territory of Arkansas 1819–1825* (XIX). Washington, D.C.: The National Archives.

Dean, E. T. (1929). *Documents relating to Detroit and vicinity, 1805–1813* (40). Lansing, MI: Michigan Historical Commission.

———. (1997). *Shook over hell: Post-traumatic stress, Vietnam, and the Civil War.* Cambridge, MA: Harvard University Press.

Forbes, J. G. (1814). *Report on the trial of Brig. General William Hull commanding the Northwestern Army of the United States by a court martial held at Albany on Monday, 3rd of January, 1814 and succeeding days.* New York, NY: Pray and Bowden for Eastburn and Kirk.

Fussell, P. (2000). *The Great War and modern memory* (25th anniversary ed.). New York, NY: Oxford University Press.

Gilpin, A. R. (1958). *War of eighteen-twelve in the Old Northwest.* East Lansing, MI: Michigan State University Press.

Goodwin, C. J. (2011). *A history of modern psychology* (Vol. 4). Wiley Hoboken, NJ: John Wiley.

Greer, A. (2010). National, Transnational, and hypernational historiographies: New France meets early American history. *Canadian Historical Review, 91*(4), 695–724.

Guice, J. D. W., & Buckley, J. H. (Eds.). (2007). *By his own hand? The mysterious death of Meriwether Lewis.* Norman, OK: University of Oklahoma Press.

Hawthorne, N. (1981). *The scarlet letter and the house of the seven gables.* New York, NY: Signet Classics.

Hemingway, E. (1929). *A farewell to arms.* London, UK: Cape.

Hindle, R. (1843). On a critical coldness supervening in the course of continued fever with cases. *Provincial Medical Journal and Retrospect of the Medical Sciences, 7,* 9–12.

Hinton, D., Hinton, S., Pham, T., Chau, H., & Tran, M. (2003). 'Hit by the wind' and temperature-shift panic among Vietnamese refugees. *Transcultural Psychiatry, 40,* 342–376.

Hinton, D., Pich, V., Chhean, D., & Pollack, M. (2004). Olfactory-triggered panic attacks among Khmer refugees: A contextual approach. *Transcultural Psychiatry, 41,* 155–199.

Hoge, C. W. (2010). *Once a warrior–always a warrior: Navigating the transition from combat to home–including combat stress, PTSD, and mTBI.* Guilford, CT: GPP Life.

Hull, W. (1824). *Memoirs of the campaign of the North Western Army of the United States, A.D. 1812 in a series of letters addressed to the citizens of the United States: With an appendix containing a brief sketch of the revolutionary services of the author.* Boston, MA: True & Greene.

Hunter, J. (1841). Course of lectures on physiology and surgery, delivered at St. George's hospital, lecture VI on disease. *Provincial Medical and Surgical Journal, 2,* 219–222.

Jones, E., & Wessely, S. (2004). Hearts, guts and minds: Somatisation in the military from 1900. *Journal of Psychosomatic Research, 56,* 425–429. doi: 10.1016/S0022-3999(03)00626-3

Jones, E., Palmer, I., & Wessely, S. (2002). War pensions (1900-1945): Changing models of psychological understanding. *British Journal of Psychiatry, 180,* 374–379.

Karl, A., Schaefer, M., Malta, L. S., Dorfel, D., Rohleder, N., & Werner, A. (2006). A meta-analysis of structural brain abnormalities in PTSD. *Neuroscience and Biobehavioral Review, 30,* 1004–1031. doi:10.1016/j.neubiorev.2006.03.004

Ledbetter, C. (1988). General James Miller: Hawthorne's hero in Arkansas. *The Arkansas Historical Quarterly, 47,* 99–115.

Lewin, K. (1917). Krieglandschaft [War Landscape]. *Zeitschrift Angewandter Psychologie, 12,* 440–447.

Lewin, K. (2009). The Landscape of war. *Art in Translation*, *1*, 199–209.

Maugham, W. S. (1946). *The razor's edge: A novel*. Philadelphia, PA: Triangle books, The Blakiston Company.

Micale, M. S. (2008). *Hysterical men: The hidden history of male nervous illness* (1st ed.). Cambridge, MA: Harvard University Press.

Micale, M., & Lerner, P. (Eds.). (2001). *Trauma, psychiatry and history: A conceptual and historiographical introduction*. Cambridge, MA: Cambridge University Press.

Miller, J. (2012). General William Hull's trials – Was this early PTSD – One possible explanation for the unprecedented surrender of Detroit, 1812. *Canadian Military Journal*, *12*, 52–57.

Ozzer, E. J., & Weiss, D. S. (2004). Who develops posttraumatic stress disorder? *Current Directions in Psychological Science*, *13*, 169–172.

Resch, J. P. (1999). *Suffering soldiers: Revolutionary war veterans, moral sentiment, and political culture in the early republic*. Amherst, MA: University of Massachusetts Press.

Rome, A. (2001). *The bulldozer in the countryside: Suburban sprawl and the rise of American environmentalism (studies in environment and history)*. New York, NY: Cambridge University Press.

Schaller, B. R. (2012). *Veterans on trial: The coming court battles over PTSD* (1st ed.). Washington, DC: Potomac Books.

Selye, H. (1973). The evolution of the stress concept: The originator of the concept traces its development from the discovery in 1936 of the alarm reaction to modern therapeutic applications of syntoxic and catatoxic hormones. *American Scientist*, *61*, 692–699.

———. (1977). *The stress of my life: A scientist's memoirs*. Toronto, ON: McClelland and Stewart.

Shay, J. (1994). *Achilles in Vietnam: Combat trauma and the undoing of character*. New York, NY: Atheneum.

———. (2002). *Odysseus in America: Combat trauma and the trials of homecoming*. New York, NY: Scribner.

Sommerville, D. M. (1995). The rape myth in the old South reconsidered. *The Journal of Southern History*, *61*, 481–518.

———. (2011). 'Will they ever be able to forget?' Confederate soldiers and mental illness in the defeated south. In S. Berry (Ed.), *Weirding the war: stories from the civil war's ragged edge,* (pp. 321–339). Athens: University of Georgia Press.

Team Red, White and Blue. (n.d.). *Team red, white and blue*. Retrieved from http://teamrwb.org/

Veterans Administration. (n.d.). *Reminders of Trauma: Anniversaries*. Retrieved from http://www.ptsd.va.gov/professional/research-bio/research/anniversary_reactions_pro.asp

Veterans of Vietnam MC. (n.d.). *History of the veterans of Vietnam motor club*. Retrieved from http://www.vovma.org/VOVMA.html

Walker, A. (1816). *A journal of two campaigns of the Fourth Regiment of U. S. Infantry in the Michigan and Indiana territories under the command of Col. John P. Boyd and Lt. Col. James Miller, during the years 1811 & 12*. Keene, NH: Sentinel press.

Wessely, S. (2006). Twentieth-century theories on combat motivation and breakdown. *Journal of Contemporary History*, *41*, 268–286.

White, L. J. (1960). James Miller: Arkansas' first territorial governor. *The Arkansas Historical Quarterly*, *19*, 12–30.

Young, A. (1997). *The harmony of illusions: Inventing post-traumatic stress disorder*. Princeton, NJ: Princeton University Press.

5

Stereotype Threat and Marksmanship Performance

Emerald M. Archer

Scholarship shows that negative stereotypes have the power to undermine the performance of various social groups in a myriad of contexts. In an attempt to understand the impact of negative stereotypes about female Marines on their military performance, a theoretical approach is adopted here to illustrate that stereotype threat (ST) is both relevant to women in the United States Marine Corps (USMC) context and may have specific negative consequences on female Marines. This chapter presents a possible model for the application of ST theory within the military domain to explain women's underperformance in tasks of marksmanship. The power of contextual cues on USMC performance is real and may have implications for the health of the American All Volunteer Force (AVF).

ST Theory and its Relevance to US Female Marines

ST is a phenomenon defined as a "social-psychological threat that arises when one is in a situation or doing something for which a negative stereotype about one's group applies." Claude Steele, the pioneer of ST research, goes on to say:

> This predicament threatens one with being negatively stereotyped, with being judged or treated stereotypically, or with the prospect of conforming to the stereotype. Called stereotype threat, it is the situational threat—a threat in the air—that, in general form, can affect the members of any group about whom a negative stereotype exists (e.g., skateboarders, older adults,

White men, gang members). Where bad stereotypes about these groups apply, members of these groups can fear being reduced to that stereotype. And for those who identify with the domain to which the stereotype is relevant, this predicament can be threatening. (Steele, 1997, p. 614)

This definition highlights the fact that for a negative stereotype to be threatening, it must be self-relevant. Steele details, as complementary to his definition, three conditions that must be present in order for ST to take place. First, there must be widespread awareness of the negative stereotype associated with one's group. A second condition, an individual's identification with the relevant domain, must also exist. Degree of individual identification, or how much an individual stakes their self-image on particular task or ability, will render individuals prone to or safe from ST. Research demonstrates that more highly domain-identified individuals are most vulnerable to ST because their self-regard is wrapped up in the successful completion of the task (Aronson et al., 1999; Leyens et al., 2000; Stone, 2002; Stone et al., 1999). Finally, ST will only occur when a negative stereotype is relevant to the individual during a domain performance situation.

All conditions of ST are satisfied for female Marines within the USMC domain. First, there is widespread awareness that service-women are generally negatively stereotyped. Literature based on the military context, for example, shows that men are perceived to be more competent leaders than women. In a study including 288 West Point cadets, Rice, Bender and Vitters (1980) constructed mixed male and female problem-solving groups and measured the attitudes of male cadets towards their female counterparts. In groups composed of males who had progressive attitudes towards female leadership, there was no difference between problem-solving group performance and "follower satisfaction" between male and female-led groups (Chemers, 1997). In groups that held traditional attitudes that were hostile towards female military leadership, the groups led by females performed less effectively than those led by male cadets. Males in the latter groups were also dissatisfied with the leadership the female cadets provided. Moreover, Boldry, Wood and Kashy (2001) inves-tigated the perceptions of ROTC cadets regarding men and women. Individual evaluations of ROTC cadets (men and women) show that more men than women are believed to have the leadership qualities for effective military performance. Women are also believed to have more feminine qualities that impair the effectiveness of military performance. These investigations demonstrate that servicemen imagine women as generally less competent military leaders, and as a consequence, women's service and membership within the community are seen as less valuable.

Second, servicewomen are also likely to be highly identified with their role in the Armed Forces since they have self-selected into the community. The AVF, as the name suggests, accepts men and women who meet the standards of a particular service branch. Women who choose to join the USMC are certainly atypical. Generally, these women are confident in their abilities, physically fit, opinionated and ready to accept the challenge that the USMC can provide. Female Marines cite many reasons for joining, but regardless of their reasoning, they knowingly join a community that mandates their minority status and promotes traditional gender roles. Women recognize these obstacles before they join, and many of them are motivated to work hard to overcome them. This *prove 'em wrong* mentality may actually render women more vulnerable to ST because every performance counts—that is, in order to be fully accepted by male counterparts, they have to continually raise the bar. This pressure, unique to servicewomen in the Armed Forces, could be at the heart of their underperformance.

Finally, stereotypes impugning the competence of female Marines apply to several domain performance situations within the Marine Corps. Women in Military Occupational Specialties (MOS) that require mechanical knowledge will have to perform their duties in the face of stereotypes that suggest women know very little about automotive and aeronautical mechanics. In deployed situations, servicewomen may engage in firefights with the enemy. Here, two stereotypes may be relevant: (a) that women are not physically and emotionally strong enough to engage the enemy, and (b) women are not as skilled at using firearms as their male counterparts. Negative stereotypes that characterize women as incompetent Marines are often relevant to the jobs and tasks women are responsible for completing. As a result, the risk of confirming a stereotype as self-characteristic rises. Equally in domains of marksmanship, the risk of confirming a negative stereotype rises considerably for female Marines.

USMC Marksmanship Qualifications: A Possible Domain of Underperformance?

Female Marines have been successful in both deployed and stateside arenas. Although many women do excel in the "most masculine of traditions" (Keegan, 1993), female Marines generally underperform relative to their male counterparts while qualifying on a firearm.[1] Every Marine, at the recruiting stage and as they progress through the enlisted ranks, qualifies annually on an M-16 service rifle. Data from Fiscal Year (FY) 1999 provided by Marine Corps Recruiting Depot-West (MCRD-West) illustrates their underperformance.

Table 5.1 shows that a much higher percentage of females do not pass their initial qualification as compared to male recruits during boot camp. Said another way, the first time they qualify on the rifle after learning marksmanship fundamentals, only 68 per cent of female Marines pass the qualification. The "initial qualification" data account for an argument that suggests women have less experience with firearms than males do when they join the USMC. Having said that, female recruits continue to underperform relative to their male counterparts after completing 13 weeks of basic training. By the final qualification during basic training, 97 per cent of female recruits pass the marksmanship qualification. The remaining three per cent do not pass, and are subsequently dropped from the programme. One hundred per cent of males in FY 2009 graduated basic training. Furthermore, there are more highly skilled males than females represented in Table 5.1. More males than females are given the grade of *expert* and *sharpshooter*. Meanwhile, over half of the female population received the grade of *marksman*, which is the lowest scoring category.

Freedom of Information Act (FOIA) data provided by the Department of the Navy in 2008 corroborates the aforementioned trends. Data analysis for 1000 Marines shows that for those ranked Private, Private First-Class and Lance Corporal, the mean rifle score for men is 207, whereas the mean rifle score for women is 195. However,

---

[1] Servicewomen generally underperform relative to their male counterparts on the Armed Services Vocational Aptitude Battery (ASVAB) examination and tests of physical strength.

Table 5.1
Marksmanship data

	Men (n = 35,000)	Women (n = 3,000)
Initial Qualification (%)	88	68
Experts (%)	23	15
Sharpshooters (%)	28	17
Marksman (%)	49	65
Total (%)	*100*	*97*

Source: Marine Corp Recruiting Depot—Parris Island.

data received for male and female Non-Commissioned Officers (NCOs) did not include actual marksmanship scores. Data were provided for each randomly selected NCO in the form of grades—marksman, sharpshooter and expert. In order to best assess average scoring, marksman, sharpshooter and expert categories were rated as one, two and three, respectively. When averages were calculated, female NCOs had a mean rifle and pistol score of 2.28 and 1.81, respectively. Male NCOs, on the other hand, have a mean rifle and pistol scores of 2.73 and 2.31, respectively. Even though actual qualification scores were not provided in the data set, it is still obvious that women underperform in marksmanship qualification. The 2008 FOIA data, along with the marksmanship data provided by MCRD-West, illustrate that female Marines generally underperform relative to their male counterparts in tasks of marksmanship.

Despite what the data suggest, many scholars posit that women are, in fact, better shooters than males. Laura Miller claims that the 'actual common knowledge is that women make better sharpshooters than men do'. She states that women typically perform better than their male counterparts because they take the time to aim carefully, and allow for breathing adjustments. Men, on the other hand, "just want to blast quickly over and over again" (Miller, 2008; personal communication). Ben Dolan, a former Marine sniper, claims that

> women can shoot better, by and large, and they're easier to train because they don't have the inflated egos that a lot of men bring to [rifle range]… Women will ask for help if they need it, and they will tell you what they think. (as cited in Wan, 2006)

Overall, evidence from several sources shows that women underperform relative to their male counterparts in the domain of marksmanship. Marksmanship, the domain under discussion here, may render female Marines vulnerable to ST because of gendered stereotypes related to firearms and the understanding that Marines are riflemen through and through.

Before one can assess whether or not the domain of marksmanship is threatening to female Marines, it is important to note that some Marines have noticed women's underperformance in marksmanship and have taken strides to improve their performance. Chief Warrant Officer William Tinney, a Marine who directed a rifle range at Parris Island, spoke about his personal project launched in December 2007: the creation of a rifle range dedicated to female recruits. He emphasized that by building a community of trainers for women, those "trainers on the range can establish an understanding for how females think". The creation of an exclusively female rifle range entailed establishing a community of coaches trying to improve women's scores and an increased coach: recruit ratio so that recruits would have more individualized instruction. Between December 2007 and May 2008, Tinney's range pushed through 1300 female recruits. More amazing than this is the fact that the rate of passed initial qualifications increased by 20 per cent. Before the range was created, only 59 per cent of female recruits passed their rifle qualification (personal communication, 2009). In 2009, 79 per cent of the female recruits who took their initial qualification at this particular range passed. Although this improvement does not meet CWO Tinney's expectations (his expectations are to have all male and female recruits a 95 per cent pass rate), this is a vast improvement and could be associated with Iecreased anxiety resulting from having no male recruits around for comparison.

Is the Marksmanship Domain Threatening to Female Marines?

Marines will repeatedly say, "Every Marine is a rifleman". Regardless of age, gender or MOS, all Marines are trained to operate the M16-A2 service rifle during boot camp. This phrase is perhaps the Marine Corps' most widely known edict. It illustrates, in the most simplistic terms, the Corps' desire to infuse into every Marine the fighting spirit and ethos upon which the organization has relied for more than 227 years (Milks, 2003).

Everything revolves around the rifleman. "Marine Aviation, Marine Armor, Marine Artillery, and all supporting arms and war-fighting assets exist to support the rifleman" (Struckly, 2001). The saliency of the rifleman is understood by the fact that every recruit repeats, memorizes and ultimately lives by the *Rifleman's Creed*. Authored by Master General William H. Rupertus after the Japanese attack on Pearl Harbor, the creed is as follows:

> This is my rifle. There are many like it, but this one is mine. It is my life. I must master it as I must master my life. Without me, my rifle is useless. Without my rifle, I am useless. I must fire my rifle true. I must shoot straighter than the enemy who is trying to kill me. I must shoot him before he shoots me. I will. My rifle and I know that what counts in war is not the rounds we fire, the noise of our burst, or the smoke we make. We know that it is the hits that count. We will hit.
>
> My rifle is human, even as I am human, because it is my life. Thus, I will learn it as a brother. I will learn its weaknesses, its strengths, its parts, its accessories, its sights and its barrel. I will keep my rifle clean and ready, even as I am clean and ready. We will become part of each other.
>
> Before God I swear this creed. My rifle and I are the defenders of my country. We are the masters of our enemy. We are the saviors of my life.
>
> So be it, until victory is America's and there is no enemy. (Sturkey, 2003)

Marksmanship, the most fundamental of Marine Corps skills, is inextricably bound to the identity of the Marine. It is so important that all Marines, spanning combat and non-combat MOSs, qualify on the rifle annually to ensure that their marksmanship skills are continually practiced and refreshed.[2] Because the identity of a Marine is bound to competent marksmanships skills, all Marines take annual qualifications seriously because their self-image as a successful Marine is contingent upon a high marksmanship qualification score.

The *Rifleman's Creed* illustrates the importance of marksmanship in the USMC. In the domain of marksmanship, women are extremely vulnerable to underperformance. Most female Marines acknowledge the statistics—specifically, they know that more men are awarded

---

[2] Note that all enlisted personnel (E1–E5) qualify on a rifle, whereas NCOs (E6 and above) and officers are issued pistols to qualify on.

grades of expert and sharpshooter than women. Women also recognize that they are an obvious minority, only making up six per cent of the USMC. Environmental cues such as a skewed gender ratio may put additional pressure on women. Anxiety may also develop from feeling the obligation to continually overperform in certain domains as a tribute to female Marines before them. When a female First Lieutenant was asked for her opinion on origins of female underperformance in terms of marksmanship, she stated:

Subject 10:   I shoot expert on the rifle and I shoot sharpshooter on pistol. In terms of the pistol, I think there is…more pressure for me to shoot better because it's my [Table of Organization][3] weapon. [And this pressure] is totally self imposed… I'll shoot expert on the pistol all week [until I get to qualification day]. Most times, when I qualify on the pistol, I miss expert by, like, ten points. Had I gotten one more round in the 10 I would've [gotten expert].

Archer: Really?

Subject 10:   It's totally self-imposed. It's like a mental block for me, like, and it's totally personal. It has nothing to do with the fact that I'm a woman or who I'm shooting next to.

Archer: Where does that come from? Where does that pressure come from do you think?

Subject 10:   Oh, for me I know exactly where it comes from. I have this horrible innate fear of failure, so sometimes I feel like, well, I don't wanna try super hard because if I give it my all and I still fail, then what, you know?

Although anecdotal, the feeling Subject 10 expresses here—fear of failure—is an issue that several Marines commented on throughout formal interviews and informal conversations (Subject 2). Moreover, a female Major also expressed feeling added pressure on the rifle range because she did not want to be responsible for giving other female Marines a bad name. Beyond all this, female Marines who want to be taken as serious, competent Marines understand that they have to excel in the face of stereotypes that depict women as incompetent riflemen. The burden they bear—the stress that results from the need to disconfirm a negative stereotype that pertains to them—could negatively affect the scores of female Marines everywhere.

---

[3] Table of Organization (TO) refers to the USMC assigned weapon. USMC officers are assigned a 9-mm pistol. Marine Sergeants and below are assigned a rifle.

Overall, the literature on performance effects of negative stereotyping is rich and related to both race and gender. Empirical support demonstrates that ST can affect members of many stereotyped social groups and has explained female academic underperformance relative to men (Steele, Reisz, Williams, & Kawakami, 2007), African-American academic underperformance relative to Caucasian students (Steele & Aronson, 1995), the underperformance of Caucasian men compared to Asian-American men on math tests (Aronson et al., 1999), and the academic underperformance of Hispanic students wherein analytical ability was measured (Gonzales, Blanton, & Williams, 2002). Beyond academic domains, ST is described as negatively affecting homosexual men in childcare domains (Bosson, Haymovitz, & Pinel, 2004), Caucasian male athletes with respect to natural athletic ability (Stone, Lynch, Sjomeling, & Darley, 1999), women in negotiation (Kray, Galinksy, & Thompson, 2002), and women with respect to their driving skills (Yeung & von Hippel, 2002). The ST literature demonstrates that negative stereotypes have the power to undermine performance for various social groups, contexts and disciplines. Moreover, specific ST consequences exist that may impact the female Marine personally and professionally.

Consequences of ST

ST is associated with a variety of negative consequences. A number of scholars have replicated Steele and Aronson's (1995) finding that invoking group memberships associated with stereotypes can harm performance on tasks in which poor performance might confirm stereotypes (Stroessner, Good, & Webster 2008). Although the pioneering studies focused on written test performance, subsequent research has been applied to a diverse set of tasks and has examined a variety of consequences. The following four consequences of ST (beyond decrements in performance) will be discussed: (a) self-handicapping, (b) reactance, (c) distancing the self from the stereotyped group and (d) altering professional aspirations.[4]

---

[4] This review of the consequences of ST is based on the website created by Steven Stroessner, Catherine Good, and Lauren Webster. The website, reducingstereotypethreat.org, was created as a tool to help individuals understand ST as a social psychological phenomenon.

Self-handicapping

Self-handicapping is thought to be a defensive strategy individuals use to provide attributions for failure on a particular task. Some individuals construct barriers that undermine performance, and when an individual underperforms on a task, they can blame the barriers for that performance rather than deficiencies in ability. If an individual performed well despite the existence of such barriers, an improved evaluation of their performance can be made on the basis of the individual's ability to overcome obstacles to performance (Stroessner, Good, & Webster, 2008). Consistent with the above remarks, research demonstrates that ST may lead individuals to adopt more self-handicapping behaviour. Keller (2002) demonstrates that girls who perform poorly on a mathematics examination under ST are more likely to invoke stress they experienced before taking the examination. Steele and Aronson (1995) illustrated similar behaviour in African–American students— specifically, the students under ST are more inclined to produce excuses for their possible failure. Overall, ST might cause individuals to reduce preparation time for examinations and expend less effort on the task.

Reactance

Contrary to the chief consequence of ST, decrements in performance, ST can sometimes increase the quality of performance for targeted individuals. Although the literature on stereotype reactance is small compared to the literature on ST, Kray, Galinsky and Thompson (2001) clearly demonstrated the phenomenon of reactance in didactic negotiations.

> Informing negotiators that stereotypically masculine traits predict performance at the bargaining table and that these traits differ by gender led to a counterintuitive outcome: Female negotiators outperformed their male counterparts. Thus, instead of confirming and validating the connection between gender and negotiation ability, women vitiated the negative stereotype by reacting against it. More specifically, women who were told that it is men who have the upper hand at the bargaining table identified with counterstereotypic traits and set higher and more aggressive goals. The

ability to marshal empowering cognitions in the face of what might seem to be the worst performance conditions is all the more surprising given that under most conditions, men do outperform women at the bargaining table. (Kray et al., 2004, pp. 399–400)

The aforementioned example clearly shows that ST does not consistently result in assimilation effects, the hallmark of ST. In the same 2001 study, Kray and her colleagues noted that explicitly recognizing that the association between "stereotypically masculine traits and effective negotiating is linked to gender differences" can produce an advantage for women at the bargaining table (p. 401). This blatant activation of gender stereotype is believed to limit the female negotiators' ability to perform, and thus, evokes stereotype reactance. Said another way, reactance is the tendency to behave and respond in ways that are contradictory to the relevant stereotype.

The work of Kray and her colleagues leads one to conclude that the way in which stereotypes are activated determines the outcome of an interaction—that is, assimilation effects (ST) and divergent effects (reactance). Based on the result of previous research (Kray et al., 2001), blatant activation of gender stereotypes leads female negotiators to act counter to the relevant stereotype. In essence, explicit masculine stereotype activation leads to disidentification with stereotypically feminine traits, and in turn, compels women to set higher aspirations for themselves as compared to implicit stereotype activation. Overall, research suggests that "reactance processes likely result from stereotypically disadvantaged negotiators' awareness that the stereotypical perception of their ability is invalid in the current context, that both negotiators are actually on a level playing field or that they themselves hold a power-based advantage" (p. 401). In terms of negotiation, power can be as simple as having the ability to walk away from the negotiation table. If individuals do not internalize the aforementioned qualities, they will likely assimilate and fall prey to ST. Also, reactance becomes more likely when individuals are high achievers and capable in terms of the relevant task (Kray, Reb, Galinsky, & Thompson, 2004; Kray, Thompson, & Galinsky, 2001; Oswald & Harvey, 2001). As female Marines self-select into the USMC, the type of personality the USMC draws might be one that defies gender roles/stereotypes. Studies that investigate this consequence in civilian settings could be applied to the military context to test claims that servicewomen sometimes outperform their male counterparts in some military tasks (e.g., Laura Miller's assessment that women are often better marksman than servicemen). Thus, this research illustrates that underperformance under conditions of ST does not apply to every individual and is not inevitable.

Distancing the Self from the Stigmatized Group

ST also affects the degree to which individuals identify with activities associated with their social group. Steele and Aronson (1995) illustrated that African–Americans who underperform relative to their Caucasian counterparts express "weaker preferences for stereotypically African American activities such as jazz, hip-hop, and basketball" (Stroessner, Good, & Webster, 2008). Distancing oneself from the stigmatized group may reflect the desire to be viewed through a different lens—one that does not incorporate racial stereotyping.

Identity bifurcation, a process that distances the self from the stigmatized group, has been demonstrated in the case of women. Identity bifurcation refers to the choice of emphasizing an unthreatened identity over a threatened one. Pronin, Steele and Ross (2004) expressed that women under ST deny their feminine traits that are strongly related to the stereotype of women's mathematic aptitude. However, these women do not disavow their feminine traits that are mildly associated with the stereotype. Most interesting is the finding that only women who highly identify themselves with mathematics bifurcate their identities in response to ST in the domain of mathematics. Overall, stereotyped individuals will sometimes distance themselves from a portion of their social identity that "bears the burden" of a negative stereotype in order to protect their identity as a competent person in a given domain. Distancing oneself from a negatively stereotyped group is a strategy many female Marines use to improve self-esteem and performance. In personal interviews, many female Marines mentioned that it was important for them to be the best Marine they could be for the trail blazing female Marines before them, while simultaneously disassociating themselves with other "piece of shit" female Marines (Archer, 2013).

Altered Professional Aspirations

The last consequence of ST reviewed here—altered professional aspirations—is clearly demonstrated in the social psychological literature. ST can redirect a stereotyped individual's career path and alter their professional identity. The literature on women in non-traditional domains is particularly interesting in terms of this investigation. Steele, James and Barnett (2002) showed that undergraduate women in male-dominated departments (e.g., engineering, mathematics) report higher levels of discriminatory behaviour, and that these women are more likely to consider changing their major compared to their female counterparts in traditionally female-dominated departments.

ST may also lead to women withdrawing from discussions relevant to their majors. Murphy, Steele and Gross (2007) found that women mathematics and science majors who viewed a discussion of mathematics and science topics in which males were numerically dominant showed lowered interest in participating in such a discussion in the future (Stroessner, Good, & Webster, 2008). Consistent with these findings, Gupta and Bhawe (2007) reported that women express less interest in entering a field that emphasizes the importance of stereotypically masculine traits. Altered performance expectations also exist for women in the military. Because women are stereotyped against in this particular environment, the standards that exist for women differ greatly than those for men. The lowered performance expectations that come from peers could negatively affect morale, lead servicewomen to question their self-worth and may even result in prompting service women to leave their subscribed service branch.

Research illustrates that stereotypes can cause targeted individuals enough discomfort that they will eventually drop out of their current domain and redefine their career goals, aspirations and identities. Where women are concerned, if they avoid domains such as mathematics and science, they ultimately lose any opportunity to enter careers in domains of engineering and technology. The literature is based almost entirely on civilian populations and domains. This research could be easily extended to the context of the Armed Forces to see how improvements in performance, retention and satisfaction could be achieved.

Conclusion and Future Directions

This chapter explores one particular application of ST theory to explain whether negative stereotypes about female Marines are powerful enough to weaken marksmanship performance. The model presented here is tested in an on-site experiment, which suggests that the threat is real: decrements in performance for women under ST conditions were subtle, yet significant (Archer, 2014). Because ST may threaten female Marines in domains of marksmanship, it might also permeate to other areas of Marine Corps performance. Stereotypes must be studied and managed to ensure that military performance in various contexts is maximized. As policymakers and military leadership investigate which combat roles to open to American servicewomen, they should be aware that social cues endemic to the institution could mask the true performance potential of servicewomen. This, in turn, could ultimately lead to the poor utilization of the fighting force.

Similarly, other developing countries such as India that are spearheading towards the inclusion of females in combat roles can take necessary cognizance of the phenomenon in order to build up the military capital for their respective nations.

References

Archer, E. M. (2013). The power of gendered stereotypes in the US marine crops. *Armed Forces & Society, 39*, 359–391.

———. (2014). You shoot like a girl: Stereotype threat and marksmanship performance. The *International Journal of Interdisciplinary Civic & Political Studies, 8*, 9–21.

Aronson, J., Lustina, M. J., Good, C., Keough, K., Steele, C. M., & Brown, J. (1999). When white men can't do math: Necessary and sufficient factors in stereotype threat. *Journal of Experimental Social Psychology, 35*, 29–46.

Boldry, J., Wood, W., & Deborah A. K. (2001). Gender stereotypes and the evaluation of men and women in military training. *Journal of Social Issues, 57*, 689–705.

Bosson, K., Haymovitz, E. L., & Pinel, E. C. (2004). When saying and doing diverge: The effects of stereotype threat on self-reported versus non-verbal anxiety. *Journal of Experimental Social Psychology, 40*, 247–255.

Chemers, M. M. (1997). *An integrative theory of leadership.* New York, NY: Psychology Press.

Gonzales, P. M., Blanton, H., & Williams, K. J. (2002). The effects of stereotype threat and double-minority status on the test performance of Latino women. *Personality and Social Psychology Bulletin, 28*, 659–670.

Gupta, V. K., & Bhawe, N. M. (2007). The influence of proactive personality and stereotype threat on women's entrepreneurial intentions. *Journal of Leadership and Organizational Studies, 13,* 73–85.

Keegan, J. (1993). *A history of warfare.* New York, NY: Alfred A. Knopf Publishing.

Keller, J. (2002). Blatant stereotype threat and women's math performance: Self-handicapping as a strategic means to cope with obtrusive negative performance expectations. *Sex Roles, 47,* 193–198.

Kray, L. J., Reb, J., Galinsky, A. D., & Thompson, L. (2004). Stereotype reactance at the bargaining table: The effect of stereotype activation and power on claiming and creating value. *Personality and Social Psychology Bulletin, 30,* 399–411.

Kray, L. J., Thompson, L., & Galinsky, A. (2001). Battle of the sexes: Gender stereotype confirmation and reactance in negotiations. *Journal of Personality and Social Psychology, 80,* 942–958.

Kray, L. J., Galinksy, A. D., & Thompson, L. (2002). Reversing the gender gap in negotiations: An exploration of stereotype regeneration. *Organizational Behavior and Human Decision Processes, 87,* 386–409.

Leyens, J. P., Desert, M., Croizet, J. C., & Darcis. C. (2000). Stereotype threat: A lower status and history of stigmatization preconditions of stereotype threat? *Personality and Social Psychology Bulletin, 26,* 1189–1199.

Milks, K. A. (2003). Ensuring 'Every marine a rifleman' is more than just a catch phrase. *Marine Corps News.* Retrieved from http://web.archive.org/web/20071224075658/http://www.usmc.mil/marinelink/mcn2000.nsf/0/b5ac3322e236c38985256feb00492f93?OpenDocument

Miller, L. (2008). Personal E-mail Correspondence.

Murphy, M. C., Steele, C. M., & Gross, J. J. (2007). Signaling threat: How situational cues affect women in math, science, and engineering settings. *Psychological Science, 18,* 879–885.

Oswald, D. L., & Harvey, R. D. (2000–2001). Hostile environments, stereotype threat, and math performance among undergraduate women. *Current Psychology: Developmental, Learning, Personality, Social, 19,* 338–356.

Pronin, E., Steele, C., & Ross, L. (2004). Identity bifurcation in response to stereotype threat: Women and mathematics. *Journal of Experimental Social Psychology, 40,* 152–168.

Rice, R. W., Bender, L. W., & Vitters, A. G. (1980). Leader sex, follower attitudes toward women, and leader effectiveness: A laboratory experiment. *Organizational Behavior and Human Performance, 25,* 46–78.

Steele, C. M. (1997). A threat in the air: How stereotypes shape intellectual identity and performance. *American Psychologist, 52,* 613–629.

Steele, C. M., & Aronson, J. (1995). Stereotype threat and the intellectual test performance of African-Americans. *Journal of Personality and Social Psychology, 69,* 797–811

Steele, J., James, J. B., & Barnett, R. (2002). Learning in a man's world: Examining the perceptions of undergraduate women in male-dominated academic areas. *Psychology of Women Quarterly, 2,* 46–50.

Steele, J. R., Reisz, L., Williams, A., & Kawakami, K. (2007). Women in mathematics: Examining the hidden barriers that gender stereotypes can impose. In R. J. Burke & M. C. Mattis (Eds.), *Women and minorities in science, technology, engineering and mathematics: Upping the numbers* (pp. 159–183). Northampton: Edward Elgar.

Stone, J. (2002). Battling doubt by avoiding practice: The effect of stereotype threat on self-handicapping in white athletes. *Personality and Social Psychology Bulletin, 28,* 1667–1678.

Stone, J., Lynch, C. I., Sjomeling, M., & Darley, J. M. (1999). Stereotype threat effects on black and white athletic performance. *Journal of Personality and Social Psychology, 77,* 1213–1227.

Stroessner, S., Good, C., & Webster, L. (2008). *Reducing stereotype threat.* Retrieved from www. reducingstereotypethreat.org

Sturkey, M. F. (2003). *Warrior culture of the US marines.* New York, NY: Heritage Press International.

Wan, S. (2006). Women's role in combat: Is ground combat the next front? *Hohonu: Journal of Academic Writing, 4*(1). Retrieved from http://www.uhh.hawaii.edu/ academics/ hoho nu/writing.php?id=112

Yeung, N. C. J., & von Hippel, C. (2008). Stereotype threat increases the likelihood that female drivers in a simulator run over jaywalkers. *Accident Analysis & Prevention, 40,* 667–674.

6

Misconduct Behaviours in Armed Forces

*Pankaj Kumar Sharma, Ashutosh Ratnam and T. Madhusudhan**

The inherent brutality of war cannot be denied—some of humankind's worst and most shameful chapters have been written in theatres of combat. While nursing on the Black Sea in 1855, the famed English nurse Florence Nightingale wrote to her family:

> What the horrors of war are, no one can imagine. They are not wounds and blood and fever, spotted and low, or dysentery, chronic and acute, cold and heat and famine. They are intoxication, drunken brutality, demoralisation and disorder on the part of the inferior… jealousies, meanness, indifference, selfish brutality on the part of the superior.
>
> (as cited in Vicinus & Nergard, 1989.)

If a historical appraisal of many major misconduct behaviours is performed, and the origins of many of them are used as a basis for conjecture, it can be theorized that perhaps there are certain peculiarities of military life which predispose towards actions fitting the definition of misconduct behaviour. Misconduct usually takes place when coping mechanisms are either flawed or inadequate. A degree of understanding regarding this does pervade amongst military circles, and it resulted in the adoption of the 'stress management briefing'. This chapter highlights certain shortcomings that such a briefing will have in a military setup. Finally, several specific interventions to help curb military misconduct are suggested. Specific emphasis is placed

---

* The authors would like to thank Martin L Freidland, University Professor and Professor of Law Emeritus at the University of Toronto for his advice and guidance, and for kind permission to source material from his landmark publication *Controlling Misconduct in the Military: A study prepared for the Commission of Inquiry into the Deployment of Canadian Forces to Somalia.*

on the unique role a military leader commands in this regard, given the degree of influence he is provided within the system.

Military Misconduct

One of only a handful social roles which is defined in superlatives is that of a soldier—he is the bearer of the 'ultimate liability', a duty-bound professional obligation to risk his own life. This vocational calling is unique because its fundamental nature isolates the soldier from the very civil society whose interests he is working to protect or further. The martial society which he then has to inhabit is again strange in how its every piece is a stringently defined evaluation. The average day is a rigid, un-malleable system framed with innumerable lines, and no line is a fine line—doing 24 push-ups when 25 are asked for is just as futile as having done 20, and will merit the same repercussions. Military society also makes an open declaration of the great emphasis it places on moral fibre and strength of character, qualities which even the most outspoken of war's critics have not been able to leave unappreciated in the individual soldier. In *A Question of Loyalties*, Massie (1989) writes, "Do you know what a soldier is, young man? He is the chap who makes it possible for civilised folk to despise war."

With such clear yardsticks of definition in place, and with moral and personal infallibility effectively being a job-requirement, it is no surprise that misconduct in the military becomes an expansive term which includes entities ranging from minor breaches of prevailing unit orders to gross violations of law (DGAFMS, 2002). Save cosmetic differences in the language used, an unmistakable consistency prevails in definitions of misconduct formulated by armies across the world. The following one is taken from the American Code of Federal Regulations (United States Office of Federal Register, 1939), but almost all definitions convey the same idea just as lucidly:

Misconduct is an act involving conscious wrongdoing or known prohibited action. Wilful misconduct involves deliberate or intentional wrongdoing with knowledge of or wanton and reckless disregard for its probable consequences.

At the outset, most military frameworks enforce respect for the civil penal code—for example, Chapter 4 of the Indian Army Act contains all sections of the penal code. However, the nature of military action requires that a framework of definition extending beyond these limits be established. Serious misconduct behaviours in militaries world over broadly fall under the following categories, and their legal ramifications are also similar.

Offences in Relation to the Enemy and Punishable with Death

Active collusion with the enemy in times of war is universally deemed the most dire and serious offence by all militaries. The maximum punishment awardable for such offenses is death (Army Act, 1955). In the United Kingdom, despite the death penalty for murder being abolished in 1969, it was still in force for assisting the enemy until 1998.

In Indian Military law, this set of offenses includes actions such as:

1. Shameful abandonment of a post
2. Shamefully casting away arms
3. Treacherously holding communication with the enemy
4. Intentionally spreading false alarm
5. Voluntarily aiding the enemy if taken as a Prisoner of War
6. Directly or indirectly assisting the enemy
7. Sleeping or being intoxicated upon one's post during times of war or alarm.

Offences in Relation to the Enemy and Not Punishable with Death

The distinction being made here is of the presence or absence of *intent* to cause harm to one's own country. These acts are performed either for want of some precautions or without authority, or embody a general laxity in duty performance (Ministry of Defence, 1961).

It includes offenses such as:

1. Holding correspondence with the enemy without due authority
2. Sending a flag of truce to the enemy without due authority
3. Being taken as a PoW by want of due precaution/neglect or failing to rejoin when able.

Mutiny

A strange disconnect exists between the lack of clarity regarding what constitutes a mutiny today, and the unmistakable and forceful nature of its punishments. Most military legal systems, such as The British Military Service Act, the United States' Uniform Code of Military Justice and Indian Military Law, have all upheld the death sentence for mutiny. Similar to assisting the enemy, the death penalty for mutiny was in force in the UK until 1998, even though it had been abolished for even murder in 1969. The Indian definition describes an act of collective insubordination and includes persons who:

1. Begin/incite/conspire to cause mutiny
2. Join such a mutiny
3. Being present at such a mutiny do nothing to suppress the same
4. Knowingly withhold information of a mutiny from senior authorities (Ministry of Defence, 1961).

However, the validity ascribed to reasons for this disobedience has, with time gathered strength. In the wake of the Second World War (WWII) Nuremberg trials, and further reinforced by the atrocities committed by American troops in the Vietnam War at Mai Lai (an American army company murdered close to 500 civilians on direct order), the US Army today deems that every soldier must obey only a lawful order. Disobedience of an unlawful order is a military service obligation. Every soldier who has made this decision to disobey what he deems an unlawful order will though almost certainly be court-martialed to determine the nature of the order, and hence the justness of the disobedience.

Threatening, Striking or Killing a Unit Leader

Military history is replete with incidents of troops resorting to murdering a squad commander who is perceived to be (Director General Armed Forces Medical Services, 2002):

1. Excessively eager to commit the unit to danger
2. Glaringly incompetent
3. Unfair in the sharing of risks.

'Fragging', the term used to describe the act is derived from how fragmentation grenades were the instrument of choice because when they were used it was near impossible to assign culpability using ballistic forensics or any other means. Although such incidents have been documented since the 18th century, fragging was particularly common in the Vietnam War—230 cases of American officers being killed by their own troops are on record. Another 1400 officers died in circumstances which 'could not be explained' (Hedges, 2003). In the Indian context, it is described under Section 40 of the Army Act, which murder aside, includes the use of criminal force/assault, and the use of threatening or insubordinate language. In times of active service, it may even result in a punishment of up to 14 years.

Desertion

The manual of Indian Military Law astutely highlights that the essence of desertion lies in the *intention* of the offender to not return to service, differentiating it from the less serious offence of Absence without Leave (AWL). Desertion during times of war or alarm is universally recognized as one of the worst forms of military misconduct, and it has always attracted severe penalty. Even today, as per Indian Military Law, when committed during active service, desertion is punishable by death.

During WWII, the German Army executed some 15,000 soldiers found guilty of desertion (Hatlie, 2005). Joseph Stalin's contempt for desertion was well known—he issued orders for deserters to be shot on sight, and to even see their families imprisoned. He also directed that each Army theatre be buttressed by a set of 'barrier troops' whose job would be to shoot deserting 'cowards' or panicking troops who retreated. By the end of his reign, Stalin had directed the execution of some 1,58,000 soldiers for desertion (Roberts, 2006).

Rape and Looting

Virtually every known historical era has seen rape accompany warfare. Even the Bible contains numerous references to wartime rape—"For I will gather all the nations against Jerusalem to battle, and the city shall be taken and the houses plundered and the women raped…." (Zechariah 14:2).

The manner and strategic ruthlessness with which it was carried out in recent conflicts in the Democratic Republic of Congo even merited rape to be considered as a military strategy, with its identified aims as: increase in military morale, decrease the military morale of the enemy, to offend the enemy, and to loot the maximum of an enemy's belongings (including women and children) (Rutagengwa, 2008). Numerous statutes and resolutions exist condemning rape, the most sweeping being the 2013 UN Resolution, demanding the complete and immediate end of all acts of sexual violence by all parties to armed conflict. The resolution noted that sexual violence can constitute a crime against humanity and a contributing act to genocide, called for improved monitoring of sexual violence in conflict, and urged the UN and donors to assist survivors.

Suicide

Although all militaries accord the utmost priority to preventing troop suicides, all acts of self-harm, including suicide, are viewed as deviant behaviour and qualify as chargeable offences in military jurisprudence. The Indian Military Act describes the attempt to commit suicide under Miscellaneous Offenses, covered under Section 64. The reason behind this universal military intolerance for suicide becomes clear from a 2010 verdict delivered by a US Navy court against Private Lazzaric Caldwell who attempted to slit his wrists while awaiting trial on charges of stealing a belt.

> Self-injury, whether it results in an intentional suicide or not, has the potential to cause tremendous prejudice to the good order and discipline within a unit. If a convening authority feels it necessary to resort to court-martial to address this type of a leadership challenge, he or she should be allowed to do so…. (US Navy Marine Corps Court of Criminal Appeals, 2010)

Intoxication

An individual is said to be intoxicated if, owing to the influence of alcohol or any other drug, he is rendered unfit to be entrusted with his designated duty, or any duty which he may be called on to perform. Behaviour in a disorderly manner or which brings discredit to the Army also constitutes intoxication, and as noted by the Army Act, it is immaterial whether the individual is on duty at the time or not (Ministry of Defence, 1961).

No Quarter—Taking No Prisoners

For all its inherent carnage, warfare is still considered by many to be an art form which continues to evolve. Each major global act of war has seen in its wake proceedings to reassess the rules governing warfare. Attempts are made to ensure that the boundaries of any future conflicts are better defined, so they have a lesser chance of disintegrating into acts of crude barbarism. A landmark event in the evolution of warfare came to being under Article 23 (d) of the 1907 Hague Convention IV—*The Laws and Customs of War on Land*. The article declared that it was now illegal for any fighting force to declare that 'no quarter' would be given to the enemy (International Committee of the Red Cross, 1907). Giving no quarter implies that no mercy would be shown to a vanquished opponent, that lives would not be spared in return for unconditional surrender, and all defeated enemies would be killed. The Nuremberg trials following WW II further made it binding on all parties entering an international armed conflict to explicitly prohibit any declarations of no quarter being given (Lillian Goldman Law Library, 2008).

Numerous origins of the term 'no quarter' have been suggested— one theory states that the term began with a declaration by the commander of a victorious army that they "will not quarter (house)" captured enemy soldiers, and hence all those defeated should be killed (Oxford English Dictionary, 1997). Another possible origin is based on an agreement between the Dutch and Spaniards during a conflict in the 18th century, by which the ransom of an officer or private was to be a quarter of his pay (Oxford English Dictionary, 1997).

Mutilation of Dead Bodies

There is an unmistakable sense of duty owed to those who have died on the battlefield, regardless of which side they fought for. Article 15 of the First Geneva Convention provides that all parties entering a conflict must "at all times, and particularly after an engagement... search for the dead and prevent their being despoiled." Any acts of mutilation or desecration of enemy corpses, or the practice of taking body parts as trophies, are indicative of a gross breakdown in the discipline and restraint of a fighting force.

Earlier, such behaviour was almost dismissively attributed to the extreme stresses of battle. However, during recent research funded by the Economic and Social Research Council, it has been observed that misconduct of this manner is usually carried out by soldiers who view the enemy as racially different from (and usually inferior to) themselves, often to a subhuman degree. They often describe their actions as similar to a 'hunt' (Harrisson, 2012). For example, in WW II, acts of corpse desecration and the taking of body parts as trophies were almost unheard of on the European battlefront. However, it was common for soldiers engaged in the Pacific theatre of operations to take body parts of Japanese service personnel as trophies.

Malingering

The deliberate faking of a physical ailment, disablement or mental illness (including PTSD) is again viewed as a serious military offense. During the Vietnam War, numerous young Americans looking to dodge conscription went to the degree of feigning homosexuality, as the Pentagon deemed homosexuality a 'moral defect' rendering someone unfit for service. Consistent problems have been faced however, in identifying malingering. Together with feigning and the production of disease of infirmity, malingering is covered under Section 46 of the Indian Army Act under the broad heading of Certain Forms of Disgraceful Conduct. Punishment awardable may extend up to seven years.

Self-inflicted Wounds (SIWs)

The majority of SIWs occur in times of active combat. The idea behind self-infliction is to have oneself removed from a hostile fighting theatre to the safety of a hospital setting. The perfect SIW is termed a 'million dollar wound'—it is serious enough to get the soldier removed from combat, but will not leave him permanently handicapped. Common SIWs include gunshots to one's non-dominant hand, arm or foot. A deliberate neglect of health, for example, 'forgetting' foot-care in damp conditions leading to a fungal infection qualifies as Self-inflicted Injury. In the military, considering the premium on bravery, all SIWs are viewed as acts of cowardice and qualify as serious military offenses. For example, the penalty for SIWs in the British Army during WWI was capital punishment, at the time death by firing squad (Duffy, 2009). Most SIWs however escape notice.

Basis of Misconduct Behaviour

Most abnormal or improper behaviours observed in military settings are not manifestations of psychiatric disorders. The majority are what fall under the broad heading of 'negative stress behaviours', and are a consequence of being exposed to the psychological effects of combat per se or to environments surrounding or suggestive of combat (Director General Armed Forces Medical Services, 2002). In light of this fact, it becomes imperative to get a broad idea of what the unique stressors affecting a soldier are (Director General Armed Forces Medical Services, 2002).

Stressors during Training

1. Loss of emotional support from family and friends
2. Newfound rigid discipline framework
3. Demanding nature of physical training
4. Erosion of privacy

5. Constant onus to compete and qualify for next stage
6. Inability to address problems at home due to preoccupation with training.

Stressors during Peace Time

1. Improper or poor interpersonal relations
2. Domestic problems related to marital life
3. Health problems of family members
4. Insecurity of family members
5. Children's education
6. Property disputes
7. Financial problems
8. Inadequate response by civil administration to the problems of service personnel.

Stressors in Field

In addition to all the above:

1. Separation from family members
2. Adverse or demanding climatic conditions
3. Isolation
4. Long tenures
5. Unknown enemy in counter insurgency areas
6. Uncertainty of life
7. Difficult living conditions
8. Fatigue.

The life of a soldier is inherently tough, and the coming together of these domestic and vocational pressures can easily become over-whelming. Frustration at fighting losing battles on all fronts often sees soldiers resorting to some of the negative coping strategies listed previously.

Contributory Factors to Misconduct Behaviour

It is valid to argue that some of the above stressors are faced by all people engaged in any difficult, physically demanding and remotely located line of work. However, it is clear that there are some unique circumstantial and cultural factors which prevail in military service, which seem to increase, and almost promote misconduct stress behaviours.

The Legal and Moral Grey Areas of War

Few workplaces are as morally and legally ambiguous settings as a theatre of war. The total suspension of systems to enforce the civilian law of the land means that with wrong enough intent and with weak enough command, effectively anything can be gotten away with. Some specific social peculiarities about war situations which are particularly conducive to acts qualifying as misconduct behaviour are (United States Army Headquarters, 1994):

1. Permissive attitudes towards, availability of and rampant use of drugs/illicit substances by civilian populations in the vicinity of garrisons or army bases.
2. Well-structured distribution networks of illicit substances usually thrive in such settings of legal ambiguity. The very price of drugs is often lower in war theatres as costs which normally have to be incurred by the drug industry to avoid the arm of the law can be done away with.
3. The victorious pursuit of a vanquished enemy has been associated with lower incidences of combat fatigue, but is usually accompanied by the commission of criminal acts (rape, plunder) and a rise in substance use, unless command assumes a tight moral control over the situation. A similar situation arises during hasty and unplanned withdrawal—in this case, disillusioned with the defeat of the cause, troops may either become deserters, or they may again resort to criminal activities. The psychological justification is that goods will anyway be taken

over by the advancing enemy, so they might as well be taken. Rape and other atrocities on civilians left behind may also be committed using a similar justification.

4. Viewing and experiencing atrocities at the hands of the enemy: providing instructions on moral conduct is well and good, but most such directive frameworks sustain serious insult when atrocities to comrades or friendlies are viewed first-hand by soldiers.

Logistical and Hierarchical Factors

The results of having to operate within the structure of military hierarchy, and the consequences of adhering to the behaviour this structure deems proper and demands are open to varied inference. On one hand, military life requires a rigid communication and discipline channel which many deem inviolable. On the other hand, a failure to communicate is often the root cause of the stress which ill-manifests as misconduct behaviour. The very nature of life in a hostile area as part of a fighting unit is besotted by its own set of unique problems and stressors, all of which can accentuate misconduct behaviour. Some such contributory factors are (United States Army Headquarters, 1994) as follows:

1. *Racial and Ethnic Tensions*: Although attempts are made to form fighting units along racial lines, such tensions are bound to arise in any large group of individuals. An even greater racial divide (and hence a greater cause for enmity) is that between troops and the local population—this is at times enough to create an almost 'dehumanising' mentality.

2. *Failure of Expected Support*: Supplies, reinforcement and relief often do not arrive at times when they are needed the most. When soldiers who have been called upon to kill and die for a cause face such basic, menial shortages, it is only natural for a sense of abandonment to emerge. Soldiers may then resort to illegal means to see their requirements fulfilled.

3. *Interpersonal Resistance*: High levels of interpersonal resistance, especially vertically between commanders and men, can have disastrous consequences with regard to the aggravation of stress.

A common negative behaviour used as a solution is the use of a substance (commonly alcohol) as a 'ticket' for inclusion into a social group.

4. ***Opposition to the War Effort at Home:*** In the absence of tangible support from the home front, soldiers often find their own belief in the validity of the war effort erodes. When they no longer believe in what they are fighting for, many soldiers often desert or openly refuse lawful orders. Others may carry out orders, but display resentment by inflicting brutalities on the local population or escaping through alcohol and drugs.

Aggression as a Job Requirement

The very nature of their vocation demands that soldiers be aggressive. As Anthony Kellet notes, "If an army is to fulfil its mission in the battlefield, it must be trained in aggression" (Kellet, 1987, p. 89). Military training is broadly a repetition-to-perfection spectrum of aggressive exercises. Several new conditioned reflexes are created which prime the soldier's response mechanisms for combative retaliation. Sadly, this up-regulation persists even outside spheres of combat, Harrisson and Laliberte (1994, p. 247) commented, "it is surprising that there is not more spill over criminal activity by members of the military than there is". The situation becomes particularly problematic when a unit trained for active combat is deployed in areas without much actual fighting, or in peacekeeping roles, for example, in March 1993 a young, unarmed Somali was tortured and beaten to death by members of the Canadian Airborne Regiment in Somalia, deployed there in a peacekeeping capacity (Crocker, 1995).

Total Absence of Social Support

Perhaps the single most significant characteristic of military life which affects the risk of violence is the removal of the military family from the social support system usually provided by the extended family, friends and neighbours. They are instead provided limited social exposure to a near-homogenous military society, whose uniformity consolidates stereotypes and weakens the ability to make social adjustments. An author commented "frequently, military couples have to live in quarters assigned according to rank. Their neighbours therefore are also young people with little more experience than they have" (Schwabe & Kaslow, 1984, p. 129). Some of the marital conflict risk factors which get accentuated in this monotonous social atmosphere are:

1. Financial problems
2. New baby in the home
3. Differences in the level of commitment to the relationship
4. Sexual problems
5. Child discipline problems or disagreements
6. Different or unrealistic expectations of marriage
7. Cultural or religious and spiritual differences
8. Poor communication and problem-solving skills
9. Chronic unresolved life stressors
10. Dual career demands

The Silent Tradition of Sexual Assault

Susan Brownmiller, the author of *Against Our Will: Men, Women and Rape*, was one of the first historians to attempt a longitudinal overview of rape in war. Brownmiller theorized that:

> the maleness of the military—the brute power of weaponry exclusive to their hands, the spiritual bonding of men at arms, the manly discipline of orders given and orders obeyed, the simple logic of the hierarchical command—confirms for men what they long suspect—that women are peripheral to the world that counts. (Brownmiller, 1975).

This objective devaluation allows rape to be viewed as an act which is no longer morally abhorrent. When such moral leeway is provided to "young males who are cut off from traditional informal controls and faced with a situation of relative unavailability and inaccessibility of females, they become prime candidates for sexual crimes" (Bryant, 1979). War rape was even once regarded as a tangible incentive to soldiers who were otherwise paid irregularly (Askin, 1997). Attempts to erase this deeply engrained aspect of the martial mind-set have achieved only partial success.

Out-of-Hand Alcohol Abuse

In June 2000, the American Forces Press Service released disturbing statistics—alcohol abuse was costing the Department of Defence (DoD) more than $600 million a year. The DoD was spending another $132 million a year to care for babies of serving women with fetal alcohol syndrome (a direct effect of their mothers' heavy drinking during pregnancy) (Rhem, 2000). This is alarming especially when US policy under Order Number 1A categorically prohibits the consumption of alcohol by any US service members stationed in Iraq, Afghanistan or Kuwait. That the use of alcohol increases the incidence of violent behaviour is a fact recognized by military hierarchy—a report published by the Major General Hewson inquiry of 1985 investigating criminal behaviour in the Canadian Army emphasized that higher intake of alcohol reduces the threshold for potential violence and acts of antisocial behaviour (Hewson, 1985). The fact that even in the face of such damning evidence, militaries across the world have been unable to reduce alcohol consumption in their troops (in 2000, 21% of American service members admitted to drinking heavily, a figure identical to that found 20 years ago) gives some idea of the complexity of the problem being dealt with. The reason is perhaps that alcohol has silently included itself as an implicit part of military service. As Bryant suggested that from the standpoint of the authorities, alcohol serves to help solve the problem of morale and boredom, and helps prevent the build-up of potentially disruptive frustrations (Bryant, 1979). The trouble is that such stopgap morale-boosting arrangements soon backfire. In the 17th century, when it was the world's premier seafaring force, the Royal British Navy used to issue half a pint of rum on a daily basis to sailors to keep ship morale high. Sailors then began hoarding away their daily liquor issues so they could get thoroughly drunk twice a week. Ship discipline deteriorated to such a degree that soon the authorities had to pre-mix all rationed rum with four parts water before issue (Barnett, 2006).

Wanting Selection Standards

With conscription no longer resorted to by most nations, the military has effectively ceased to be an institution and is now an occupation. Retired US Vice-Admiral J. B. Stockdale stated:

> With the closing down of obligatory military service, the armed forces lost the strength of a cross-section of the nation's youth. Now they must make do with the least highly qualified segment of the nation's young people. They have to deal with illiteracy, drug abuse, alcoholism, as well as with an increasing rate of desertion and criminality. (as cited in Gabriel, 1987, p. xv)

Foremost, such a compromised strata of society is already morally suspect. Secondly, having joined mainly for financial reasons, such soldiers neither have any natural affinity for nor deemed it necessary to truly imbibe and embody the character qualities the military ideal strives towards—it is, after all, just a job.

A common call is for the implementation of psychological tests to aid selection during recruitment. However, two problems are faced:

1. As illustrated by the Hewson report (Hewson, 1985), the institution of these tests could qualify as a possible human rights violation and would also not be cost effective. The sheer volume of tests, added screening procedures and increased staff requirements would make added psychological fitness testing of questionable value.
2. The present system of psychological testing can easily spot those with serious mental illness, but persons with personality disorders can manage to slip through undetected. Further, Hewson report (1985) showed that although the incidence of serious mental illness was lower in the military than in the general population, the rate of personality disorders was significantly higher.

Question of Leadership

Military leadership is set apart from that of other organizations by it having two opposing functional requirements—a strict maintenance of discipline and control on the one hand, and the need to allow for flexibility in the field on the other. The Canadian Forces Military Training manual, *Leadership in Land Combat*, mentions that a military commander is a commander by virtue of the legal authority he holds. He becomes a leader when his men accept him as one. Good leadership demands a willingness to lead by example, and it is common knowledge that few things boost troop morale as directly as the extent to which an officer is willing to risk his life in battle. For example, the Israeli Army is particularly noted for the sacrifices made by its officers. In the 1967 Six Day War, almost half of the Israeli fatalities were officers. "There is no doubt," a study of the Israeli Army concluded, "that the fact that so many commanders, proportionally, fell in battle had a salutary effect on the morale of the troops…they were not being asked to give their lives for something for which the commander would not give his own" (Rolbant, 1970, p. 166). The Hewson report (1985) observed how the relationship between men and their immediate leaders has been increasingly eroded. One of the primary reasons for this in their view was the temporary absence of personnel from the unit to undertake other tasks or attend courses. In the absence of constant and effective leadership, prolonged stress may lead to low morale and disciplinary infractions.

Good leadership also requires that clarity regarding the war effort be transmitted through the ranks. A glaring example to the contrary is the infamous taking of Hill 937 by US forces during the Vietnam War. The hill was cynically dubbed 'Hamburger Hill' by troops as the high casualty rate reminded many of a meat grinder. The operation lasted for 10 days and comprised 10 assaults, and almost 100 Americans were killed and 400 wounded with a staggering 70 per cent casualty rate (Zaffiri, 1969). It later dawned that Hill 937 had no strategic importance, and was taken primarily as a diversionary tactic. Eight days after its conquest, the hill was abandoned. This much publicized failure in leadership crippled military morale, and effectively ended support for America's war effort.

There are few organizations in which a leader is provided this much ability to improve not just the output, but the very lives of his men where so much is at stake on his being able to do so.

The Dilemma of Youth

War has always been deemed a young man's game, and is often viewed as a rite of passage to manhood. Most armies have well-demarcated age cut-offs for troop enrolment, and active combat is usually an exclusive purview of an army's youngest, freshest troops. However, a study to identify the psychosocial predictors of military misconduct showed results pointing towards a disturbing truth—the two variables most strongly associated with misconduct-related discharge from the military were receipt of a psychiatric diagnosis, and age at first combat deployment (Booth-Kewley, Highfill-McRoy, Larson, & Garland, 2010). Troops deployed to a combat zone at a relatively older age (22 years and above) were at significantly less risk of receiving a bad conduct discharge from service or a demotion compared to younger troops (aged 19 or less) exposed to a theatre of war. The study states emphasized that younger individuals may be less able to cope effectively with the traumatic experiences and pressures associated with being in a war zone, perhaps because they have less life and/or military experience. Although very little can possibly be done about the broad age-demography of combat soldiers, a realization of this vulnerability could perhaps lead to interventions which could lessen such unfortunate negative fallouts of active service.

Negative Stereotypes

Critics of war argue that viewed objectively, war is an act of murder on a scale which would not be acceptable to anyone with a functioning moral compass. Throughout history, this problem has been overcome by skewing this compass through the propagation of negative, dehumanising stereotypes regarding the enemy. It is at times the only way soldiers can be convinced to perform acts of the nature war demands, and a nation's people be convinced to support the war effort. The most obvious example is, of course, the Nazi portrayal of Jews just prior to and during the holocaust—anti-semitism was taught in public schools through textbooks with racist themes. For example, a mathematics problem taken from a textbook for school children during the reign of the Third Reich stated, "The Jews are aliens in Germany—in 1933 there were 66,060,000 inhabitants in the German Reich, of whom 499,682 were Jews. What is the per cent of aliens?" Jews were deemed evil and wantonly corrupt, and an alternative history was fabricated which held them responsible for all of Germany's troubles. The US media helped propagate a similar hatred for the Japanese during WWII, often describing them as "yellow vermin" (Ferguson, 2007). An official US Navy film went to the extent of describing Japanese troops as "living, snarling rats". Such use of negative stereotypes and the resultant 'dehumanising' of the enemy makes it far easier to perform acts of brutality violating the martial code on them.

Why Just 'Coping Training' Isn't Working?

At the centre of the majority of episode of misconduct behaviour in the army is a soldier whose mental coping resources have been overwhelmed by environmental demands. Coping training is a concept excluded from most military training programmes, whose onus is on relevant skill acquisition. Military training teaches usually by a systemic process of 'overlearning' a task until it can be approached with familiarity and confidence, and so interference from competing responses can be blocked out (Thompson & McCreary, 2006). The vagueness about the implicit psychological lessons means any learning is at best implicit, and the individual is left to his own devices in learning to control thoughts and emotions. Yet, it is evident that emotions and thoughts can affect behaviour and may be elements critical to the acquisition of proficiency (ibid.). Individuals vary in their degree of ability to teach themselves this control, which can at best delay military skill acquisition, and at worst place them and their teammates in a constant, impending danger.

In recognition of the adverse effects of ill-managed stress reactions, militaries have begun to address the issue by use of the 'stress management briefing'. The normal template is one or a series of lectures, usually mental-health professionals covering issues such as the stress–strain relation, the general mechanisms of stress generation, stressors specific to the military and information regarding good and bad coping mechanisms (ibid.). The intent cannot be faulted, but the effectiveness of these lectures seems limited as follows:

1. A strong stereotype exists that psychiatric illness reflects an inherent character weakness. This will naturally be heightened in military culture that places high premium on fitness, courage and toughness.
2. Use of mental-health professionals to deliver these lectures may further fortify the above natural resistance, as mental-health professionals are seen as treating those already injured, not providing training which may enhance operational efficiency. Important lessons are likely to be rejected if not given by military personnel who are perceived to have had applicable operational experience.

3. Soldiers' response to the conventional lecture format has consistently seen to be poor.
4. The lecture/briefing is usually limited to 'a talk' which does not provide specific training on the techniques that could be put to effective use during stressful situations. If any techniques are featured, they are done so only as a 'demonstration'. This limits a person's ability to generalize the practical techniques to any real-world setting.

In light of all the above, perhaps the most important step which can be taken to allow soldiers to better cope with the stressors of hostile deployment is an integration of coping principles into active military training curriculums, and seeing skill impartation performed by trainers with recognized operational experience and credibility. *Stress management should not be seen as distinct from 'normal' military training.*

Specific Interventions to Control Misconduct

Controlling Selection

It is imperative that the military at the outset makes clear to any prospective applicants that its identity as an institution is paramount, and that certain required standards of honour and morality are expected from every soldier. It is commonly observed that persons with problematic personality disorders will show visible resistance or indifference to such concepts. The manner in which this is implemented may perhaps be along the lines of the US Army's 'Soldier's Rules', an integral part of basic US military training.

1. Soldiers fight only enemy combatants.
2. Soldiers do not harm enemies who surrender. Disarm them and turn them over to your superior.
3. Soldiers do not kill or torture enemy prisoners of war.
4. Soldiers collect and care for the wounded, whether friend or foe.
5. Soldiers do not attack medical personnel, facilities or equipment.
6. Soldiers destroy no more than the mission requires.

7. Soldiers treat all civilians humanely.
8. Soldiers do not steal. Soldiers respect private property and possessions.
9. Soldiers should do their best to prevent violations of the law of war. Soldiers report all violations of the law of war to their superiors.

Establishing a Concrete Code of Ethics

Gabriel (1987) suggested that one needs a very clear statement of the ethical obligations that one ought to observe if one is to be expected to behave ethically. He suggests a one-page code of ethics which contains provisions such as a soldier will never require his men to endure hardships or suffer dangers to which he is unwilling to expose himself. Every soldier must openly share the burden or risk and sacrifice to which his fellow soldiers are exposed and no soldier will punish, allow the punishment of or in any way harm or discriminate against a subordinate or peer for telling the truth about any matter. Moral decisions are marked by social or peer-group deterrence. Creating and then putting up on clear display a legally protected environment in which moral uprightness is protected and valued will encourage more soldiers to take these difficult decisions correctly. Having a code brings the issues of unethical behaviour to the forefront, and stimulates conversations about right and wrong.

Early Identification of At-Risk Soldiers

It is quite possible using an objective assessment to identify the individual soldier atrisk for misconduct behaviour. The main factors which contribute to the likelihood of such behaviour are:

1. High levels of ambient stress
2. Poor coping mechanisms which lead to unresolved residual stress
3. A predisposition to resort to negative mechanisms as an outlet for this stress.

Following are the broad baseline set of unit factors, progressing to individual risk factors which affect each soldier uniquely, and lastly aberrant behaviour traits that can identify the at-risk combatant with a significant degree of accuracy (United States Army Headquarters, 2009):

1. **Unit Risk Factors**

    i. A high incidence of soldier and civilian deaths occurring in the same area of operation and over a short period of time.
    ii. A high operation tempo with little respite between engagements.
    iii. Rapid turnover of unit leaders.
    iv. Manpower shortage.
    v. When there is overly and unreasonably restrictive or confusing set of rules of engagement.
    vi. When there is an enemy that is indistinguishable from innocent civilians.
    vii. If there is a perception of lack of support from higher command.

2. **Individual Risk Factors Affecting Soldiers**

    i. Poor social support.
    ii. Home front or unit problems.
    iii. History of reacting impulsively in past.
    iv. History of disciplinary actions and military disciplinary proceedings.
    v. Suffering a combat loss (friend or a team member who was wounded in action or killed in action).
    vi. Personally witnessing the injury or death or being involved in the medical evacuation of friend/unit member.
    vii. Witnessing a particularly gruesome or horrific loss of life.

3. **Individual Behaviours of Soldiers At-Risk**

    i. Verbalization of thoughts about:

        a. Anger toward or lack of support from higher command
        b. Indiscriminate revenge.

ii. Appearance and/or behaviour changes which may include:

    a. Lax military dress/bearing,
    b. Appearing on edge,
    c. Being subject to angry outbursts,
    d. Taking excessive and/or intentional risks,
    e. Appearing to be depressed and having minimal or no contact with others.

iii. Changes in sleep patterns and appetite.
iv. Alcohol use or substance abuse.

Knowledge of unit risk factors, individual risk factors and individual behaviours of soldiers at risk can easily help the command and control to identify and reach the soldier at risk. Same has been depicted in the form of a pyramid (Figure 6.1).

Figure 6.1
Identifying and reaching the soldier at risk

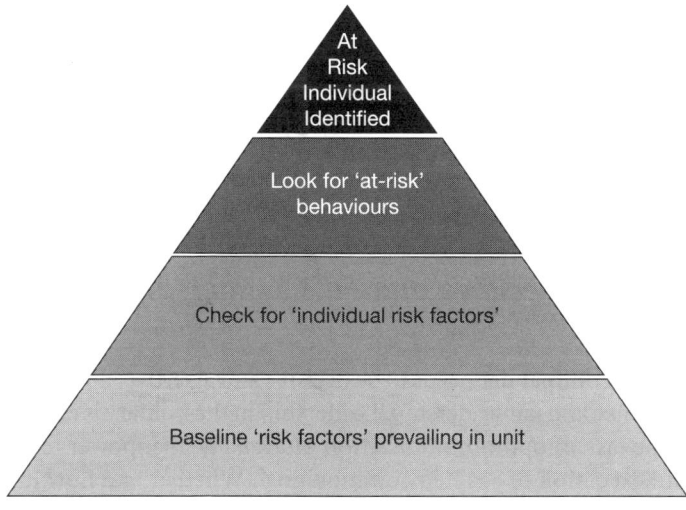

Source: Authors.

The Role of Leaders in Misconduct Behaviour Control

A leading military encyclopaedia reflects that "the 'secret' of good leadership continues to elude explanation" (Dupuy et al., 1993). The qualities of good military leadership require little deliberation upon. With regards to controlling misconduct behaviour, the following specific interventions, especially when viewed in light of the degree of influence a military leader commands, will certainly control misconduct behaviours.

Creating Ethical Environment

It is imperative for a commander to create an environment in which unethical behaviour is first clearly defined, then not just discouraged but also dealt with according to established norms at face value. The system of military jurisprudence has effectively sealed most gaps in the definitions of unethical behaviour, and continues to evolve, for example, rape and sexual crimes were even not prosecuted at the Nuremberg Tribunals, but today rape is considered amongst the most heinous of war crimes.

As highlighted, few organizations exist in as much of a legal grey-area as military units involved in active combat. A commander will be consummating his role if he ensures that the framework's influence is allowed to reach forward areas, and to function there without interference.

Practising Decisive and Fair Leadership

A military commander needs to be decisive and assertive; demonstrate competence and fair leadership. Leadership in the military is entrusted with the task of optimum utilization of available manpower for successful execution of a stated common goal. Whether and how this is achieved in the face of prevalent shortcomings and obstacles is, in large part, the measure of leadership. However, a ruthless desire for achievement despite wanting resources often sees men being overburdened by responsibility. Pervasively, such task mismanagement erodes confidence in a leader as he is viewed as a poor decision-maker. Feeling poorly led, a unit's morale can dip significantly, placing an entire mission in jeopardy. Thus, it is essential that the foremost thing to be done is proper task allocation to maximize efficiency and minimize stress. This proper allocation includes (Hewson, 1985):

1. Adequately assessing an individual's talent, training and ability for a particular task well to see that the right man is assigned the right job.
2. Duplicating critical tasks—tasks requiring complete accuracy and behavioural alertness are best assigned to two people at once. Though working on the same project, it is seen they check each other's work by actually performing the same task independently.
3. Cross-training—save special technicians such as medics, ideally in addition to his own duties, each soldier should also be trained in the task profile of another, increasing available manpower and creating a backup in times of exigency.
4. Using performance enhancers—the execution of critical tasks during times of inattention can be simplified if SOPs, checklists and guidelines are kept handy.

Honest Efforts and Concerted Actions

A leader must ensure that every effort is made to provide for the soldiers' welfare, such as (United States Army Headquarters, 2009):

1. **Diet:** An inadequate diet degrades performance and judgement, reduces resistance to disease and hampers the ability to recuperate from stress. For a successful mission, it is imperative that the requisite calorific requirement be provided.
2. **Sleep:** While at times essential for operational excesses, it is not possible for a deprivation of sleep to leave an individual

unaffected. Although a sleep debt may be 'paid off' partially by obtaining the same quantum of sleep later, prolonged deprivation can even result in florid nervous breakdowns.

3. **Avenues of Recreation:** The military has always had high regard for all forms of sport. This stems not just from a commonality between the principal tenets of sport and combat, but also because organized games serve as an excellent releaser of stress and promote cohesion in the unit.

Thoroughness in Training

Thoroughness in training (United States Army Headquarters, 2009) is another requisite for commanders to follow. Unit leaders, having experienced the stressors and hardships of field life and peacetime activity beforehand, will recognize the need to impart training of a relevant quality and of a sufficient toughness to optimize the sense of readiness in their troops. Training of this manner serves many benefits, such as:

1. Realistic training sensitizes and prepares soldiers to the stressors of a combat theatre, and allows them to prepare beforehand their own unique, positive coping mechanisms. This will not just enhance performance, but also instil confidence as soldiers first develop and then become aware of an inherent capacity to deal with what active duty will offer.

2. Rigorous training reinforces the positive image that a soldier will harbour about the general preparedness of his unit personnel and equipment. Not only will this allay a considerable degree of stress related to the combat experience, but will also add to unit cohesion and espirit de corps.

3. It has been observed that when involved in the purpose they believe ordained (i.e., training for active combat) soldiers are far more receptive to information and directives regarding the war effort at hand. Training is a wonderful opportunity to prime troops about the capabilities of their enemy, and the limitations which they will be hampered by when they go forth to meet him. A clear and fair idea of where they stand and the honesty with

which this is conveyed is usually appreciated, and does away with any misconceptions or senses of deception which a lack of clarity can propagate.

Assistance in Personal and Familial Matters

Although it is an area which needs to be treaded on with caution, military commanders should recognize that troops are often unaware of the nature of preparations they should be making prior to active deployment. The normal rigid nature of unit communication channels means that they have no-one more experienced to actually turn to for this nature of advice. A sense of leaving family and personal obligations unaddressed, if persistent, can seriously hamper the focus of the soldier on the job at hand. It is highly beneficial if unit leaders encourage soldiers to:

1. Generate or update wills
2. Update insurance policies timely
3. Finalize power of attorneys for spouses
4. Provide spouses staying alone with a specific list of people to contact if faced with a specific problem (e.g., healthcare, mechanics)
5. Ensure a means of transportation, if required is made available
6. Resolve major legal issues prior to deployment, for example, property dealings, major purchases (United States Army Headquarters, 2009).

Encouraging Social Support within the Unit

The encouragement of a 'Joint Family Model' within barracks can help provide a degree of social support which separation from the normal family structure takes away. In this setup, an elder/senior member of the unit takes up a paternal role, and becomes instrumental in providing guidance to younger less experienced military couples. The model is particularly useful because the problems experienced by troops are unique, and a true understanding of them comes only with first-hand experience. Following are some suggestions:

1. All personnel deployed in sensitive/stressful areas should be granted regular and frequent spells of leave. Turnover/rotation of duties should also be timely ensured.
2. The promotion of informal forums between unit hierarchy and troops may be encouraged, where general domestic problems can be brought out into the open, and where matters of general benefit such as information regarding housing schemes, admissions and scholarships for children of troops and awareness regarding monetary benefits may be shared.
3. Senior members of the unit should actively be directed to mentor and guide new recruits. These newcomers to the establishment are less likely to become disillusioned if they are at the outset, provided a realistic picture of what military life entails. They should also be told clearly about the larger goals towards which their actions will be directed.

Discouraging Substance Abuse

Few things can have as pervasive and detrimental an impact on the general cohesiveness and efficiency of a unit as a lax attitude towards substance abuse, or worse the promotion of substance abuse as a coping mechanism. Alcoholism's effects on the military has been well documented, and a leader needs to serve both by example and by making examples of what is to be strived for and what cannot be tolerated with regard to substance use.

Appreciation of the Larger Cause

Without proper reigning in, any large group can easily turn into a mob and its behaviour can collectively become reckless or morally suspect. The reinforcing to troops that their larger goal is to complete the mission on hand and leave with their moral integrity beyond reproach should be a continuous activity. It is to be made very clear that acts such as rape and plunder will be dealt with harshly, and are grounds for ostracism from a cohesive unit.

Conclusion

There has always been a scope for great heroism and nobility in combat, which implies that there is also a pure and true manner in which to wage war—a combat ideal. Misconduct behaviours both find their roots in, and by themselves become direct violations of this combat ideal. However, they can be contained with little awareness at the personal level and lot of concern at the unit level. Undoubtedly, unit commander shoulders loads of responsibility for ensuring the same at both the levels.

References

Army Act. (1955). *Discipline and trial and punishment of military offences*, UK Statute Law Database. Retrieved from http://webarchive.nationalarchives.gov.uk/20110805174918

Askin, K. D. (1997). *War crimes against women: Prosecution in international war crimes tribunals*. Hague: Kluwer law International.

Barnett, G. (2006). The blood of Nelson. *Military History, 23*, 36.

Booth-Kewley, S., Highfill-McRoy, R. M., Larson, G. E., & Garland, C. F. (2010). Psychosocial predictors of military misconduct. *The Journal of Nervous and Mental Disease, 198*, 91–98.

Brownmiller, S. (1975). *Against our will: Men, women and rape*. New York, NY: Simon & Schuster.

Bryant, C. (1979). *Khaki-collar crimes*. New York, NY: Free Press.

Crocker, C. A. (1995). The lessons of Somalia: Not everything went wrong. Foreign Affairs, 74, 2–8.

Director General Armed Forces Medical Services. (2002a). *Mental health programme for the armed forces*. Nashik, India: Govt. of India Press.

———. (2002b). *Psychiatric disorders in the field* (1st ed.). Nashik, India: Govt. of India Press.

Duffy, M. (2009). *Self-inflicted wounds (SIWs)*. Retrieved from www.firstworldwar.com / atoz/siw.htm.

Dupuy, T. N. (Ed.). (1993). *International military and defence encyclopaedia* (Vol. 3). Washington, D.C.: Brassey's.

Ferguson, N. (2007). *The war of the world: History's age of hatred*. London, UK: Penguin Books.

Gabriel, R. A. (1987). *To serve with honour: A treatise on military ethics and the way of the soldier*. Westport, CT: Greenwood Press.

Harrisson D., & Laliberte, L. (1994). *No life like it: Military wives in Canada*. Toronto, ON: James Lorimer.

Harrisson, S. (2012). *Soldiers who desecrate the dead see themselves as hunters*. Retrieved from http://www.esrc.ac.uk/news-and-events/press-releases/21182/soldiers-who-desecrate-the-dead-see-themselves-as-hunters.aspx

Hatlie, M. R. (2005). *Memorial to deserters in ulm*. Retrieved from http://sites-of-memory. de/main/ulmdeserters.html

Hedges, C. (2003). *What every person should know about war* (1st ed.). New York, NY: Free Press.

Hewson, C. W. (1985). *Mobile command study: A report on disciplinary infractions and antisocial behavior within FMC with particular reference to the special service force and the Canadian airborne regiment* (Hewson Report). Ottawa, ON: Department of National Defense, Canada.

International Committee of the Red Cross. (1907). *Convention (IV) respecting the laws and customs of war on land and its annex: regulations concerning the laws and customs of war on land.* Retrieved from http://www.icrc.org/ihl.nsf/INTRO/195

Kellet, A. (1987). *Combat motivation: The behaviour of soldiers in battle.* Canada: Springer.

Lillian Goldman Law Library. (2008). *Judgment: The law relating to war crimes and crimes against humanity.* Retrieved from www.avalon.law.yale.edu/imt/ judlawre.asp

Massie, A. (1989). *A question of loyalties.* London, UK: Hutchinson.

Ministry of Defence. (1961). *Manual of Indian military law* (3rd ed.). Delhi, India: Manager of Publications, Ministry of Defence.

Oxford English Dictionary (1997). New York, NY: Oxford University Press.

Rhem, K. T. (2000). *Alcohol abuse costs DoD dearly.* American forces press service. Retrieved from http://www.defense.gov/News/NewsArticle.aspx?ID=45284

Roberts, G. (2006). *Stalin's wars: From world war to cold war, 1939–1953.* London, UK: Yale University Press.

Rolbant, S. (1970). *The Israeli soldier: Profile of an army.* Cranbury, New York, NY: Yoseloff.

Rutagengwa, C. S. (2008). *Rape as a weapon of war.* Retrieved from http://www.author-me. com/nonfiction/rapeasaweapon.html

Schwabe, M. R., & Kaslow F. W. (1984). *The military family: Dynamics and treatment.* New York, NY: Guilford Press.

Thompson, M. M., & McCreary D. R. (2006). *Enhancing mental readiness in military personnel.* Retrieved from http://ftp.rta.nato.int/public/PubFullText/RTO/MP/ RTO-MP-HFM-134///MP-HFM-134-04.pdf

United States Army Headquarters. (1994). *Leader's manual for combat stress control* (Field Manual 22–51). Washington, D.C.: Headquarters, Department of the Army.

———. (2009). *Combat and operational stress control manual for leaders and soldiers* (FM 6-22.5).Washington, D.C.: Headquarters, Department of the Army.

United States Office of the Federal Register. (1939). The code of federal regulations of the United States of America. Washington, USA: G.P.O.

US Navy Marine Corps Court of Criminal Appeals. (2010). *Special Court Martial - Lazzaric T. Caldwell* (NMCCA 201000557). Retrieved from www.jag.navy.mil/courts/documents/ archive/2011/CALDWELL,L.T.201000557.pdf

Vicinus, M., & Nergaard, B. (Eds.) (1989). *Ever Yours, Florence Nightingale: Selected Letters.* London: Virago

Zaffiri, S. (1969). *Hamburger Hill.* San Francisco, CA: Presidio Press.

7

Psychological Operations in Warfare

Ron Schleifer

Looking at the emergent psychological warfare, this chapter examines Iran's client regime of Hamas in Gaza, specifically the way Hamas conducted its Psychological Operations (PsyOps) campaign against Israel in Operation Cast Lead (27 December 2008 to 18 January 2009). Israel can anticipate that bombing Iran's nuclear depots may result in Iran retaliating by instructing its clients, Hamas and Hezbollah, to attack Israel simultaneously (Hamas, Hezbollah Would Run Riot, 2012). A growing element of the conflict revolves around persuasion, namely PsyOps.[1] Therefore, a look at how Israelis and Hamas have conducted their PsyOps campaigns during battle may foretell an important aspect of a future armed conflict in the region.

Since its foundation, the Israeli Defense Forces (IDFs) have not invested substantial efforts in psychological warfare. In Operation Cast Lead, however, an abrupt change occurred. For the first time in its history, the IDF launched a military operation with a psychological warfare strategy prepared in advance—devised by a specialized unit and coordinated with its operational forces. The IDF's operational unit for psychological warfare is the Mercaz L'Mivtzaei Toda'a (MALAT, The Center for Consciousness Operations), which is subordinate to the army's operational and intelligence branches. The earlier unit, known as Lohamat Modi'in (LOM, Intelligence Warfare), was dismantled at the end of the Second Intifada (2005) and then reassembled shortly before the Second Lebanon War (2006) broke out (Harel, 2005).

This chapter describes the basic theoretical assumptions of military PsyOps and its execution in the field in Operation Cast Lead, and seek to extrapolate from this to the Israeli–Hamas PsyOps battle in the scenario of a possible armed Israeli conflict with Iran.

---

[1] Current Pentagon terminology for PsyOps is 'Military Information Support Operations' (MISO), which attempts to dodge the tradition of propaganda insinuation, yet it is unlikely to last long due to criticism from professional levels (Paddock, 2011).

Theoretical Principles of PsyOps

Psychological warfare is a doctrine of warfare that aids an army in achieving its objectives via essentially nonviolent persuasion (Taylor, 1998). The characterization of PsyOps has gone through several renderings over the years (Taylor, 1998). Since the First World War (WWI), its techniques have been increasingly implemented; as the intensity of physical warfare decreases, the use of psychological warfare increases. Ironically, psychological warfare has developed a malicious image, as if its main enterprise is the exploitation of lies and deception, and disseminating paper leaflets from the air. This is a popular misconception. Psychological warfare is an extremely advantageous battle technique, particularly in clashes in the Arab–Israeli Conflict since 1982, when battle against Israel took the form of unconventional warfare (Frisch, 2003).

Psychological warfare's main objective is to transmit information to designated groups during wartime in order to support military and political objectives (United States Department of the Army, 1994 [This was followed by Field Manual 3–13]). This has been particularly true in recent decades, during which radical Islam has achieved political victories against the armies of the industrialized West, such as in the US withdrawal from Iraq and its gradual pullout from Afghanistan. Consequently, there has been much effort to develop and implement the doctrine in the armies of the United States, Great Britain and countries that comprise the North Atlantic Treaty Organization (NATO) (Tatham, 2008).

PsyOps messages transmitted during a war or conflict can be divided into two main categories: persuasive messages and vital information messages. Persuasion aimed at changing attitudes need to be carried out long before any battles begin; once the first shot is fired, this stage of PsyOps has ended. At this point, the initiator of psychological warfare typically aspires to change the conduct and behaviour of the enemy's soldiers and/or civilians through transmitting vital information— updates on the situation or news articles containing information that their leaders would, of course, prefer to remain concealed.

As noted, paper leaflets containing a few lines of written text or some relevant illustration are unfortunately the generally accepted image of this type of warfare. Although this method has been proven to be the most practical over the last century, other transmission methods—such as loudspeakers, electronic mail or the Internet—are also used, as long as the enemy can receive the information during the course of the conflict.

Traditional psychological warfare doctrine distinguishes between two means of transmission (United States Department of the Army, 1994). One is traditional media, such as TV, radio and the press. The second is alternative media—used by those with limited resources— which includes leaflets, fax, graffiti, loudspeakers, rumors, etc. The new-media age has blurred the distinction between these two types of transmission, and the Internet is now an arena that integrates TV, radio and the press, together with electronic mail, blogs, Twitter, and the like, all at extremely low costs. Cell phones have turned each carrier into a broadcast station, as well as an accessible way to receive information through Short Message Service (SMS), texts and news video clips.

Nonetheless, even though a century has passed since armies first exploited the massive use of leaflets dropped from the air into the battlefield; this is still the most accessible medium for transmitting messages during conventional warfare.

The Psychological Warfare: Operation Cast Lead

IDF's PsyOps

OBJECTIVES

The following psychological warfare objectives of the IDF during Operation Cast Lead can be identified by analysing the MALAT's messages and transmission methods:

1. To damage Hamas's psychological warfare capability, its transmission channels, and the credibility of its contents.
2. To damage the credibility of Hamas as a governing organization.
3. To amplify Israel's achievements, in contrast to Hamas's failures, and to display the demoralization among Hamas's activists and its various supporters.

TARGET AUDIENCES

During the war, the MALAT was involved in imparting vital information to three key groups: Hamas soldiers, Hamas's civilian supporters and Palestinian residents in Gaza.

MESSAGE CONTENT

The messages to Hamas soldiers were of the standard demoralization approach: "You who are dying in battle: You have no chance against the IDF's special units and their weapons. Your leaders are hiding. You are alone in the battlefield", and so forth (Friedman, 2013).

During the operation the press reported that the IDF had taken control of the enemy's tactical communications network, but Hamas also managed to break into the IDF's communications network (Feffer, 2012). Nonetheless, one may assume that the IDF's direct communication with Hamas soldiers during battle gave Hamas soldiers the disagreeable impression that there is nothing the enemy does not know about them.

Messages to civilians mainly attacked their leadership: "Your leaders have fled and abandoned Gaza's civilians. They were completely wrong about Israel's response and are consequently unable to function. Hamas exploits civilians as human shields and steals for itself the aid designated for Gaza residents". The concluding message was: "The IDF is completely prepared to enter the Gaza Strip".

Messages of this type were transmitted through leaflets. One of these leaflets was entitled, *Inform on Them*. It called upon the residents of the Gaza Strip to report (without compensation) hiding places used for weapons or booby traps, thereby saving their own lives and those of their families, as well as protecting their property. The notice included a telephone number, which indeed received thousands of calls—most of which were abusive (Ya'ari, 2009).

Hamas made numerous comments about these particular leaflets. From this, one may infer that the matter was quite troubling to the organization—they feared that those who did not support them in Gaza would exploit the opportunity and indeed report information to Israel. Consequently, this action on Israel's part was considered a success; in fact, it was a classic case of "driving a wedge"—one of many psychological warfare techniques also known as the Roman maxim of: *divide et impera* (divide and rule) (Tzu, 1963). The objective in such a case is to cause a division between various elements of the population—in this case between Hamas and Palestinian society in general—in an attempt to challenge the legitimacy of Hamas's rule.

The message regarding the IDF's complete preparedness for a ground invasion was aimed at undermining the feeling Hamas had cultivated about having built a comprehensive booby-trap network against Israeli soldiers, and that the population need not worry about Israel possibly launching a ground invasion of the Gaza Strip. Messages such as these are abundant in the Iranian PsyOps campaign designed to deter Israel from bombing Iran's nuclear depots (`Iran Commander', 2012; 'Iran Says', 2012).

TRANSMISSION CHANNELS

In Operation Cast Lead, both sides exploited all transmission means to an extent unprecedented in Israel's previous wars. During the operation, millions of leaflets were dropped on the Palestinians, even though the conflict took place in the winter, when strong winds might have carried the leaflets out of the targeted areas. To accommodate the interests of the Israeli and foreign media with representatives concentrated on one of the local hills overlooking the Gaza Strip, they were also included in the drop zone.

The media factor in what the US Army calls "Information Theater", is a most complex issue in democracies. The latest doctrinal document of the US Army does its best to sidestep the sensitive issue of including the media within a country's PsyOps efforts. For instance, the manual on handling the media is neutrally called *Public Affairs* (United States Department of the Army, 2000).

Most of the leaflets were directives relating specifically to the Palestinians' behaviour. The IDF also transmitted humanitarian messages that dealt with when and where food would be distributed. In the final analysis, the number of drops and the quantity of leaflets were limited; presumably, this was in order not to diminish the dramatic effect of information falling from the sky by diluting it, thus rendering it commonplace.

In the electronic arena, the method of transmitting messages was mainly through commandeering Hamas communications channels. In as much as broadcasts of the Israel Broadcasting Authority are not received in the Gaza Strip, the IDF broke into and took control of the Palestinian broadcasting media, exploiting them to transmit messages of the MALAT. In fact, since the start of the war the Palestinians had complained that the IDF had taken control of Hamas's TV stations and was broadcasting videos in Arabic (Israel Interrupts Its Broadcasts, 2009). Consequently, an alternative Hamas news programme was instituted on the radio, and video clips were broadcasted on Hamas TV. On a tactical level, messages were broadcasted on Hamas's internal communications network, SMS messages were transmitted to the cell phones of many Gaza Strip residents and recorded messages were transmitted to their home phones.

TYPES OF MEDIA EMPLOYED

Every message transmitted to the enemy must cut across the psychological barrier of it being a message from *their* enemy, whose purpose is to exert influence. Therefore, above all, the message must be perceived as vital and relevant, and must also elicit interest. The format proven effective in every war in the 20th century has been news reports, insofar as following the heat of battle the demand for information grows exponentially. At the same time, as noted, humanitarian content is also an effective means for eliciting consumers' interests.

In most cases of transmitting messages to Palestinians, it is possible to distinguish an unchanging line of penetration. The message is, in fact, pushed into Palestinian space and consciousness, although this says nothing about its reception. The printed leaflets occupy the physical space, and the SMS messages also achieve certain penetration. Perhaps this is because, following the initial irritation, the consumer becomes accustomed to the method; this is also the case with recorded phone calls to a person's home.

However, when an unexpected frequency takes control of radio broadcasts or one's TV, the sense of one's privacy being invaded is overwhelming. Messages in the form of news broadcasts were transmitted on TV, including an opening signal confirming that a broadcast was about to begin. In order to increase the programme's appeal, it opened with vital information about humanitarian aid, and matter-of-fact continued with reports on developments in various battles. The printed announcements were written in spoken Arabic, and radio and TV broadcasts were communicated in the local high language [The Arab language consists of two distinct levels, the literary (*al fusha*) and the colloquial (*al amiyah*). Speaking at the familiar level or dialect would be a primary requirement].

Hamas's PsyOps

Hamas's psychological warfare operated mainly before the fighting and afterward, and less during the fighting itself. Before the outbreak of hostilities, the organization formulated its deterrent, which was based on the message that the Gaza Strip had been turned into a death trap of hidden tunnels and explosive devices.

This technique very much resembles the Iraqi psychological warfare tactics used in the First Gulf War, on account of which the Americans deployed a force much larger than required to execute the mission. This also applies to the IDF; the pressing need to send IDF soldiers into harm's way is a factor that comes from a long-range perspective on future confrontations, and takes into account the burden on Israeli society. This burden includes the amount of time spent in reserve service, a decrease in the public's readiness to enlist and the economic cost involved in pushing back the confrontation to the extent possible.

Hamas reserved the bulk of its psychological warfare effort for the stage following the war, with the greatest political achievement culminating in the Goldstone Report. This report was named for the head of the UN Human Rights Council, Richard Goldstone, a jurist who served as the prosecutor of the United Nations International Criminal Tribunal for the former Yugoslavia and Rwanda. The Israeli government abstained largely from assisting the delegation, taking the position that the very title of 'International Criminal Tribunal' is tantamount to a conviction. As expected, the report condemned Israel for committing war crimes. Only months after, a closer inspection of the testimonies revealed serious irregularities in the accusations. As it turned out, most of the testimonies were arranged by a network of hostile NGOs financed by EU and US sources. Although Goldstone (unlike his tribunal colleagues) finally acknowledged he was misled by the testimonies, by then the political damage had already been achieved, and Israel was labeled as war-criminal state. During the fighting itself, although Hamas's efforts in the psychological warfare front were limited due to its inferior manpower and technology, it did apply itself to influence the local community—the Palestinian population—during the fighting.

TARGET AUDIENCES

Hamas's priorities were different from those of Israel. Hamas's highest priority was to reach the local community—Palestinians—followed by the West, and finally Israel. This stemmed from the gap in their physical means relative to Israel, a gap that also influenced how messages were formulated.

Following Hamas's taking control of Gaza and the violence with which it established its rule there, Palestinian society split into a number of segments whose loyalty to Hamas's regime, especially in the face of Israeli pressure, needed to be verified. Israel's PsyOps efforts, it seemed, were directed towards maintaining and even deepening that split.

The West, from the perspective of the conflict in question, was essentially a neutral target audience that included powers not tied directly to the conflict but whose influence could be advantageous to either side. Hamas's objective, then, was to urge Europe and the United States to apply pressure on Israel to withdraw from Gaza and halt the military force.

As mentioned, the Israeli population was of the lowest priority for Hamas during the battles, although Hamas's messages about innocent Gazans suffering injury and property damage were directed at the Israelis nevertheless. However, once the fighting stopped, Hamas stepped up this campaign in order to direct a split within the Israeli society after the level of wartime patriotism had declined and the average Israeli citizen was inclined to examine how the war had affected his own life, the IDF and Israel's position in the international arena.

MESSAGES

Before the operation, Hamas transmitted two types of messages: those that cultivated the image of the Gaza Strip as a giant death trap (with respect to the IDF and Israeli society), and those that publicized the damage from Israel's siege on Gaza and, in particular, the siege's deleterious effect on noncombatants (to the neutrals).

Once the shelling and ground operation had begun, Hamas moved to assert its position on the Palestinian society so that its political influence there would not falter. Its messages turned inwards and dealt mainly with national unity and Israel's offensive. At the conclusion of the war, Hamas leaders hurriedly left their bunkers, and the battle over the effect of victory started immediately. In Israel, this move was considered absurd, but this was not the case for Hamas, as perception of victory is a psychological matter, not necessarily linked to the physical reality on the ground.

Transmission Channels

The media's finest hour is during war. The craving for information is extremely strong and therefore each side aspires to harness the media to itself and its interests. Privately owned media outlets are generally perceived as reliable and therefore sending messages through them is a main objective for each side. In order to approach the Israeli public, Hamas exploited Israel's openness and fed Israeli and world media through the phone and Internet, as Israel forbade journalists to enter Gaza from its side. The approach to Israelis was mainly concerning the theme of injustice being committed against noncombatant Palestinian civilians. The messages were aimed at increasing the public's concern and feelings of guilt, with the anticipation that these feelings would ultimately reach IDF soldiers, a technique already highly developed during the First Intifada (Schiff & Ya'ari, 1990).

At first, Hamas operated a satellite TV station (Al Aqsa) and a ground station (Al Quds), newspapers, radio stations and various websites, such as http://www.aqsatv.ps and http://www.palestine-info. info. A few of the websites are official sites that are at risk for breaches by hackers (organized and voluntary); most of the Islamic sites are not directly identified with the organization, but feed off its content and disseminate its messages. Likewise, Hamas has in its possession a public relations centre that serves as a clearinghouse for its psychological warfare programme and is headed by Fathi Hamad, a Hamas member of parliament. A number of correspondents remained in the Gaza Strip and broadcast photos of the IDF offensive to various news agencies, such as Ramattan (http://www.ramattan.tv). After broadcast stations signals were hijacked by the IDF, websites, SMS and cell phones remained, using interpersonal communications and broadcasts at times when the Palestinians were able to do so without interference (The dilemma in imposing a block on or breaking into an enemy channel is that the enemy will avoid using that channel for broadcasting or receiving messages). It was impossible to deal with news websites, blogs and social media that were fed by organization activists using satellite phones, due to the large number of websites and their ease of connectivity. The IDF deduced several lessons following the war, establishing a department within the IDF Spokesperson's Unit for the purpose of employing these social networks and supplying them with information (Haaretz, 2009).

Hamas's other communication methods were mostly meager, but one should note that all of this activity was undertaken confronting a military offensive of superior military strength. Yet, Hamas managed to hijack the signal of the IDF Radio regional frequency and caused a considerable communications disturbance for the Israelis. SMS messages were transmitted to cell phones of civilians in the southern region (Balousha & O'Loughlin, 2009). Hamas used the medium of rumors effectively. Some rumors were disseminated regarding the numbers of the wounded, but the IDF spokesperson had a reliable reputation in these matters, and no serious damage was caused following these fabricated reports. In one case, Hamas took control of the situation and managed to penetrate the Israeli psychological barrier when a rumor it spread claimed that the abducted soldier held by Hamas, Gilad Shalit, was wounded in a bombing (Yahav, 2008). This is a clear example of how sophisticated equipment is not a prerequisite for engendering significant psychological damage to the enemy.

Assessing the PsyOps' Effectiveness

In conventional warfare, it is relatively easy to assess the influence of the measures taken against the enemy and make needed subsequent improvements. Regarding psychological warfare, however, this is far more difficult. Few citizens—soldiers and officers alike—will acknowledge (even to themselves) that they have been influenced by some campaign or message. As is the case in commercial advertising, PsyOps consumers may be aware of the enemy's intent to influence opinion, but the actual impact is often imperceptible to them. This means the standard tools used to measure effectiveness, such as surveys and focus groups, are naturally not applicable. Therefore, psychological warfare is compelled to find indirect ways to evaluate its influence.

The first criterion in assessing PsyOps influence is looking at whether the population followed directives, that is, modified their behaviour. The general impression is that there was Palestinian compliance with IDF orders. Out of the millions of printed leaflets that were disseminated, it is difficult to assess how many reached their targets and were read by the population, and how many people viewed the videos broadcast on the Hamas TV frequency hijacked by the IDF. But, it is easier to monitor the prerecorded phone messages. In some cases, the party receiving the call simply hung up, and in some cases he or she stayed on the line and listened to the message. From Palestinian complaints on blogs and websites, it appears there was a significant percentage of Gazans who did indeed listen to the prerecorded message (Major [ret.] M., personal communication, August 2010).

Tracking Hamas's message content reveals the oft-repeated idea that the Palestinian people supported the Hamas government. Another claim, repeated day and night, was that Israeli psychological warfare had no influence on Palestinians. This is an indirect indication that Israel's psychological warfare was possibly effectual, leading Hamas to devise countermeasures.

Likewise, comparing messages sent by both sides reveals the dynamics of a sort of indirect dialogue, in which Israel initiated and to which Hamas responded. Each time the MALAT issued a statement, Hamas messages were immediately broadcasted to the Palestinians for the purpose of refuting the Israeli reports (Lt. Col. [retd.] Rami, personal communication, December 2011). In his visit to his patron Khamenei of Iran, Khaled Mashal, head of the Hamas political bureau, flattered the Palestinian people for bravely withstanding the IDF's divisions and massive psychological warfare campaign (Khamenei, 2009). MALAT staff may have taken his statements as a compliment, especially in light of the wide gap between needs and allocated Israeli resources.

Israeli–Hamas PsyOps: Lessons to Learn

In general, although there were no surprises in the IDF campaign during Operation Cast Lead, none were necessary. Psychological warfare operates according to fixed principles of self-empowerment while wearing the enemy down: weakening him by encouraging desertion and falling captive, cultivating apathy and disinterest, and lowering the morale and motivation of the opposing army and its civilian support. This is the general outline—there are endless details, of course, and there is much room for creativity. The basic content of psychological warfare messages has repeated itself throughout history. A conventional army will tell the enemy, "You don't have a chance". Relatively weak armies or guerilla movements will stress their determination and willingness to sacrifice, along the lines of "It isn't the tank that wins…".

What changes throughout the ages is primarily the means of transmitting the messages. In the distant past, the means of transmission were mere shouts and written notes. In our present Information Age, means include Twitter, beepers, and lasers. In this area, the IDF is making significant strides. The challenge faced in the current Information Age is that the cost of using these media has dropped significantly, thereby making them more accessible to terrorist and guerilla organizations. They employ creativity and imagination while doing so, and the themes that come up are not likely to be stalled at any transition point by a long chain of command.

Based on Hamas's reactions, the MALAT's PsyOps execution was admirable. Its messages managed to infiltrate every layer of the population, sometimes including the leadership itself, diminishing their confidence. The MALAT almost succeeded in creating the impression that the Hamas regime in Gaza was collapsing, and this obligated Hamas leadership to invest much effort in reestablishing its ruling authority. There is as yet no clear indication whether the collapse of Hamas rule was indeed in Israel's interests, but it appears that this was not a guiding objective of the MALAT. The unit was particularly focused on stressing Israel's humanitarian side, although under ideal conditions this function may be more suitable for the IDF Spokeperson's Unit, were it to have the means to broadcast to the Arab public. Since Israel's High Court's Tzoran ruling (in the wake of an environmental damage suit, mandating the dismantling of a transmission antenna), however, Israel has lost the ability to broadcast to the Arab public through radio, particularly through the Voice of Israel Arabic Service.

Hamas, whether deliberately or under duress, in accordance with rules of guerilla warfare, refrained from direct military confrontation with the IDF, and its psychological warfare operated at low intensity. In terms of physical fighting, the organization concentrated on enticing IDF soldiers into traps and attempting to kidnap them.

Hamas's psychological warfare was mainly reduced to responding to the messages of the MALAT, reserving its main PsyOps efforts after the war. In other words, it focused on the political battlefield, which, according to the Clausewitzian maxim, is the principle battlefield in the final analysis. Immediately following the fighting, Hamas began efforts to create a consciousness of victory—that of spirit over substance. It fabricated a general demonization of the Israeli enemy (as a satanic killer of children) and also stressed the irrefutable fact that the Hamas regime was still firmly in control. Hamas's victory speech was written in advance and presented to the public at the first opportunity when there was no imminent peril to the organization's leaders. Of course, there is no relationship between consciousness of victory as it was marketed to the Palestinian public and the reality on the ground, and here Israel failed in its understanding of the situation. Although in Israel photos of the destruction in Gaza symbolized the IDF's indisputable victory, for Hamas this same destruction represented *its* victory.

The battle for awareness was on a relatively small scale. It focused on the populations of Gaza and Judea and Samaria, and was targeted mainly at Western audiences. This was due to Hamas's perception that Israel depends for its survival on the West's mercy, and if the moral and political support of the West is suspended, Israel's military and economic support will also be suspended, resulting in Israel's total collapse before the sword of Islam. The Goldstone Report and Hamas's successful conduct of the Shalit affair are examples of links in a long chain of psychological warfare tactics orchestrated by Hamas.

Extrapolating PsyOps Programmes in Future Conflicts

What would the PsyOps perspective of a future conflict look like? If a coordinated attack took place, which included missiles from Iran, Hezbollah in the north, Hamas in the south, and possibly the Palestinian Authority (PA) in the centre, the Palestinians would have to take a risk of enormous consequences.

If Israel survives the first strike from Iran and its accomplices, it will respond with an armed entry into the Gaza Strip and Samaria that could result in an unplanned 1948-type exodus into Egypt and Jordan. In an armed invasion, Hamas is likely to repeat the essence of its successful PsyOps campaign of 2009, such as stressing Israeli cruelty against noncombatants. Its messages will highlight the devotion of the Palestinian people to sacred values as opposed to Israeli brutality. The methods of delivery are likely to remain the same, as there has been no substantial novel technological breakthrough that can be exploited. If Hamas and its Iranian backer have acquired any lesson since 2009, they will attempt to maintain an interruption-free TV channel and expand the popular use of smartphones as an alternative medium, whose reception and broadcast will be enhanced by satellite communication. Following Hezbollah's 2006 success in maintaining the activity of the Al Manar TV station (through the preplanned use of alternative transmission equipment in Lebanon), Hamas might do so, perhaps in cooperation with Egypt (after a hurried reconciliation). The classic paper leaflet droppings are most likely to continue as a delivery channel in a future conflict. As for message content, Hamas would require strong visual images of noncombatant casualties in order to force Israel to discontinue its attack. Again, based on the Hezbollah experience, it will initiate rocketing Israel from Gaza residential areas, preferably but not necessarily populated by Hamas opposition, and absorb a large death toll despite Israel's attempt at surgical-precision bombing. These actions will produce PsyOps-worthy materials for Hamas. In case of a prolonged military operation, civilians fleeing the battlefield can always fall back on the successful sixty-five-year-long Palestinian refugee campaign and provide enduring historical familiar images of suffering.

In a future conflict, the humanitarian message appeal is likely to be minimal. Israel may try to use its counterattack/preemptive strike to topple the Hamas regime entirely and help the PA regain control, or fully strip the Gazans off the offensive weapons. Israeli messages are likely to be more coercive, as Israel will be confronted in more than one arena. The messages may direct the population to abandon their own residential area for their safety and assemble on the rural southern region of the Gaza Strip, to serve as an impediment to the entrance of Egyptian regular or volunteer forces. Alternatively, residents may be instructed to block Gaza City and demand protection from Hamas, which apparently it would be unable to provide and thereby engendering political havoc, which could then lead to Hamas's downfall. At the same time, Israel could continue its humane-focused efforts in order to deliver one of the oldest messages in the practice of PsyOps: "We don't have anything against you; we only oppose your leadership".

For those questioning how Hamas could plausibly initiate rocket attacks against Israeli civilians and then blame Israel for barbarous acts, Hamas's PsyOps campaign throughout Operation Cast Lead, following a five-year shelling campaign—provides the answer.

As the balance of physical power is still against Hamas, it is quite likely it will continue its use of psychological warfare in future conflicts with Israel. Based on its past successes, it will enhance its communication capabilities in terms of delivery channels, such as satellite communication (phone and Internet), journalists (local and foreign), and PsyOps teams that will produce, at any cost, images of civilian suffering. The Iranian input may include technical equipment to break into Israeli cell phones for purposes of gathering standard intelligence as well as data on population morale, and as a channel for prerecorded phone messages to diminish it. Based on experience gathered during Cast Lead, these means are relatively simple to operate and contribute greatly to the most important phase of the war—the political one.

Conclusion

Detailed understanding of the HAMAS-Israeli PsyOps can provide enough inputs for designing future campaigns in the event of nonconventional warfare worldwide. Undoubtedly, such psychological warfare shall become the reality of tomorrow, haunting every nation while safeguarding its frontiers from internal and external aggression. Lessons need be learnt to equip the forces well with the techniques of psychological warfare to keep an edge over the opposing forces irrespective of its physical prowess.

References

Balousha, H., & O'Loughlin, T. (2009, January 3). Text messages and phone calls add psychological aspect to warfare in Gaza: Hamas fires threatening text messages at Israeli mobile phones while Israel bombards Palestinians with menacing phone calls. *The Guardian*. Retrieved from http://www.theguardian.com/world/2009/jan/03/Israelandthepalestinians-middleeast

Feffer, A. (2012, November 19). The psychological campaign: Israel breaks into broadcasts, Hamas threatens to reveal officers' personal information [Hebrew]. *Haaretz*. Retrieved from http://www.haaretz.co.il/news/politics/1.1868001

Friedman, S. A. (2013). *Israel vs. Hamas (2008–2009)*. Retrieved from http://www.psywarrior.com/GazaPsyOps.html

Frisch, H. (2003). Debating Palestinian strategy in the Al-Aqsa Intifada. *Terrorism and Political Violence, 15*(2), 61–80. doi:10.1080/09546550312331293037

Hamas, Hezbollah would run riot under Iranian nuclear umbrella, general warns. (2012, January 18). *Israel Hayom*. Retrieved from http://www.israelhayom.com/ site/newsletter_article.php?id=2724

Harel, A. (2005, January 25). The IDF decides to reestablish the psychological warfare unit. *Haaretz*. Retrieved from http://www.haaretz.com/print-edition/news/idf-reviving-psychological-warfare-unit-1.148134

Iran commander says Islamic Republic could launch pre-emptive strike on Israel. (2012, September 23). *Haaretz*. Retrieved from http://www.haaretz.com/news/diplomacy-defense/iran-commander-says-islamic-republic-could-launch-pre-emptive-strike-on-israel-1.466411

Iran says World War III may erupt if attacked by Israel. (2012, September 23). *Xinhua News Agency*. Retrieved from http://news.xinhuanet.com/english/world/2012-09/23/c_12 3750479.htm

Israel interrupts its radio, TV broadcasts in Gaza: Hamas. (2009, January 3). *AFP*. Retrieved from http://www.google.com/hostednews/afp/article/ ALeqM5hFj HD4 gSPfb4lPEcyIy XQclnnTbg?hl=en

Khamenei, S.A. (2009, February 2). Gaza people conferred honor on all of us. *The Office of the Supreme Leader Sayyid Ali Khamenei*. Retrieved from http://www.leader.ir/langs/ en/?p=contentShow&id=4735

Paddock, A. H., Jr. (2011, February 15). The future of MISO: A critique. *Small Wars Journal*. Retrieved from http://smallwarsjournal.com/blog/journal/docs-temp/677-paddock.pdf

Schiff, Z. & Ya'ari, E. (1990). *Intifada: The Palestinian uprising—Israel's third front*. New York, NY: Simon and Schuster.

Tatham, S. A. (2008). *Strategic communication: A primer*. Shrivenham, England: Defense Academy of the United Kingdom. Retrieved from http://www.aco.nato.int/resources/9/ Conference%202011/08(28)ST[1].pdf

Taylor, P. M. (1998). Foreword. In R. Cole (Ed.), *International encyclopedia of propaganda* (pp. xix–xxiii). Chicago, IL: Fitzroy Dearborn Publishers.

Tzu, S. (1963). *The art of war* (S. B. Griffith, Trans.) Oxford, UK: Oxford University Press.

United States Department of the Army. (1994). Psychological operations, techniques and procedures (Field manual 33-1-1). Washington, D.C.: Headquarters, Department of the Army. Retrieved from http://arcdc.org.il/attachments/article/24/fm33-1-1.pdf

———. (2000). *Public affairs, tactics, techniques and procedures* (Field manual 3-61.1). Washington, D.C.: Headquarters, Department of the Army. Retrieved from http://armypubs.army.mil/doctrine/DR_pubs /dr_a/pdf/fm3_61x1.pdf

———. (2013). *Inform and influence activities* (Field manual 3-13). Washington, D.C.: Headquarters, Department of the Army. Retrieved from http://www.globalsecurity.org/ jhtml/jframe.html#http://www.globalsecurity.org/military/library/policy/army/fm/3-13/ fm3-13-2013.pdf

Ya'ari, E. (2009, January 7). Write it down, Ehud Barak is a terrorist [Hebrew]. *Channel 2 News*. Retrieved from http://www.mako.co.il/news-military/security/Article-9640dabc182be11004.htm

Yahav, N. (2008, December 29). Report: Gilad Shalit wounded in bombing [Hebrew]. *Walla News*. Retrieved from http://news.walla.co.il/?w=/9/1406802

8

Future Warfare and Mind Control

Swati Johar and Updesh Kumar

The perspective planning for the defense of the nation brings to focus the issues of changing nature of warfare. Tremendous changes in socio-economic milieu and geopolitical systems on gross or subtle levels since the Second World War (WWII) have resulted in changing nature of social order across the world. Short, high-tech wars along with challenges of dealing with various low-intensity conflicts are expected to change the nature of warfare in the future. The locations of war might vary from deep sea to the expanse of the outer space. The soldiers would be required not only to fight to protect their territorial integrity, but are more likely to fight for peace building, peace keeping and protecting the natural resources; most of the time operating as a part of multinational coalition forces. Besides the traditional role of the military, the soldiers can also be foreseen performing disaster mitigation and management, negotiating hostage situations and engaging non-state enemy forces in non-traditional conflicts.

Research on psychological aspects of war started since the beginning of modern psychology but it is yet a commonly overlooked aspect of warfare. With the beginning of WWII began a new style of war. The dropping of paper-filled bombs from bomber planes during WWII to target the morale of the troops provides a perfect example of the power of psychological warfare. For the first time, psychology was used to deteriorate the motivation and sentiments of the enemy forces. In recent times, the weaponry and violence of war itself has spread its influence to such a large extent that the weapon of 'psychological warfare' has been completely ignored. The focus on suffering rather than death makes this concept more justifiable and provides answers to the following questions:

1. How the growing trend of technology and machines can be arrested and diverted towards non-conventional systems?
2. What are the means by which fears can be manipulated?

A vision for the future defence forces must include prospective research areas that would enable the soldier for operations directed at the enemy's mind rather than his body. The significance of psychological dimension of a conflict needs to be broadened in this information technology (IT) age. Clearly, the British military analyst and historian prophesied that

> the so called traditional means of warfare might be replaced by purely psychological warfare, wherein weapons are not used or battlefields sought... but dimming of the human intellect, and the disintegration of the moral and spiritual life of one nation by the influence of the will of another is accomplished. (Fuller, 1920)

The overdependence on technological superiority has left insufficient attention on the importance of the study of enemy's psychology. Undermining enemy resistance by the planned use of communications is a long process. It influences human attitudes and behaviour by deconstruction of the enemy and creates in target groups, behaviour, emotions and attitudes that support the attainment of national objectives. The role of Psychological Operations (PsyOps) in shaping the moral and intellectual environment of the battlefield and its use as an influence weapon at tactical, strategic and operational levels in manoeuvre warfare is known, but not well understood and implemented. Post (2005) illustrates that terrorism is a war for hearts and minds and cannot be won with smart bombs and missiles. It is a psychological warfare waged through media and PsyOps should be the primary weapon in the war against terrorism.

Given the widening difference between growth of technology and physical limits of human capability, technology is likely to replace and retard human capability. Therefore, any future perspective for the defence forces must enable the soldier to effectively defend oneself from techno-savvy attacks. Some options that come visible are, for example, personality structuring for optimizing human resources for creating militarily designed soldiers to fit in specified roles which would involve predisposing the human mind. Penetrating the human mind using technological tools in the form of cognitive hacking for causing information paralysis, disruption and distortion might leave the adversary in an ineffective state. Brain washing by using wireless internal voice transmission, blocking thought streams using ultrasonic sound as carrier and controlling the mind by remotely altering brain-wave technology would be another way of either aggressing the enemy or as a defence against it. Mind-enabled tools, such as mind readers, meta-cognitive tools and decision-making tools might prove to be another option in shaping the future environment in which the soldiers need to operate.

Need for developing special forces and evolving adaptability to changing face of technology and changing nature of the theatre of warfare has necessitated the change of focus in Military Psychology from retroactive management to proactive preparedness and restructuring of the military endeavours of nations. The challenge before psychology is to enhance the efficiency of new scientific and technological advances, by ensuring effective application of the inevitable human and psychological components. The explanation of how mind, the brain, other biological systems of the body and human environment interact to produce behaviour in the context of changing situational variables and varied human capabilities moderated through the use of technology needs to be broadened. Certain specific areas that pose a challenge to the science of Military Psychology can be identified. Threat detection systems, prediction and forecast models may be developed to enhance capabilities for proactive preparedness. Advanced psyops and psycho-social immunizations may be considered as potential areas of research in the near future. Keeping this in mind along with the urgency to handle and control the present world situation, it is important to explore these aspects of warfare as it will aid not only in matters of critical importance but also have far-reaching implications for the civilian populace.

Information Age Opportunities

Advancements in information and communication technologies, together with increased availability of information, have changed the ways in which societies work. Today, more and more information is available to people, thus creating a knowledge–based society surrounded by high-tech global economy. This makes the task of decision making even more challenging. Some of the information-age developments have contributed greatly and are likely to influence future wars.

Fast Computing and Knowledge Processing

Computers have made the task of managing ever increasing amount of information at a time simple. Computation speed has increased exponentially over the past few decades. The proliferation of computers, together with the miniaturization of machines, has made it possible to solve complex problems in a matter of seconds. The information is largely centralized making processing of data easier and faster. Representation of knowledge and optimal searching of knowledge base systems using artificial intelligence and simulation have given way for development of automatic trackers and self-engaging weapons to minimize human aspect. Human–computer interaction is an obvious development that has taken place in this era. The intelligence and real-time decision-making power of man is incorporated in the machine to create a symbiotic relationship of man and computer in order to facilitate formulative and intelligent thinking. Keeping in view these emerging trends and innovative technologies, rationalization of Armed Forces and effective utilization of unprecedented computing speed and time would form the vital part of battle scenario. The time required for precise decision making by the central authority will reduce to a few seconds and automatic software development for real-time processing will reshape future wars.

Networking

Networks have existed since people started interacting within groups. Later, email and Internet led to the exchange of data between systems. The emergence of Internet and cyber space has transformed the approach to handle information proliferation worldwide. The introduction of network as an organizational scheme in business, government and society only recently found utility (Bowdish, 1999), especially when the wars have gone global.

Advanced Mass Media

Today, mass media comprising communication services ranging from satellites, TV, radio to print media such as newspapers and magazines envelops the entire globe. The technologically advanced mass media has a great potential to influence future wars. The way information is interpreted and transferred are the crucial aspects of media. The ability of media to influence governments, society and military can be seen in various instances of the past. During Operation Uphold Democracy, radio and TV programming was used in carefully crafted interagency campaign "to prepare Haitians for democracy restoration and the imminent arrival of US forces". The mission was a booming success (Bowdish, 1999). In another instance, images brought war to an entire society.

Due to the prevailing unstable global environment, it is difficult to predict the nature of future wars. Mass media, specifically news and radio, can assist in routine tracking of movements and unusual happenings. It can act as the intermediary between the forces and the public to act on the spur of the moment. The effect and effectiveness of media plays a major role in formulating planned military operations. Squire (1995) illustrates application of mass media as a principle of war and refers it as a high-stake player in military operation and planning equation. It goes without doubt that media in the form of miscommunication and propaganda can be used by the adversary and can have a negative impact. Thus, when properly applied, it is a potent tool against the enemy and if neglected, can easily be used by the enemy to his advantage.

Impact of it on Future Wars: Information Warfare

The world has transformed from the industrial age to the information age. Persistent technology innovation, growth of worldwide networks, escalation in the reach of mass media and ever-increasing power of computers have all contributed in cyber revolution and have tremendous impact on warfare. The effective utilization of technology in various phases of war will be a critical aspect in the future due to the increasing dependence of military forces on IT. Due to sweeping influence of technology, future wars will be characterized by ambiguity and uncertainty owing to increasing lethality, stealth and complexity.

Information refers to the data and events that are identified and interpreted. It is a strategic resource which is now available gradually more in a digital format. Information becomes a potent weapon and extremely vulnerable when effectively managed and exploited by the enemy. As a result, nations seek to obtain and protect information in support of their objectives. This use of knowledge for destruction and immobilization of the enemy forms the core of future wars. The powerful impact of IT has opened gates for a completely new form of war structure called information warfare which would require innovative policies, skills and re-organization for its effective implementation. Making use of enemy's information to enhance our own capabilities, while protecting our own, is not a novel approach. Information warfare (InfoWar) can be defined as any action or class of techniques used to destroy and exploit adversary's information to the extent of degrading his will to fight, regardless of the means. More technically, it may also be defined as actions taken to achieve information superiority by affecting adversary information, information-based processes, information systems and computer-based networks while defending one's own information, information-based processes, information systems and computer-based networks (Bowdish, 1999). InfoWar builds a battlefield environment in space and time combining the power of land, sea and air. When IT is used to feed intelligence into the operation, it is called Intelligence-Based Warfare. Command and Control Warfare attacks the central command of the enemy organization and is considered most important. Warfare over the Internet is gaining momentum these days due to expansion of Internet networks, and this type of warfare forms the Cyber warfare. Psychological Warfare is a non-lethal IW weapon system which directly affects the psyche of soldier. The unstable and uncertain future battlefield is expected to be too demanding on man, and psychological warfare seems to be the most revolutionary to deal with the problems of this arena. Like traditional warfare, information warfare has various weapons to debilitate the targets without physically damaging it.

Weapons of Information Warfare

To be involved in a war weapons as they are the backbone of any war. Information Warfare also has its own weapons of information. Information elicitation is the foremost requirement to know details about the enemy. The greater the information, the higher is the advantage of decisiveness over the adversary. It helps in better planning against the enemy. Once the information is collected, it should be transported securely to the destination through networks. Transporting information in a timely manner is a weapon so that it becomes usable against the enemy. In modern warfare, networks are not only used to transmit information but also are the core to the entire command and control structure.

In military, networks are used for intelligence operations. Any network-based environment is characterized by the effective linking and coordination of its elements. Network warfare is a new style of fighting in the information age. Identified during the 1970s by US Air Force strategist John Boyd, Observation-Orientation-Decision-Action (OODA) is an abstraction which describes the sequence of events which must take place in any military engagement. The opponent is observed to gather information and the attacker must orient himself to the situation or context, and then decide and act accordingly. The OODA loop is thus fundamental to all military operations, from strategic down to individual combat.

Arquilla and Ronfeldt (2001) describe this phenomenon in Networks and Netwars. Network Warfare is the "use of network forms of organization, doctrine, strategy and technology attuned to the information age". While the Industrial Age favoured state-run military hierarchies, the Information Age favours non-state actors in social-networks. It consists of a shared awareness of the battle space that can be exploited by various network-centric operations. Internet, cell phones and fax machines are the common networks in use. A large network is broken down into isolated clusters of dispersed forces which are linked to each other through leaders. This structure provides the benefits of evasion and asymmetry to the troops so that if one of the clusters is destroyed the other remains safe from the adversary. Thus, a mechanism to rapidly acquire and distribute target information will make the operations even more complex and protracted and focus more on the human dimension.

Information Manipulation and Degradation

Another very important weapon that prevents the enemy from getting correct information is information manipulation and degradation. The data are altered so that reality is not presented to the opponent. Use of guided virus into enemy's information infrastructure, micro robots may be used. An information-dependent environment can be critically exploited if these weapons are engaged. But, at the same time, all these weapons can be attacked by the enemies using various techniques such as spoofing and jamming. Virus intrusion and semantic attacks are other kinds of threats which hamper information integrity and credibility.

To ensure that information that is reaching the user is true and consistent, various encryption mechanisms, firewalls and antivirus software are employed on the data. As the defence forces become more technologically sophisticated, the greater is the challenge of adapting it and dealing with threats they pose.

From Conventional to Non-Conventional Warfare

Wars are witnessed since time immemorial; however, today's war seems to have integrated itself with the issues of the society. It has become a part and parcel of life. The nature of warfare has been changing and evolving over these years. It started with animal-based weapons in which people were fighting. Then, machines took over in the era of mechanized warfare and 21st century marks the beginning of automatic warfare or non-conventional warfare wherein biological, chemical, nuclear wars will be fought along with space warfare. This age has already started and still in its nascent stage but considering the rapid technological progress it will not be long when traditional warfare will be completely replaced by non-conventional warfare. Gone are the times when infantry weapons such as tanks, guns and bombs formed the basis of wars. Although the tremendous growth of science and technology in the fields of communication, information, infrastructure and transportation has made life easier, but at the same time it has made the task of national security a matter of great challenge and effort. Future warfare will be a digital theater of war. There is a need to revolutionize our technology management, transform strategies and philosophies and reengineer our thoughts to face the changing nature and tools of warfare.

The tremendous growth of science and technology has shrunken the world. The increasing impact of technological changes has brought revolution in world bringing critical elements of complexity, uncertainty and swift decision making in the battleground. There has been a gradual shift from physical to technological strength, and today the time has come when psychological forces play an important role in warfare. Changing socio-economic milieu, nuclear proliferation, trans-national terrorism and cyberspace battle are threatening the security balance of the country. The world has become a much more dangerous place. As discussed above, the revolutionary developments of net-centric warfare, human–computer symbiosis and robots have enhanced our capabilities and we are progressing from a human-centric warfare towards a technology-intensive environment.

Technology has not only altered the way of living but its employment in warfare is also very crucial. Man in future warfare will have to be fully proficient and self-contained. He will possess high decision-making power and will continue to be at the centre stage, but his role as a soldier will be minimized. The most important trend contributing to this development is the rise of technologies such as artificial intelligence, man–machine interaction and VR. The systems in the future will be small and too fast creating a complex environment beyond the ability of human's reaction time. There will be information overload that will make it practically impossible for humans to participate directly in decision making. The weaponry would perhaps include bio-genetically engineered systems producing physiological changes and affecting emotional and psychological aspects of humans. Therefore, the focus should shift towards human factor research and exploitation of human potential, training system and leadership styles. As a Chinese saying goes "If you are planning for one year, sow grain, if for two years, plant trees, but if planning for a 100 years, grow men". The future ideal of Armed Forces should be an agile leader who can plan innovatively in response to new challenges.

Future Warfare Technology

Advances in military technology driven by innovation and army have managed to use technology in new and creative ways to gain an edge over the enemy. Today, many agencies are working on programmes and missions to tackle future battlefield. Lighter infantry equipment, ultra-light and faster vehicles, crowd control weapons and swift and efficient battlefield communication are some of the basic requirements of future warfare that are being researched upon. It is clear from the above discussion that achieving victory in modern high-tech warfare should be the basis for war preparation and weapons that do not kill, but temporarily incapacitate adversaries seem to be effective elements of war in the given scenario.

Non-lethal weapons can range from rubber bullets and pepper spray to lasers, high-powered speakers and lights that cause disorientation. These can not only be used on the battlefield but also play a role in evacuations and other control operations. Directed energy weapons emitting laser, heat, etc., deliver enough power to destroy and burn materials and enemy, thus distracting them. Similarly, a grenade hitting the ground with a loud thud and producing flash can distract and alert the enemy (Beidel, Erwin, & Magnuson, 2011). There will be a need for high-speed continuous communication in the non-linear battlefield. 4G spectrum will provide reliable network and would facilitate the use of smart radios which make use of unused spectrum automatically. Night-vision technology is a new dimension to facilitate the need to identify targets at night and under poor visibility conditions. Goggles, pocket viewers, gunner sights and anti-tank missile guidance are some of the tools which make use of IR technology to enhance night vision.

Future warfare scenario requires multiple functionality unmanned technology with human intelligence interface (Chander, 2013). The challenge is to move machines from tele-operated mode to self decision makers that can work automatically on their own. Since the brain can adapt and function in dynamic environment, it is ideal for a rapidly changing battlefield. Brain–computer interface (BCI) technology has shifted the focus from imaginary fiction to scientific reality. Various critical technologies such as sensors, robotics, advanced materials and electronics are paving the way for development of unmanned warfare systems. These unmanned systems perform human tasks by mimicking the human brain. Unattended sensors, micro Unmanned Aerial Vehicles (UAVs), robot sentry and autonomous underwater vehicles are some of the unmanned systems under development. These systems may include mobility, intelligence, scene awareness, manipulation and friend–enemy identification capabilities. An agile four-legged robot may be developed for carrying military gear over long and high distances easily. These systems may be further refined to recognize individuals and interpret visual and vocal commands for more efficient performance.

Miniature robots called mesh worms (Sammon, 2012) are one of the tiniest robots in development. It can move silently using its artificial muscles to tiny, secret places and gather information such as temperature and even record audio and video of the place. This technology stealthily reports data and finds immense application in surveillance and secret missions. Moreover, very small flying robots are being built which can be sent on reconnaissance missions to areas inapproachable by soldiers. A certain type of highly mobile robots could also assist soldiers in reliable communication by autonomously moving and acting as nodes in a wireless communication network.

Due to the existence of cyber space, satellites, radars, etc., the deployment is bound to be much more diverse and mobile. The battlefield will be largely non-linear contrary to the old sequential field (Bhushan & Jain, 1999). There will not be a clear enemy and enemy forces will be small, independent mobile units. In order to tackle this kind of situation, there is an urgent need for a paradigm shift from hard killing to soft killing environment. Lethal weapon systems and explosions would fail to serve the purpose. The thrust needs to be more on incapacitating the enemy by attacking his mind and making him ineffective. The 2050 battleground is likely to remain largely non-lethal though highly technical in nature with unique manpower requirements. In this kind of scenario, psychological warfare or mind-controlled warfare is one strategy that may prove to be a potential solution for the concerned issues.

Psychological Warfare and Mind Control

The advancements in technology will present new opportunities and risks for future combat. The trauma of war may be further heightened, creating increased anxiety among the soldiers. The battlefield is bound to be ever more demanding of human potential and will add new facets in the methodology of warfare. The science of psychology to conduct war is as old as war itself but its significance is not acknowledged. This field has grown slowly and its application has still not been exploited. Advent of technologies has made its application widespread and necessary. Psychological Warfare is the oldest strategic and tactical weapon system that may prove to be most useful in the given future scenario. Even the great epics of Mahabharata and Ramayana involve various illustrations of Psychological Warfare. The most famous among them being the victory of Pandavas over Kauravas is an excellent depiction of such kind of warfare. When the Pandavas were about to face defeat against the Kauravas due to the fear that Dronacharya, the commander-in-chief of the Kauravas, would deploy the weapon of *Brahmastra* (a weapon created by Brahma) against them, Krishna advised that somehow Dronacharya be made to leave the battlefield. And, he also suggested that Drona would lay down his arms if told that his son, Ashwatthama, has been killed (due to existential concerns Dronacharya's desire and will to fight would cease to exist). The Pandavas carried out this propaganda by killing an elephant named Ashwatthama and made a proclamation that Bhima, a Pandava, had killed Ashwatthama. As anticipated, Drona was deeply hurt by this announcement and sought confirmation from Yudhishthir, the eldest Pandava, who was known for his moral uprightness. The Pandava said that it was true but added an under breath rider that it could be either a man or an elephant (Mahabharata, CLXL: 55).

WWII also has instances in which the psyche of men was being attacked for attainment of objectives. The US instrument of choice for producing strategic effects has been the air attack. During WWII and the Korean, Vietnam and Persian Gulf conflicts, the US conducted air attacks against strategic targets located within the enemy heartland to degrade both the enemy's physical capacity to wage war and his will to do so. A major psychological objective common to these strategic attacks was to convince enemy leaders that they could expect to pay a heavy price for their continued refusal to agree to allied peace terms. In addition, the US also attempted to use strategic air attacks to demoralize and frighten enemy civilian populations and thereby deny labour to an enemy's war industry (a primary objective of allied bombing in WWII); foment indigenous opposition to an enemy government's war policies; and, in the case of Iraq, prompt an enemy government's overthrow by a coup or popular uprising (Hosmer, 1999).

Psychological Warfare is a technique that affects the psyche of man influencing his emotions, attitudes and behaviour. According to Paddock (1982), Psychological Warfare may be defined as the dissemination of propaganda designed to undermine the enemy's will to resist, demoralize his forces and to boost friendly morale. It is essentially a non-violent, non-lethal type of warfare resulting in the surrender of parties and not death. It is not confined to time and place. It is a continuous process that begins much before the actual war and continues even after the last bullet is fired. Like any other warfare, psywar is also designed for certain objectives. Mobilizing hatred against the enemy, undermining enemy morale and sustaining fighting spirit are among the major intentions and aims of psywar.

Propaganda is the most dominating element of psywar. It is any kind of planned communication that affects the mind, emotions and beliefs of other people to achieve a specific purpose. It is the content that eventually affects the target. Military role of propaganda is not obscure and has been used since World Wars against the enemies. Paper bombs and leaflets were used to solve the problem of accuracy in air defense. Thousands of leaflets carrying messages were stuffed in a cylinder which when fused would release those bundles of paper and the messages would scatter over large distances (Bhatt, 2006). Target audience in denied areas can be informed using propaganda. Also, propaganda enables overcoming illiteracy and disrupted communication. A propagandist must be trained personnel and must think objectively keeping his emotions and feelings covert. Propaganda is characterized by its source, intention and selected target audience. It is generally perceived as spreading of inaccurate and incomplete information to deceive the target population. On the other hand, if used correctly it acts as a strong weapon. The effectiveness of propaganda is determined by its delivery medium. A propagandist must use a suitable media at the right time and place. A suitable media is the one which is available timely and suits target conditions. Leaflets and pamphlets, for example, are ineffective for those who cannot read. Propaganda may use means of radio, TV, Internet, interpersonal communication, etc. It uses non-violent means for persuasion.

In the absence of proper dissemination of information the propaganda is futile. Thus, psywar needs experts in various fields for able target analysis and planned execution. Propaganda may be defensive or offensive depending on its proposition to sustain or interrupt a social action respectively (Stranglove, 1998). Psychological warfare is a propaganda used to convince the enemy. It reinforces existing attitudes, making one believe and act under the propagandist control. Mind is the central organ for carrying out psyops, and various mind-control tools have been developed that direct an individual's behaviour.

Mind-enabled Tools

The BCI technologies connect the human brain to devices and may involuntarily penetrate and compel the mind. Brain is connected to robotic systems for manipulating one's thoughts or even actions. Various technologies have been developed on the basis of this paradigm. The most famous among these are the drones. Drones are UAVs either controlled by pilots or fly autonomously following a pre-programmed mechanism. These are increasingly used for remote sensing and ultra-high speed aerial surveillance. It will not be long when mind-controlled weaponized drones will enhance the power and lethality of our soldiers. Another technology exploits the power of microwaves to create loud sound discharge directly into a person's head leaving him practically inactive for a considerable period of time. Ultrasonic brain wave clusters have been used by the Middle East countries during the suicidal attack by Iraqi troops on the deserted city of Al-Khafji (ITV News, 1991). According to reports, a new kind of high-tech subliminal messages of ultra-high frequency was used that were completely silent to the ear. The negative voice messages placed on the tapes alongside the audible programming by psyops psychologists were clearly perceived by the subconscious minds of the

Iraqi soldiers and the silent messages completely demoralized them and instilled a perpetual feeling of fear and hopelessness in their minds. Thus, it can be seen that on one hand, BCI allows injured soldiers to remain active and on the other hand it can manipulate soldier's mind. A disrupted BCI may be hacked by the enemy and force the victim to behave indifferently. He may share secrets with the enemy voluntarily or may introduce inaccuracies in his own computer system and exploits the system from within.

The association of neuroscience with military systems has altered the ways in which BCI functions, especially in the area of threat warning and detection. An innovative step towards threat warning systems is the use of cognitive visual processing to monitor the subconscious signals of a soldier's brain. It enables detecting a threat a soldier has perceived even before he is consciously aware of it. Nanoparticles that can percolate through the brain and induce a desire in the enemy to become calm is another area of research in the field of neuroscience.

Remote neural monitoring is another mind-control technology having immense potential in future war settings. It is a form of functional neuroimaging that employs satellite-delivered extra low-frequency (ELF) waves to communicate voice to skull transmissions producing schizophrenic symptoms. These frequencies can locate a target anywhere on Earth and can penetrate water, rock, concrete and other dense matter (Finney, n.d.). It allows access to a person's thoughts without his knowledge. This technique can track individuals in any location or country. Infrared detectors, X-ray viewers and directed energy weapons are other more commonly used spying and stalking weapons. Also, EMF waves are widely used in surveillance operations. They can tap into computers wirelessly and can track persons with electric currents in their bodies. It then becomes possible to monitor suspects from a distance. Information is gathered through electromagnetic waves.

These strategies affect the soldier in ways he may not realize consciously. Thus, the battlefield of guns and bombs will slowly perish away and digital or computer battlefield will take over. The gap between technology supremacy and human competence will become so large that future planning of defence forces must not only cater for defence against tech-savvy attacks but psychological harassment as well. The innovation gap needs to be shortened as far as possible.

Limitations of Psychological Warfare

Psychological Warfare is not liberated from factors restricting its effectiveness. It is an old concept that has not been exploited to the fullest and the lack of expertise, and knowledge of its application will become a cause for its limited application to enhance military effectiveness in the future. Much research in this field is required to be done. Dearth of trained personnel and proper use of manpower in understanding political and social conditions and devising innovative procedures may reduce performance and output. Tactical psychological warfare needs extensive planning and deployment time for execution. Existing methods of mass communication and information dissemination for propaganda might get in the way of security and some secret information might be revealed to the enemy in the process. The credibility of media is also a serious concern in order to carry out a successful PsyOp.

Political compulsions and legal hurdles usually come in way of successful propaganda by restricting the reach of media, its means of proper delivery and timely availability. Paucity of accurate, accessible and complete information is a bottleneck for an effective operation. Coordination malfunction between army units and civilian population may open avenues for counterpropaganda and lead to severe reliability concerns. We must continue to upgrade our skills and professional knowledge in order to overcome the stated limitations and make psychological warfare a valuable asset for future war scenario.

Conclusion

Wars shall continue in newer forms with emerging technologies. New technologies including biological weapons, cyber weapons and mind enabled tools which can affect several people in one attack are compelling the forces to transform the techniques of war. Wars of the future may be decided through manipulation of soldier's behaviour and countering enemy propaganda. New targets of brain may be recognized for incapacitating the enemy. Operations would focus on affecting the perceptions and attitude of the target. The battleground would envelop entire societies instead of a specific enemy group. Effective use of such technologies will require sophisticated command and control and rapid and close interaction between different systems. Although direct human participation seems to be less likely, but human role will be critical at certain levels of such a psychological battle.

In this state of affairs, collapsing the enemy's will to fight and inducing troops to surrender and abandon their equipment would be considered a critical capability that must be employed. A fully evolved model of warfare must be formulated to deal with situations in which application of non-lethal force would be the tactical preference. Thereby, issues surrounding information warfare, mind control and behaviour modification need serious attention. However, the potentials of technology in this new realm of intelligence gathering and warfare may be constructive or destructive. On one side is the idea of enhancing mental and physical performance but on the other side the danger of misuse of this science by the creators themselves is a grave possibility. Hence, research, development and application of mind-invasive technologies must be promoted but with care.

References

Arquilla, J., & Ronfeldt, D. (2001). *Networks and netwars: The future of terror, crime, and militancy*. Santa Monica, CA: RAND.

Beidel, E., Erwin, S. I., & Magnuson, S. (2011, November). 10 technologies the U.S. military will need for the next war. *National Defence Magazine*. Retrieved from http://www. nationaldefensemagazine.org/archive/2011/November/Pages/10TechnologiestheUSMil itaryWillNeedFortheNextWar.aspx

Bhatt, A. (2006). *Psychological warfare and India*. Delhi, India: Lancer Publishers & Distributors.

Bhushan, N., & Jain, R. K. (1999). Mathematical and computer modeling of future battlefields. In P. N. Chaudhary & W. Selvamurthy (Eds.), *Battle scene in year 2020* (pp. 259–272). Delhi, India: DESIDOC.

Bowdish, R. G. (1999). Information age psychological operations. *Military Review*, (Dec 98–Feb 99), 28–36. Retrieved from http://www.c4i.org/bowdish .pdf

Chander, A. (2013, July 08). Unmanned technology will be the focus of future warfare: DRDO. *The Economic Times*. Retrieved from http://articles.economictimes.indiatimes. com/2013-07-08/news/40443738_1_warfare-chander-systems

Finney, D. (n.d.). *Remote neural monitoring*. Retrieved from http://www.greatdreams.com/ RNM.htm

Fuller, J. F. C. (1920). *Tanks in the great war*. London, UK: Murray.

182   Swati Johar and Updesh Kumar

Hosmer, S. T. (1999). The information revolution and psychological effects. In Z. M. Khalilzad
& J. P. White (Eds.), *Strategic appraisal: The changing role of information in warfare* (pp.
217–251). Santa Monica, CA: RAND.

ITV News. (1991, March 23). *Military use of ultrasonic brain wave clusters.* ITV News Bureau
Ltd. Retrieved from http://whale.to/b/mc_brain.html

Paddock, A. H. (1982). *US army special warfare: Its origins.* Washington, DC: National
Defence Univeristy Press.

Post, J. M. (2005). Psychological operations and counterterrorism. *Joint Force
Quarterly, 37.* 105–110, Retrieved from http://www.thefreelibrary.com/Psychological
operationsandcounterterrorism.-a0141576784

Sammon, R. (2012, September 20). *9 amazing technologies of the future.* Retrieved from
http://m.kiplinger.com/slideshow/business/T043-S001-9-amazing-military-technologies-
of-the-future/ index.html

Squire, L. (1995). *Mass media: The tenth principle of war?* Newport, RI: Naval War College.

Stranglove. (1998). *Propaganda101.* Retrieved from http://www.mindcontrol101.com/ pdfs/
Propaganda101.pdf

SECTION II

Psychological Interventions in Military Context

9

The Secret Weapon of Optimism

Eyal Lewin

When armies struggle against all odds, the source of the soldiers' motivation to fight remains an unsolved puzzle. According to scholarly literature, comradeship and social cohesiveness are key factors in the conduct of any struggle. However, persistent fighting and determined resistance seem to occur also when soldiers are totally unacquainted with each other. This phenomenon calls for another observation, recognizing that social cohesion may not necessarily be the only explanation for human persistence on the battlefield.

Doubting Some Common Knowledge

Why soldiers fight is a puzzle that bothers students of human behaviour particularly when confronted with historical evidence of struggles in which people have fought until the bitter end even when all, for them, had been lost. Facing fatal casualties and overrun by overwhelming military forces, one would expect soldiers belonging to disadvantaged armies to give in. Yet as Somerset Maugham has taught us, passion does not know the cost, and the heart has its reasons that reason takes no account of (Maugham, 2000).

One common explanation for a continuous struggle even when the chance of succeeding seems slim lies within the social bonds of military units. Social cohesion as a human motivator has been examined time and again with numerous case studies of armies, starting far back with the Spartan military tradition (Sekunda & Hook, 1998), continuing through the poor but socially cohesive Confederate army of the American Civil War (Frank, 1991) as well as accounts of the *Wehrmacht's* social community structure (Shils & Janowitz, 1948), and concluding with records of current reserve units in the Israeli Defense

Forces (IDFs) (Ben-Dor, Pedahzur, & Hasisi, 2002). On the other hand, in contrast to common wisdom, there is evidence that sometimes persistence in combat exists even when army units are destroyed and reduced to individual soldiers who are not necessarily acquainted with each other. The social capital theory of networks and group cohesion fails to explain why troops continue to fight even when their units have been dismantled and their comrades have all died.

Perhaps an extreme example of that is the phenomenon of the Japanese holdouts. In the aftermath of Second World War (WWII), tens of thousands of Imperial troops were bypassed by the advancing American forces, and many of them were left stranded in isolated islands in the pacific. Those Japanese soldiers went into hiding, waiting for attacks that would never come and commands from military authorities that no longer existed. Devastatingly short of supplies and cut off from their homeland, thousands of them hid from the American patrols in the thick jungles and mountains of the islands, scattered in isolation and unaware that the war had ended. Like satellites that are lost in space, remote and alone among the numerous islands of the Philippines, Indonesia and the Chinese sea, many perished quietly without their officers or comrades around them, alert for a battle that would never take place. Even when eventually located and finally persuaded that the war had long been over, their surrender proved sometimes impossible to achieve without assistance from their former long-ago commanders in Japan (Ononda, 1999).

Another possible explanation for people's uncompromising struggle might be the factor of collective fear. Collective fear results from a threat perception according to which the community's fate is in danger and the social group is contested by a challenge that risks the social order in such a way that the group might fail to recover (McCann, 1997). It is the collective fear that encourages one's activity for the sake of social survival even at the price of high personal costs (Higgins, 1999; Stephan & Stephan, 2001). Under threat, the social group sees itself as one under siege and becomes a community in which it is all for one and one for all; social solidarity and patriotic feelings are bound to flourish and prosper (Sherif & Sherif, 1969). This certainly has proved to be the case of American citizens who immediately demonstrated pro-social activity and volunteered to give aid right after the 9/11 events (Allen, Sargeant, & Bradley, 2003). Supreme emergencies, during which the very existence of a state is threatened from the outside, raise extreme feelings of social solidarity and encourage patriotic actions, drawing attention away from domestic disputes and thus contributing to the consolidation of the political community (Canetti-Nisim, Arieli, & Halperin, 2008; Evrigenis, 2008; Gordon & Arian, 2001; Kinder & Sears, 1981; McConahay, 1982). This corresponds, of course, to the rally-round-the-flag syndrome (Mueller, 1973).

However, these theories too fall somewhat short of enveloping a comprehensive explanation to the behaviour of highly motivated patriots risking their lives against all odds. In fact, one can find historical proof for cases in which fear was rather a paralyzing factor that prevented collective action. Such, for example, was the case of the Czech president Edvard Benes who on 30 September 1938 decided that fighting Germany was hopeless, and therefore submitted to the Munich agreement, allowing Germany to annex the Sudetenland; with this surrender, the Czech government practically put an end to its country, since the incorporation of the industrial Sudetenland into Nazi Germany left the rest of Czechoslovakia weak and powerless, unable to resist any subsequent occupation. In March 1939, meeting no resistance, the *Wehrmacht* crossed mainland Czechoslovakia and marched into Prague, finally terminating the short-lived First Republic of Czechoslovakia (Vital, 1967, 1971).

Proponents of the Czech leadership claim that the small country that was betrayed by its Western patrons who preferred appeasing Hitler stood no chance against the mighty Third Reich; a militant reaction would have led, not only to useless bloodshed that would not change the political outcome, but would also put the historical blame on Czechoslovakia for having been the state that caused a World War everyone had been so eager to prevent. By refraining from struggle the Czech leaders saved the lives of their people, whose country had been doomed to be conquered by powers far beyond its strength.

Yet historic accounts, particularly updated ones, reveal an altogether different version. Military and strategic reports show that the balance of forces in 1938 left the Germans with no advantage whatsoever. Comparing the size and quality of armies, the highly protected Czech defense lines, the Soviet alternative for foreign aid and the devastating winter that would not enable an effective *Luftwaffe* activity—all these considered, the Czech capitulation surprised even the German leadership (Angelucci, Matricardi, & Pinto, 2008; Ben-Arie, 1990; Lewin, 2012; Lukes, 1993; Vital, 1966).

What, then, led the Czech leadership to succumb? Why did they concede to hostile demands and political pressure until finally committing an act of national suicide? One possible answer lies within the effects of collective fear. In Czechoslovakia, just like in other countries, the memories of First World War (WWI) boosted anxiety; the Western political and military establishments were practically possessed with the emotion that such a war should never again take place. Above all, a widespread fear of an immediate fatal blow from the *Luftwaffe* greatly affected Western foreign policy and has come to rule the customary comprehension of Europe's peoples (Showalter, 2003). This collective fear was produced through the outstanding implementation of an extremely aggressive airfare doctrine of the German army during its intervention in the Spanish Civil War. On 26 April 1937, in what would later be known as the Guernica raid, the *Luftwaffe* led an aerial attack on the Basque town of Guernica, destroying the city and killing hundreds of its inhabitants. The strategic weight of the devastated city was marginal, and the whole event was taken out of its objective historical proportions, but the tremendous psychological and political impacts of the air assault went far beyond its local context. In Britain and the United States, the major newspapers generated a common shock from what was later to be named terror bombings; indignation and dismay over the destruction of Guernica echoed around the world. More importantly, the reports of the Guernica raid propelled a public fear that in any coming war the air forces of the warring nations and those of Germany in particular would be able to wipe whole cities off the map. The cumulating shocking impression throughout the West was that the *Luftwaffe* was an aggressive force whose violent activity could not be blocked or restrained (Bialer, 1980; Patterson, 2007).

Other competing rationalizations of outstanding patriotic struggles refer to ideology, particularly—national ideology, as the leading cause for the people to fight stubbornly. The national ethos of a country is the array of the particularistic shared values and traditions from which the people's image of its future and past is envisioned. The ethos integrates the community into feeling a common mutual destiny and forms the foundations for its unique identity as a distinctive social group. The integrative ethos is also the moral source for the national community's informal social controls; it enforces commitments upon society and drives its members into a largely voluntary social order. Thus, the ethos of a nation holds, in fact, one of the most important keys for a people's ability to unite into a cohesive society (Etzioni, 2009).

The use of national ethos in political science goes back to the German romanticism of the late 18th century with philosopher Johann Gottfried Herder speaking about the cultural, ethical and political climate in which a nation evolves and crystallizes (Barnard, 2003). These ideas project a strong association between the ethos and the representations of a long history that the nation claims for itself. The features of a community, some scholars claim, originated in the historical stages when the mental maps of the people, their prevailing culture, norms and ideas were cultivated (Rothstein, 2000). This attitude corresponds also to the writings of sociologist Maurice Halbwachs, who spoke of a group memory that was shared by its members, passed on, and constructed by the social group. Rejecting Freudian and other purely psychological approaches, Halbwachs (1992) claimed that it was impossible for an individual to remember in any coherent and persistent fashion outside his group contexts; memories were not preserved in one's brain or in one's mind, they were rather external recollections controlled and constructed by the group.

Whether judging nationalism and its ideologies as positive or destructive factors of our society, decades after Halbwachs, the basic understandings are that a social memory shapes images of the past and draws, by doing so, the lines of political cultural profiles (Fentress & Wickham, 1992; Olick & Levy, 1997). According to international studies theorist Benedict Anderson, the idea of a collective memory resides deep in comprehension of the nation as an imagined community; nationalism has a symbolic and constructed nature, and by utilizing the communicative media it is capable of reaching dispersed populations (Anderson, 1983). The collective identity of a nation, as a unique combination of a public that shares mutual values and beliefs, lies in its common narratives, that is—its constructed collective memory, and in the united role that its members believe that fate had destined for them in this world.

There is no doubt that national ideology has the capacity to mobilize people and to lead them to join forces and reach unachievable accomplishments or endure unbearable hazards. Perhaps, the historical meeting point of the 20th century's two colossal ideology-driven forces is Stalingrad, where between 23 August 1942 and 2 February

1943 over two million soldiers and civilians held one of the bloodiest battles in human history. Neither the soldiers of the Nazi *Wehrmacht* nor the Soviet troops of the Red Army or the men and women who fought alongside with them were willing to give in; each party, in its turn, suffered its painful disasters inflicted by close-quarters combat, thirst and starvation, medicine shortage and eventually—the cruel Russian winter.

First, the German offensive was supported by intensive air-raids that reduced large parts of the city to rubble. But since the Soviet defenders were willing to endure casualties and refused to leave their posts the *Wehrmacht* was driven to door-to-door bayonet-range fighting. Even when most of the city was eventually conquered, the Russian forces clung tenaciously to the west bank of the Volga River, refusing to break down. Three months into the campaign, the Red Army launched a counter-attack and through heavy fighting managed to surround the German 6th Army inside Stalingrad. The Nazi predator was now encircled, trapped in a lethal enclave, cut off from any supply.

In an attempt to break the lock, the *Luftwaffe* launched an aerial operation to supply what, to begin with, had never exceeded more than a small portion of the daily 500 tons of food and ammunition that the 6th Army needed for its survival. The result was that instead of relieving the besieged army, as many as 500 German aircraft were lost to bad weather, and to a sophisticated Soviet defense system combining rings of guns and ground-controlled fighters. The superior Soviet forces were determined in their constant pressure on the ground, and only a fraction of the required supplies made it to the doomed German forces. Even at this point in time, desperately isolated and standing no chance to win this hopeless battle, the German commander, Friedrich Paulus, turned down Soviet offers to negotiate and refused to surrender. The battle, as well as the whole 6th Army of the *Wehrmacht*, was to be entirely lost.

By January 1943, the Germans had already gone through months of living on rations measured by dozens of grams, supplemented irregularly by horsemeat and by occasional rats. Conditions in their military hospitals were practically medieval. Thousands of kilometres away from home, often, literally, freezing to death, with hardly any ammunition, they were determined to choose bravery over life. In the final assault, thousands of Soviet guns and mortars were firing in every corner of the German pocket. Tanks and infantry were ruthlessly advancing simultaneously in all sectors. Within a week, the German remaining forces were reduced to half their size. Once again Paulus was summoned to surrender; once again he refused. The German soldiers fought back against all odds and, like their commander, refused to submit. German die-hards fell back into the city's ruins, using tactics learned from the Russians several months before, to prolong the end as ammunition ran out and men sought terms at bayonet point.

Even after Paulus was captured, fire did not cease, and German resistance was active, literally, to the last bullet. In all, some 400,000 German soldiers were killed in this battle. In late summer 1942 Soviet warriors driven by faithfulness to Mother Russia rejected any option of collapsing even in the dark moments when all seemed to have been lost. In the winter of early 1943, it was the German troops loyal to the Third Reich who remained loyal to their mission, fighting literally until the very end. Hungry, cold, sick and wounded with no treatment at hand, they continued to lead a lost war, refusing to accept their fate (Beevor, 1999; Roberts, 2006). Yet, the inspiration of national ethos, performed by Russians as well as by Germans, seemed to be far less effective when speaking of other nationalities. It was no coincidence that the besieged forces in Stalingrad were purely German; their allies, the Romanians, Hungarians and Italians, who were also driven by highly inspiring national spirits, surrendered almost immediately, not necessarily facing any better fate than the German fighters. Surrender was, in fact, the general trend amongst most European countries throughout WWII, even though some of them maintained centuries-long national ideologies. In spite of their national traditions, when called to struggle for their very existence, Denmark, Belgium, Norway and France, just to name a few, demonstrated mainly defeatism (Lewin, 2012).

Hence, the enigma of the soldiers' motivation remains, after all, veiled and obscure. Notwithstanding the importance of social bonds, what drives soldiers to exert their willpower when chances are negligible? What enables troops, even when isolated and remote from their comrades, to fight stubbornly and to insist on changing reality even when the probability of positive outcomes is clearly low? In quest of an adequate answer, this chapter has taken the theoretical foundations of positive psychology as an alternative explanation for soldiers' persistence in fighting.

Optimism

Optimism is a topic that belongs to a broad field of psychological thought often referred to as *positive psychology*. Positive psychology is an umbrella under which areas of concern such as happiness, talent and personal empowerment are studied and the results implemented (Peterson, 2006; Seligman, 2000; Snyder, 2000). Optimism relates to the sense of control that resilient individuals demonstrate in times of hardship. It is a constructive motivational state, relying on a sense of a possible future success. Group solidarity and social cohesion are often achieved through the influence of leaders who are capable of inspiring positive shared interpretations of stressful events, of restoring a sense of control and of holding up a vision of a positive future (Bartone, Barry, & Armstrong, 2009; Bartone & Bartone, 2011; Ben-Dor, Canetti, & Lewin, 2010).

Martin Seligman defines optimism as the tendency to interpret reality in a manner that (a) considers the role of external factors in producing bad outcomes; (b) views the bad event as a one-time episode and (c) confines the bad outcome to just one performance area (Seligman, 1991, 1994, 2002). Charles Snyder's definition describes optimism as goal-directed thinking in which people implement *pathways thinking* (i.e., their perceived capacity to find routes to the desired goals) together with *agency thinking* (i.e., the necessary motivations to use those routes). Pathways thinking, according to this point of view, is the production of alternate paths once original ones have been blocked; agency thinking is the positive emotional mindset that stimulates high hopers to maximize their pathway thinking (Snyder & Lopez, 2007).

Optimism, then, is considered to be a personal human resource for enabling one to achieve success. If one is an optimist, he can meet a challenge successfully. He plans how to cope with it productively and tell oneself that he can do it; upon an initial success his self-confidence grows, he tries new approaches and finds improved pathways, gradually recognizing oneself as a winner—as an individual who can make things happen. Hence, the personal process of optimism is based on one's articulating his goals, identifying the routes he can use to achieve these goals and finding the personal self that will encourage him to maintain his efforts until he finally completes the required accomplishment. If any of the stages in this process is lacking, optimism will no longer form a human resource for change, but will rather remain wishful thinking, daydreaming or even a form of escapism (Snyder, 2000).

In search of the applied outcomes of psychological theories, optimism has been referred to as a possible purposely acquired trait, a learned human pattern of thinking and certainly not a genetically endowed one (Bandura, 1986; Barone, Maddux, & Snyder, 1997). Indeed, some scholars of positive psychology have gone the extra mile and translated their research into useful manuals on how to become more optimistic. The very titles of some of their books convey the concept that positive psychology is an attainable trait one does not necessarily have to be born with—*Making Hope Happen* (Snyder, 1999), *The Psychology of Hope: You Can Get There from Here* (ibid.) or *The Great Book of Hope: Help your Children Achieve their Dreams* (McDermott & Snyder, 2000).

In this study, collective hope and optimism reflect the core ideas of positive psychology: self encouragement, led by deep faith that affirmative outcomes are achievable provided that the members of the group maintain their belief in success. It is the profound belief that the human spirit is capable of changing reality; the approach according to which will-power is stronger than physical barriers.

The Weapon of Optimism: Case Study of Arab–Israeli 1973 War

On 6 October 1973, in the middle of the Day of Atonement [*Yom Kippur*], the most solemn and strictly observed holy day on the Jewish calendar, Egypt launched a surprise attack on Israel in the Sinai Peninsula, coordinated with a simultaneous assault by Syrian forces in the Golan Heights. Because of the tremendous surprise effect that characterized this attack, Israel failed to mobilize its army in time; as a result, its defenses were fragile and inadequate.

The Egyptian army successfully crossed the Suez Canal, easily over-running the 450 Israeli soldiers who manned the fortified waterline. The Egyptian invasion, with 100,000 soldiers deployed, 1,400 tanks and 2,000 heavy mortars, was met by a vastly outnumbered defense;

in all of Sinai the Israeli army counted no more than 290 tanks; only about a third of them were close to the canal when war broke out, and the rest were scattered some 50 kilometres to the east. Masses of advancing Egyptian soldiers carrying portable anti-tank missiles created an overwhelming military formation that the Israeli strategists were unaware of in advance. As early as 24 hours after the initial attack, two thirds of the Israeli tanks had been lost and the remaining 90 tanks had to cope with hundreds of Egyptian tanks that had already managed to cross the canal.

The Israeli fortified points along the Suez Canal had been easily bypassed; the soldiers in each small fortification were now under siege. Within a few days some of them would surrender, of whom several would be immediately executed by victorious Egyptian forces who by now had regained the canal line. The Israeli air force's superiority that had proved so important for Israel's victory six years earlier was now eliminated by Soviet surface-to-air missile batteries that the Egyptians had posted on the western bank of the Suez Canal. Within a day's fighting, over 40 Israeli airplanes had already been shot down.

On the northern front, things were not any better for Israel. The Syrians invaded the Golan Heights and were virtually threatening the whole northern part of the country. The Syrian army consisted of 45,000 soldiers, 1,600 tanks and 950 heavy mortars, enormously outnumbering the Israeli defenses that counted 4,000 soldiers, 177 tanks and 44 cannons. Aside from the huge numerical advantage, Israel was caught totally unprepared for this war, none of its units were properly equipped and the reserve forces, the spinal cord of the IDF, were called into action far too late.

With these unfortunate opening conditions, with the routes to mainland Israel on the verge of being breached, Israeli reserve forces congregated for a final struggle for their homeland. Shocked, and overwhelmed by outnumbering forces and new Soviet technologies, the IDF fought with its back to the wall. Against all odds, however, within a week of fierce fighting and heavy losses, Israel recovered and launched counter-offensives in the south and the north. Immense tank battles took place in the Sinai as well as on the Golan heights, some even greater in scope than the WWII desert battle of El Alamein. Starting on 15 October 1973, the IDF organized a remarkably daring counterattack in the south. An Israeli division broke through the enemy's forces and crossed the Suez Canal in order to advance towards Cairo; other IDF divisions cut off and encircled the whole Egyptian 3rd Army that was now trapped in the Sinai desert without water, food or ammunition supplies.

Three weeks into the war, when a ceasefire was achieved, Israel had already recaptured almost all of the territory that it had lost in the first few days of fighting. In the north, most of the Syrian tanks, along with thousands of their vehicles and hundreds of guns, remained scattered in the battlefields; the route to Damascus was almost open. In the south, IDF troops gained power over a large stronghold on the western bank of the Suez Canal, and a large portion of the Egyptian army was locked in, urgently needing humanitarian supplies (Ezov, 2011).

Methodology: Inquiring Optimism

In order to establish the theory of positive psychology as a key to understanding how and why people struggle, at times, against all odds, the testimonies of several Israeli soldiers and officers who had experienced the hardships of war and had coped with what might count as unbearable situations were tested. This research is, therefore, based on a qualitative analysis of testimonies relating to what is considered one of Israel's most traumatic war experiences, the 1973 war.

The testimonies in this research have been extracted from books and pamphlets, in most cases within a military framework, as well as from magazine sections of Israeli daily newspapers. The technique applied in this study is the framework approach that was developed in the 1980s by policy researchers at the National Center for Social Research, considered to be Britain's largest independent social research agency, as a method to manage and analyse qualitative data in applied policy research. This method is particularly suited for cases in which content and thematic analyses describe and interpret participants' views and opinions (Ritchie, Spencer, & O'Connor, 2003).

Producing themes from data is common in qualitative studies and forms a widely used analytical method. Thematic analysis is an interpretive process in which patterns are systematically searched within the data and provide an illuminating description of the inquired phenomenon (Tesch, 1990). The development of meaningful themes, without necessarily generating any theory at this stage, provides rich insights into complex observable occurrences and can be applied across a range of theoretical approaches (Ritchie et al., 2003). In addition, the framework approach process is based on a series of interconnected stages that enable the researcher to move back and forth across the data until a coherent account emerges. Thus, a constant refinement of themes is achieved, which eventually enables the development of a conceptual framework (ibid.).

Because it is suited to analyse descriptive data and enables different aspects of the investigated phenomenon to be captured, the framework approach was chosen in this research. In addition, the researcher's interpretations of the testimonies' contents can be kept transparent through this approach. Three stages form the procedure in which the framework approach is implemented. (a) Data management: becoming familiar with the various texts through multiple readings; identifying central themes; developing a coding matrix and assigning it to the themes. (b) Descriptive accounts: refining initial themes in order to summarize and synthesize the diversity of the data; identifying associations between the themes so that more abstract concepts could be developed. (c) Explanatory accounts: finding meanings, interpreting and explaining the concepts and themes and searching for coherent concepts (Ritchie et al., 2003). In this research, altogether 217 testimonies of soldiers and commanders from the case study of the Arab–Israeli 1973 War have been reviewed. Consequently, five themes were found to be the leading ones (Table 9.1):

1. **Description of Military Movements:** This theme can be found in every one of the testimonies, which are naturally dedicated first to reports about the advance or retreat of IDF and Arab forces. These descriptions include accounts of the specific sceneries of battlefields.
2. **National Causes of the War:** This theme was found in 11 testimonies and refers mainly to the deep belief of soldiers that they were practically forming a last-defense line that would save the whole

Table 9.1
Themes of the testimonies

Themes Mentioned in Testimony	*Number of Testimonies*
Description of military movements	217
National causes of the war	11
Fear of death	8
Thoughts about home	5
Positive psychology	18

Source: Author.

country. References are made to the Holocaust and to the 1948 war, usually with a fundamental conviction that losing the battle would put an end to the State of Israel and to the Jewish people.

3. **Fear of Death**: This theme was found in eight testimonies and refers to one's notion that the specific incident is the one which he would not survive. In those testimonies soldiers testify that they were sure that their death was very near.

4. **Thoughts about Home**: This theme was found in five of the testimonies and refers to thoughts about home and particularly letters that soldiers wrote to their loved ones, expressing deep concerns not about the war that was going on but rather about the personal future of those whom they may never meet again.

5. **Positive Psychology**: This theme was found in 18 of the testimonies and its reference to optimism and hope will further be detailed. In fact, the testimonies that include this theme form a new sample, or a sub-sample, that enables us to elaborate on a specific theory, in this case—the theory of positive psychology. This strategy of sub-sampling is common when developing grounded theory, but is also frequently used in qualitative investigations that are based on interpretation (Marshall, 1996).

Results: Interpreting Optimism

In order to provide a variety of cases, the examples that were chosen here reflect the two major geographical fronts in which the IDF had to cope with some devastating conditions: the Golan Heights, where the Syrian army was on its way to conquer the northern part of Israel, and the Sinai Peninsula, where the Egyptian army had to be stopped from invading Israel from the south. Four illustrations are brought here from every one of these large battles, representing perspectives of commanders as well as soldiers.

The examples from the northern front include the following: (a) Avigdor Kahalani, who with poorly armed 40 tanks led a successful counter-offensive against an army of 500 Syrian tanks; (b) Ori Orr, the Brigade Commander who personally convinced the tank crews that they were going to win what at that point in time seemed to be a lost war; (c) Zvika Greengold, who had only one tank under his command but managed to fool the enemy into thinking that he was controlling a sizeable defense force; (d) Arad, who was a young soldier in an infantry company that was left to fight on its own from the top of a hill against thousands of Syrian troops.

The examples from the southern front include the following: (a) General Ariel Sharon, who refused to admit failure and decided instead that the IDF should cross the Suez Canal and head towards the Egyptian capital of Cairo; (b) Division Commander Amnon Reshef, who was determined to reflect confidence in his troops although outnumbered they were fighting for their lives; (c) Eliashiv Shimshi, who, like his commander, insisted on speaking calmly on the wireless even when soldiers and officers were being killed in front of his very eyes; (d) Yoel Nachshon, the Military Rabbi who inspired wounded soldiers with optimism during some grave moments in the battlefield. In addition, the testimony of Air Force Commander Benny Peled is brought here, particularly because it became well-known and often quoted in the form of a poem in Israeli national ceremonies.

These examples exhibit, each in its own way, the three major attributes of optimism: (a) an articulation of goals; (b) an identification of the routes to achieve these goals; (c) fostering of the self-perception that would encourage the optimist to maintain his efforts.

Starting with examples taken from the northern front, the war caught Battalion Commander Avigdor Kahalani in the Golan Heights, where disjointed and overwhelmed Israeli troops were, during the first day of the war, at a loss. Kahalani assembled groups of tanks and crews from the dismantled and destroyed units, some of them mentally shocked, and talked them into forming an armored unit. Kahalani's force counted no more than 40 tanks and soon they were about to face 500 attacking Syrian tanks in a site that would later be named the Valley of Tears [Emek HaBacha], in the Northern Golan Heights. For four days Kahalani managed to hold back one assault after another; each Syrian attack that was halted had its costs in dead and wounded commanders and soldiers and damaged tanks. At one point Kahalani's forces decreased to about a dozen functioning tanks that were constantly running out of ammunition. In fact, when at a certain point in the battle one of his officers had to pull back in order to reload his gun, Kahalani asked him to postpone his resupply and to stick around for a while so that the Syrians would not notice how few the Israeli forces were becoming. The Israeli forces managed to maintain the struggle until the Syrians finally gave in and retreated. When the Syrians at long last withdrew, the whole valley was littered with hundreds of their burned tanks, and the Golan Heights were once again under Israeli control.

One of the company commanders, Amy Planet, described the peak of the final battle.

> [...] At a certain point we [the commanders] started to shout "the Syrian tanks are retreating! They are running away!" This was not true, though perhaps some of them did turn away; but this cry boosted our forces with sudden new energies. This was how we managed to overcome one of the major breaking points. (As cited in Kahalani, 2004, p. 270)

As for Kahalani, who later received the Medal of Valor for his part in the war, once he adopted the self-perception that he was going to succeed, the possibility of being overrun by the enemy forces was no longer an option. He was well aware of the fact that he remained the last battalion commander alive in the northern part of the Golan Heights, but he had the gut feeling that it was up to him to create the change (ibid., 1992).

Another hero of the northern arena was Ori Orr, who was appointed slightly before the war to command a newly formed reserve brigade. Concentrating mainly on the southern and central parts of the Golan Heights, his forces, still in the process of gathering and organizing their equipment, fought against the superior Syrian forces. At the end of a day's struggle, Orr realized that in spite of heavy losses on his part, the Syrians were only warming up; the next morning they would start with the advantage of controlling the ridge, with the sun at their back. The self-perception that encouraged Orr to maintain his efforts was the decision that he was a winner. He ordered the troops to evacuate injured and dead soldiers from all the armored vehicles and to form new tank crews. He felt it was his duty to speak with each of the soldiers, most of whom he had never met before, and to assure them that the next morning they were going to win. He was certain that it is going to happen, and he sensed that talking to the people would make it happen (Orr, 2003). One of the soldiers recalled the night time meeting with his formerly unknown brigade commander—Haim Sabato—who would years later become a Rabbi and author, remembers:

> We arrived at Naffach [an Israeli central military base attacked by the Syrians]. Bombs lighted the darkness and we could see that tanks were on flames everywhere. Wounded soldiers were scattered all over, lying on the ground, shouting so that we would not accidentally run over them. On the

tanks that had been hit we could notice injured and dead crewmen situated in strange positions.

At around midnight someone signaled us to stop. A robust looking man with a big face mounted the tank-turret and we all gathered around him. He had a quiet and soft voice and he asked that we pay attention. We were tired and shocked. He then silently caressed the hand of each of us and said: "Hello, I am the division commander".

We were astonished; we had never been addressed in such a manner by such a high ranked officer. The division commander pulled out a chocolate pack, and handed us the broken pieces. Then he said: "I know, it is very hard on you, you are young. It is hard for me too. I have already fought one tough war, but this one is different; totally different. The brigade has lost a lot of tanks. The whole commanding chain has been crippled. You have lost your regiment commander and your company commander. It is hard, very hard. But I am entirely certain that we will win. Whoever holds on shall win. If we do not let ourselves fall down we shall win. We will win because we have no choice. We simply have to win. The whole Jewish people counts on us, so we have to win… We shall reorganize the brigade from whatever we have still got, and with dawn we shall attack […]".

Before he left us he shook our hands and said again "Guys, I love you"; then he approached another tank. That was how he spoke with every crew in every tank that night. (Orr, 2003, p. 64)

Ori Orr's optimism was turned into a collective one, the kind of group hope that gives a sense of meaning and coherence to life. It was this hope, transmitted through his direct contact with each crew member, that encouraged the soldiers' readiness for sacrifice, so crucial for the upcoming events that were awaiting all of them.

Perhaps a unique example of a junior officer who changed the course of the whole war is Zvika Greengold, a 21-year-old lieutenant who was at home on leave when his story began. He was not attached to any unit, but once he realized that war had broken out, he sensed deep in his heart that it was up to him to play a major role in any war. Zvika Greengold hitchhiked to the Golan Heights in search of any formal duty he could fulfill. In the midst of chaos, he took command of two tanks that had retreated after their crew members were either killed or badly wounded. After having gathered new crews, Zvika headed with the tanks to the Petroleum Road, a key route running along the Golan Heights, where the Syrian invasion force was bound to pass sooner or later. Transmitting through the communication military net, he was coded by the name 'Zvika Force'.

Arriving at the Petroleum Road, Greengold spotted Syrian tanks that had broken through the Israeli crumbling defense lines and were now advancing unopposed. In spite of his inferiority as a force of just two tanks, he immediately engaged the enemy's forces and destroyed six of their tanks. As darkness fell, he lost contact with his other tank, but more and more Syrian tanks were closing in. Zvika also realized that the military radio network was being tapped, so he insisted on calling himself 'Zvika Force,' deliberately forming the impression of a large unit; even his superiors had no idea that the whole force consisted eventually of one lonely tank. He would move and shoot constantly, fooling the Syrians into thinking that they had run across a sizable defense force, and throughout the night he succeeded in destroying about a dozen of the enemy's armored vehicles. Recalling that night he would later say:

> I had no choice. I was alone; there was nobody else who could defend the whole area [so] I took responsibility over the mission of halting the Syrian forces. […] I got the feeling that it was up to me to determine the fate of my relatives, my family, and the fate of the whole Jewish people.
>
> […] It was clear to me that a scenario of failure was out of question, and deep in my heart I silently repeated that over and over again.
>
> […] I knew that I had to dredge from the energies that I hardly possessed, to dig in for personal physical as well as spiritual forces, or else I would not be able to complete my mission. (as cited in Greengold, 2008, pp. 23-25, 116).

Zvika, one ought to remember, had no preliminary acquaintance with any of the crew members, let alone any form of a committed friendship with any of them; but he was led by an optimistic drive to change the catastrophic situation. At dawn, the Syrian brigade commanders decided that it was too risky to encounter the strong Israeli force by daylight and halted optimistically, waiting for further orders. By the time their assault was renewed, larger Israeli forces that had managed to organize in the meantime were already on their way to fight them. Thus, had it not been for one junior officer, off duty, whose attitude had led him to search for alternate routes, the results of the whole campaign might have proved even more tragic (ibid., 2008; Rabinovich, 2005).

On the Southern Golan Heights, near the Syrian-Israeli border, stands a small but strategically positioned hill called Tel Saki. It was used as a military reconnaissance post, and during the first three days of the war a handful of IDF soldiers who were posted there stood their ground, fighting off thousands of Syrian assaulting troops. Outnumbered 100 to 1, short of weapons, ammunition, food and water, the Israeli fighters of Tel Saki never surrendered. At a certain point in time, the Syrian soldiers managed to mount the hill, and then—practically at the last moment—IDF enforcement forces reached the place. In total, 35 out of 80 Israeli soldiers who fought in Tel Saki died, and practically everyone else was wounded. Arad, one of the survivors, a twenty-year old soldier, explains how victory was achieved at the battle of Tel Saki.

> [...] It was all about making possible something that seemed to be simply impossible. We knew that we were holding the place until the rest of the [IDF] forces would get organized and join us. [...] I sometimes ask myself what made them [the Syrians] retreat. I think that the answer is that it is like in Tug of War [rope-pulling game]; the two teams are locked, determined not be pulled for even one centimeter. Both sides are running out of energy, but the winners are those who have this internal spiritual force telling them all the time to hold on, reminding them that the other team is not doing any better; and then it is just one more effort and you win! (as cited in Laskov & Arad, 2003, pp. 144–145)

In the southern front things were certainly not any better than in the north at the first day of the war. In spite of the buffer zone that the Sinai Peninsula formed, a loss at the southern front was just as dangerous for Israel's existence as losing the Golan Heights. During the first two days of the war, even the reserve troops that started a counter-offensive failed to halt the massive Egyptian troops that now flooded the whole western part of Sinai. When the high command of the IDF was occupied with finding a way to stop the enemy from storming mainland Israel, Ariel Sharon, by October 1973, a reserve general, decided that the IDF should cross the Suez Canal and head towards the capital of Cairo. Speaking of crossing the canal at that point was out of the question, to say the least. It is true that thoughts of such an operation had been brooded over before within the IDF's command as theoretical options for future wars, but the circumstances were now those of attempting to survive an unprecedented attack in the face of a disaster. The IDF had lost control of the western part of the Sinai, all of its forces were either organizing for combat or under fire, and crossing the water was something that had never been practiced. The high command of the IDF totally rejected Sharon's ideas.

Yet, Sharon was immovable; he sent out some patrolling units that discovered there was a breach between the 2nd and the 3rd Egyptian Armies. It was through this breach, leading to the town of Kantara on the bank of the Suez Canal, that Sharon decided to instigate what would later become one of the turning points of the war. He located the appropriate crossing equipment and gave orders to take the equipment westward as quickly as possible. Motorized large rafts were hence mobilized across the war zones, and a cylinders bridge 200 metres long and weighing 400 tons was dragged by 24 tanks. The possibility that all this would eventually work seemed at the time like a desperate wish for a miracle.

All this time, Sharon's forces were engaged in some of the fiercest battles of the whole campaign, but in addition to managing the battle-field Sharon persuaded the high command that it was time to initiate the operation. The major dispute was between most of the army leaders who agreed that ensuring a clean corridor for Israeli forces to move westward was vital before any crossover was implemented, and Ariel Sharon who insisted on a plan that meant altering reality.

The arguments continued even once Sharon had managed to form a bridgehead and to land a force of paratroopers and 24 tanks on the Egyptian west bank of the Suez Canal. Those forces immediately stormed a 25 kilometres long front, and with no losses at all caused severe damage to the anti-aircraft Soviet missiles that had practically paralyzed the Israeli air force during the first day of the war. The IDF's high command was disinclined to share Sharon's optimism, and his suggestion to expand the bridgehead was rejected for fear that the Israeli forces on the Egyptian territory would be cut off. Although most of the time Ariel Sharon was overruled by his superiors, his firm belief in the optimistic scenario and his political shrewdness enabled him to trick the high command step by step into what has eventually become one of the brilliant operations in military history. Thus, it was due to positive thinking, as well as personal stubbornness, that the war ended with IDF forces on the western bank of the Suez, merely 101 kilometres from Cairo (Ezov, 2011; Sharon & Chanoff, 2001).

In the testimonies, the most prevalent description of the combat zones is one of chaotic situations. This is typical particularly concern-ing the large desert landscapes of Sinai, where everywhere immense numbers of Egyptian army forces were in motion. Whereas the IDF deployed infantry and armored troops separately, the Soviet military doctrine, in which the Egyptian army had been trained, combined the forces; infantry soldiers carrying *Sagger* anti-tank missiles were mingled with the tanks and would move between them. Encountering those massive amounts of manpower, the relatively few Israeli units that were sent to halt the invaders would soon unwillingly split and often lose any access to their necessary wireless communication (Asher, 2009).

On the very night when Sharon's forces were getting organized for the crossover, one of his tank divisions was sent to outflank 3 Egyptian armored divisions. An unanticipated alignment of the enemy forces caused the outnumbered Israeli division an additional disadvantage. In the course of several hours, 54 out of 97 Israeli tanks were hit and 122 soldiers were dead. Almost dueling, Egyptian and Israeli tanks were forming a field of dozens of large bonfires. Both Sharon and the division commander, Amnon Reshef, were aware of the casualties, they knew that their best officers were being killed by the minute; but they spoke calmly on the wireless, lest they should startle their forces. Reshef testifies, recalling that night:

> You know exactly who got hit; sometimes it is a young officer whom you have educated for years. You know them personally. In some cases you are even acquainted with the family. And you have to speak on the radio coolly, without even the slightest tone of hysteria. (*Halochem [The Warrior]* October 6, 2007, pp. 21–26)

Gaining control over the situation by imposing calmness was indeed a technique used by field officers of all ranks, from the commander downwards and from the bottom rank to the top. Eliashiv Shimshi, a company commander whose forces were on the verge of being overwhelmed, recalls:

> I did not report about the situation to the division commander because I heard on the radio that he was himself occupied with enemy soldiers that surrounded his tank. There was no point in distracting his mind now. I also heard that a commander of one of the regiments was wounded and I did not think it was necessary to burden the division commander with additional worrying accounts. (Shimshi, 1986, p. 45)

It was, therefore, not just a manner of speaking or shallow flattery when division commander Amnon Reshef told some of his soldiers, moments before they went into battle, "[…] I had a chance to speak with the Chief of Staff and with some of the Generals. I was strengthening their confidence in us. I could do so only after having seen you in the battlefield" (Ezov, 2011, pp. 82-83).

It was during those battles in the South that Military Rabbi Yoel Nachshon decided that his destiny was to pour optimism where for long hours everything seemed to be lost. He was ranked a lieutenant, but he had hardly any combat training and he was anything but a fighting commander. He was the Military Rabbi of a mechanized division, and his duties were therefore merely to provide religious services to soldiers. Caught in the midst of the fierce battles, he joined the medical forces, where he was expected to be occupied mainly with identifying the dead in order to make sure that they were properly buried. However, he soon volunteered to rescue wounded soldiers from armored vehicles and to aid the doctor and the paramedics under fire. During one of the Egyptian assaults, when shells were falling all over, he calmly went from one injured soldier to another, confidently telling them that he knew that they would survive.

> I took every opportunity to sing happy songs. Sometimes soldiers would join me [in singing], usually very shyly. […] I wanted to cheer them up, so I kept speaking, often telling them things I had no information about. I told them that I heard from the commanders that our forces were advancing in all fronts; I informed those who were badly wounded that their situation was not as bad as they thought and that they would soon be evacuated. […] I simply boosted them with optimism because I knew it would be part of their recovery. (As cited in Segal, 2007, pp. 287-289)

According to military historian Amiram Ezov who had been studying the course of this war for years, the fate of the crossover of the Suez Canal was not determined by fighting techniques, but rather by the determination of the fighters. Even when coping with obstacles that were virtually impenetrable, even when absorbing unexpectedly painful losses, even when whole units were suddenly reduced and dismantled—the IDF's forces kept pouring westward. Hence, it is no wonder that Ezov described the spirit of the Israeli field soldiers and officers with terminology approximating that of collective positive thinking (Ezov, 2011).

Perhaps one of the commanders who best verbalized a deep sense of positive psychology was General Benny Peled, who was the commander of the Israeli air force in a situation when failure and loss was evident. At the first 24 hours of fighting, 40 Israeli air planes were already shot down; by the end of the war over 100 aircraft were lost and 53 pilots were dead. When the Israeli 44 pilots who were prisoners of war (POWs) were brought back home, Peled said in a ceremony honoring their return:

We are like a violin with a thousand strings. Once in a while one of the strings is torn and we have to manage with one string less; and then another string and another one… Some of the strings shall never return, and others could be fixed. Overall, it is a melody that must never be stopped. […] There are two ways to play the melody. One is with written notes. Sometimes there are no notes and we still have to play, like we did this time. And although we lost many strings, the melody was a good one, and it can be vividly heard all over the region, and we shall keep playing it! (Peled, 2004, p. 417)

Peled's words were transformed into a poem and then into a song that is usually recited in independence day ceremonies, inspiring Israeli citizens with the notion that they shall always win their wars.

Discussion and Implication: Establishing Optimism

The data on which this research is based has from the start some basic drawbacks that should be honestly mentioned. Autobiographical accounts should not be automatically referred to as purely authentic, but rather as speech acts that may not necessarily be genuinely true; they often construct personal and collective experience, bound by social conventions and aimed at justifying actions and evaluating other protagonists (Plummer, 1995; Scott & Lyman, 1968). One should bear in mind that the testimonies were given publically and voluntarily (Berg, 2009); specifically in the case of former commanders or soldiers who are telling their versions of the war, therefore, one can hardly expect, for example, any evidence for cowardice, malfunction or personal failures. In order to cope with some of these reliability issues, testimonies were checked by means of triangulation and cross-examined in comparison to other historical accounts. However, at the end of the day, to some extent all of the testimonies should be taken with a grain of salt.

This being mentioned, the study proves how the importance of positive psychology as a component of resilience of an army unit originates from some of its basic attributes. Optimism is considered to be based on high cognitive processing that requires mental representations of positively valued abstract future situations; it entails setting goals and planning how to accomplish them; it also calls for a use of imagination and creativity. In fact, optimism is a state of mind that prescribes a development of new scripts and of programmes for future actions; it ordains conviction about the not yet proved. Perhaps more than anything else it forms the source of power to resist the temptation to compromise a view of present reality for a better future (Breznitz, 1986; Clore, Schwarz, & Conway, 1994; Isen, 1990; Lazarus, 1991; Snyder, 2000).

This calls, of course, for contemplation about the nature of a better future. A particular disturbing scenario occurs when one strives for an improved upcoming situation which he might never personally have the chance to experience. Such state of affairs is typical, of course, for altruistic deeds, but it is also characteristic, of course, for the case of soldiers whose very fate is at stake in the theater of the battlefield. There is rich evidence for the willingness of soldiers to demonstrate acts of optimism wherein they have no chance to ever participate in the better future that results from their action. For example, 54 of the 99 medals awarded in the United States during the Vietnam War were presented to the families of soldiers who had thrown themselves on hand grenades or other explosives to prevent fellow soldiers from being killed or wounded (Foley & Goudreau, 1996). These cases, in which soldiers enthusiastically sacrifice themselves, seem—on the face of it—to contradict the positive psychology explanation for motivation that is suggested in this study: how could optimism, for that matter, explain the motivation of the Japanese Kamikaze who deliberately crashed their aircrafts, heavily loaded with explosives, into American warships, knowing that they were putting an end to their lives? How could positive psychology relate to the incentives driving suicide bombers throughout centuries whose actions would ultimately deprive them of better personal futures?

Bridging the gap, a partial answer might be found within the term *spirituality*. The various definitions of spirituality spin around the idea that the word describes an overall sense of belonging, of wholeness, of connectedness and of openness to the infinite (Kelly, 1995). Ralph Piedmont uses the term *spiritual transcendence* to describe the capacity of people to step outside their immediate sense of time and place and to view life from a larger perspective. According to this line of thought, people experience—to one extent or another—a feeling that they are a part of a larger human reality that cuts across generations and social groups. Piedmont's research demonstrates that spiritual transcendence proves to be present in religious as well as in secular societies; it is a general human trait across cultures and within differing spiritual traditions (Piedmont, 1999; Piedmont & Leach, 2003).

Once a social group has identified a spiritual goal as sacred, it will strive to protect and preserve it; doing so lends deep significance to human existence. Thus, spiritual transcendence has a unique empower-ing function and serves as a source of motivation for human action in various areas of life (Pargament, 1997, 2002). The spiritual sacred goals that stand far above daily life and sketch the landmarks of a desired future are the framework of positive psychology. In the case of the 1973 war, Israeli soldiers at all fronts had the deep feeling that they were fighting for the very existence of their country for future generations. As mentioned earlier in the methodology clause of this study, one can find within the various testimonies the theme of national causes for the war, wherein the soldiers deeply believed that they were forming a last-defense line that would save the State of Israel and prevent a second Holocaust. Thus, the different themes are intertwined, with collective hope and optimistic attitudes forming the major motivating factors.

When and how does positive psychology prevail? If it is not inherent in one's personality—as scholars claim may be the case—how can it be encouraged? The good news is that under different names for the same phenomenon and using a variety of terms that eventually refer to the same behavioural occurrence, the idea of training for positive psychology has in fact been studied. For example, based on field stud-ies of army personnel, covering deployments ranging from the Gulf War to Bosnia, military psychologist Paul Bartone recognized positive psychology as a key factor in soldiers' response to stressful conditions; consequently, he called for developing training programmes in which officers, as part of their military schooling, would be trained to build up an ability to control or influence events (Bartone, 2006). Likewise, during recent years, professional trainers of army units tend to foster techniques for the acquisition of optimism just the way they do for other military proficiencies (Maddi, 2007; Zach, Raviv, & Inbar, 2007).

Although the specific strategies for fostering positive psychology are somewhat beyond the scope of this study, the recognition of the great value of optimism is an important step forward.

Conclusion

Although comradeship has long been counted as a major motivating factor for soldiers' actions in combat, a thorough review of the tes-timonies clearly points out how collective hope is a factor that must not be ignored. Collective hope, stemming not only from the positive psychology of the commander but also reflected by his officers and soldiers, seems to be the tool with which a military leader can excite his troops, as well as his superiors, into action. The comprehension that hope leads to better performance under stress is bound to lead wise military commanders to adopt the notion of optimism as an integral factor in the training system. Realizing that optimism is controllable will turn positive psychology into an important part of the professional military preparations and will consequently equip the forces with the additional significant weapon of optimism in the soldier's arsenal of resilience.

References

Allen, B. C., Sargeant, L. D., & Bradley, L. M. (2003). Differential effects of task and reward interdependence on perceived helping behavior, effort, and group performance. *Small Group Research, 34,* 716–740.

Anderson, B. (1983). *Imagined communities: Reflections on the origin and spread of nationalism.* New York, NY: Verso.

Angelucci, E., Matricardi, P., & Pinto, P. (2008). *The complete book of World War II combat aircraft.* Italy: White Star Publishers.

Asher, D. (2009). *The Egyptian strategy for the Yom Kippur War.* Jefferson, NC: McFarland & Co. Inc. Publishers.

Bandura, A. (1986). *Social foundations of thought and action.* New York, NY: Prentice Hall.

———. (1997). *Self-efficacy: The exercise of control.* New York, NY: Freeman.

Barnard, F. M. (2003). *Herder on nationality, humanity and history.* Montreal and Kingston, Canada: McGill-Queen's University Press.

Barone, D., Maddux, J. E., & Snyder, C. R. (1997). *Social cognitive psychology: History and current domains.* New York, NY: Plenum.

Bartone, P. T. (2006). Resilience under military operational stress: Can Leaders influence hardiness? *Military Psychology, 18,* (supplement), s131–s148.

Bartone, P. T., Barry, C. L., & Armstrong, R. E. (2009). To build resilience: leader influence on mental hardiness. *Defense horizons, 69,* 1–8.

Bartone, P. T., & Bartone, J. V. (2011, October). *Raising resilience in social groups: The example of grief leadership.* Paper presented at the Inter-University Seminar on Armed Forces and Society Biennial Meeting, Chicago, IL.

Beevor, A. (1999). *Stalingrad.* London, UK: Penguin Books.

Ben-Arie, K. (1990). Czechoslovakia at the time of Munich: The military situation. *Journal of Contemporary History, 25,* 431–446.

Ben-Dor G., Canetti, D., & Lewin, E. (2010). *The social component of national resilience.* Haifa, Israel: National Security Studies Center, Haifa University.

Ben-Dor, G., Pedahzur, A., & Hasisi, B. (2002). Israel's national security doctrine under strain: The crisis of the reserve army. *Armed Forces and Society, 28,* 231–257.

Berg, B. L. (2009). *Qualitative research methods for the social sciences.* Boston, MA: Allyn & Bacon.

Bialer, U. (1980). *The shadow of the bomber: the fear of air attack and British politics 1932-1939.* London, UK: Royal Historical Society.

Breznitz, S. (1986). The effect of hope on coping with stress. In M. A. Appley & R. Trumbull (Eds.), *Dynamics of stress: Physiological, psychological and social perspectives* (pp. 295–306). New York, NY: Plenum.

Canetti-Nisim, D., Arieli, G., & Halperin, E. (2008). Life, pocketbook or culture: The role of perceived security threats in promoting exclusionist political attitudes toward minorities in Israel. *Political Research Quarterly, 61,* 90–103.

Carver, C. S., & Scheier, M. F. (1998). *On the self-regulation of behavior.* New York, NY: Cambridge University Press.

Clore, G. L., Schwarz, N., & Conway, M. (1994). Affective causes and consequences of social information processing. In R. S. Wyer & T. K. Srull (Eds.), *Handbook of social cognition* (vol. I, pp. 323–417). Hillsdale, NJ: Lawrence Erlbaum, 323–417.

Etzioni, A. (2009). Minorities and the national ethos. *Politics, 29,* 100–110.

Evrigenis, I. J. (2008). *Fear of enemies and collective action.* New York, NY: Cambridge University Press.

Ezov, A. (2011). *Crossing* or *Yehuda.* Israel: Kinneret, Zmora-Bitan, Dvir

Fentress, J., & Wickham, C. (1992). *Social memory.* Oxford, UK: Blackwell.

Foley, R. F., & Goudreau, D. A. (1996). Consideration of others. *Military Review,* (Jan–Feb), 25–28.

Frank, J. A. (1991). Profile of a citizen army: Shiloh's soldiers. *Armed Forces & Society, 18,* 97–110.

Gordon, C., & Arian A. (2001). Threat and decision making. *Journal of Conflict Resolution, 45,* 196–215.

Greengold, Z. (2008). *Zvika force.* Ben Shemen, Israel: Modan.

Halbwachs, M. (1992). *On collective memory.* Chicago, IL: University of Chicago Press.

Higgins, T. E. (1999). Promotion and prevention as a motivational duality: Implications for evaluative process. In S. Chaiken (Ed.), *Dual-process theories in social psychology.* New York, NY: Guilford Press.

Isen, A. M. (1990). The influence of positive and negative affect on cognitive organization: Some implications for development. In N. L. Stein, B. Leventhal & T. Trabasso (Eds.), *Psychological and biological approaches to emotion* (pp. 75–94). Hillsdale, NJ: Lawrence Erlbaum.

Kahalani, A. (1992). *The heights of courage: A tank leader's war on the Golan.* Westport, CT: Praeger Publishers.

———. (2004). *The Yom Kippur War - Fighters' Stories.* Jerusalem, Israel: Keter.

Kelly, E. W. Jr. (1995). *Spirituality and religion in counseling and psychotherapy.* Alexandria, VA: American Counseling Association.

Kinder, D. R., & Sears, D. O. (1981). Prejudice and politics: Symbolic racism versus racial threats to the good life. *Journal of Personality and Social Psychology, 40,* 414–431.

Laskov, H., & Arad, A. (2003). *Stories of soldiers who participated in the Yom Kippur War.* Jerusalem, Israel: Arad.

Lazarus, R. S. (1991). *Emotion and adaptation.* New York, NY: Oxford University Press.

Lewin, E. (2010). *Patriotism: Insights from Israel.* Amherst, NY: Cambria.

———. (2012). *National resilience during war: Refining the decision-making model.* Lanham, MD: Lexington books.

Lukes, I. (1993). Stalin and Benes at the end of September 1938: New evidence from the Prague archives. *Slavic Review, 52,* 28–48.

Maddi, S. R. (2007). Relevance of hardiness assessment and training to the military context. *Military Psychology, 19,* 61–70.

Marshall, M. N. (1996). Sampling for qualitative research. *Family Practice, 13,* 522–525.

Maugham, W. S. (2000). *The razor's edge*. London, UK: Vintage Books.

McCann, S. J. H. (1997). Threatening times, strong presidential popular vote winners, and the victory margin. *Journal of Personality and Social Psychology, 92*, 131–138.

McConahay, J. B. (1982). Self interest versus racial attitudes as correlates of anti-bussing attitudes in Louisville: Is it the buses or the blacks? *Journal of Politics, 44*, 692–720.

McDermott D., & Snyder, C. R. (2000). *The great book of hope: Help your children achieve their dreams*. Oakland, CA: New Harbinger Publications.

Mueller, J. (1973). *Wars, presidents, and public opinion*. New York, NY: Wiley.

Ohayon, B. (2008). *We are still there*. Or Yehuda, Israel: Kinneret.

Olick, J. K., & Levy, D. (1997). Collective memory and cultural constraint: holocaust myth and rationality in German politics. *The American Sociological Review, 62*, 921–936.

Ononda, H. (1999). *No surrender: my thirty-year war*. Annapolis, MD: Naval Institute Press.

Orr, O. (2003). *These are my brothers*. Tel Aviv, Israel: Yedioth Ahronot.

Pargament, K. I. (1997). *The psychology of religion and coping*. New York, NY: Guilford.

———. (2002). The bitter and the sweet: An evaluation of the costs and benefits of religiousness. *Psychological Inquiry, 13*, 168–181.

Patterson, I. (2007). *Guernica and total war*. Cambridge, MA: Harvard University Press.

Peled, B. (2004). *Days of reckoning*. Ben Shemen, Israel: Modan.

Peterson, C. (2006). *A primer in positive psychology*. New York, NY: Oxford University Press.

———. (2009). Minimally sufficient research. *Perspectives on Psychological Science, 4*, 7–9.

Piedmont, R. L. (1999). Does spirituality represent the sixth factor of personality? Spiritual transcendence and the five-factor model. *Journal of Personality, 67*, 985–1014.

Piedmont, R. L., & Leach, M. M. (2003). Cross-cultural generalizability of the spiritual transcendence scale in India: Spirituality as a universal aspect of human experience. *American Behavioral Scientist, 45*, 1888–1901.

Plummer, K. (1995). *Telling sexual stories: Power, change and social worlds*. London, UK: Routlrdge.

Rabinovich, A. (2005). *The Yom Kippur War*. New York, NY: McGraw-Hill.

Ritchie, J., Spencer, L., & O'Connor, W. (2003). Carrying out qualitative analysis. In J. Ritchie & J. Lewis (Eds.), *Qualitative research practice: A guide for social science students and researchers* (pp. 219–262). London, UK: SAGE Publications.

Roberts, G. (2006). *Victory at Stalingrad: The battle that changed history*. Evanston, IL: McDougal Littell.

Rothstein, B. (2000). Social capital in the social democratic state: the Swedish model and civil society. In R. Putnam (Ed.), *The decline of social capital? Political culture as a condition for democracy*. Princeton, NJ: Princeton University Press.

Scott, M. B., & Lyman, M. (1968). Accounts. *American Sociological Review, 33*, 46–62.

Segal, M. (2007). *Voices across the dunes – The paratroopers' bloody battle in the Chinese Farm*. Ben Shemen, Israel: Modan.

Sekunda N., & Hook, R. (1998). *The Spartan army*. Oxford, UK: Osprey Publishing.

Seligman, M. E. P. (1991). *Learned optimism*. New York, NY: Knopf.

———. (1994). *What you can change and what you can't*. New York, NY: Knopf.

———. (2000). Positive psychology. In J. E. Gillham (Ed.), *The science of optimism and hope*. Philadelphia, PA: Templeton Foundation Press.

———. (2002). *Authentic happiness: using the new positive psychology to realize your potential for lasting fulfillment*. New York, NY: Free Press.

Sharon, A., & Chanoff, D. (2001). *Warrior: the autobiography of Ariel Sharon*. New York, NY: Touchstone.

Sherif, M., & Sherif, C. W. (1969). Ingroup and intergroup relations: Experimental analysis. In M. Sherif & C. W. Sherif (Eds.), *Social Psychology* (pp. 221–266). New York, NY: Harper & Row.

Shils, E. A., & Janowitz, M. (1948). Cohesion and disintegration in the Wehrmacht in World War II. *The Public Opinion Quarterly, 12,* 280–315.

Shimshi, E. (1986). *Storm in October.* Tel Aviv, Israel: Ministry of Defense Publications.

Showalter, D. E. (2003). Plans, weapons, doctrines: The strategic cultures of interwar Europe. In R. Chickering & S. Forster (Eds.), *The shadows of total war: Europe, East Asia and the United States, 1919-1939.* Cambridge, UK: Cambridge University Press.

Snyder, C. R. (1994). *The psychology of hope: You can get there from here.* New York, NY: The Free Press.

———. (1999). *Making hope happen: A workbook for turning possibilities into reality.* Oakland, CA: New Harbinger Publications.

———. (2000). Hypothesis: there is hope. In C. R. Snyder (Ed.), *Handbook of hope: theory, measures & applications* (pp. 3–21). San Diego, CA: Academic Press.

Snyder, C. R., & Lopez, S. J. (2007). *Positive psychology: The scientific and practical explorations of human strengths.* Thousand Oaks, CA: Sage Publications Inc.

Snyder, C. R., Simpson, S. C., Ybasco, F. C., Borders, T. F., Bavyak, M. A., & Higgins, R. L. (1996). Development and validation of the state hope scale. *Journal of Personality and Social Psychology, 70,* 321–335.

Stephan, W. G., & Stephan, C. W. (2001). *Improving intergroup relations.* Dubuque, IA: Brown & Benchmark.

Tesch, R. (1990). *Qualitative research: Analysis types and software tools.* Abington, UK: Routledge-Falmer.

Vital, D. (1966). Czechoslovakia and the Powers, September 1938. *Journal of Contemporary History, 1,* 37–67.

———. (1967). *The inequality of states.* Westport, CT: Greenwood Press.

———. (1971). *The survival of small states: studies in small power – great power conflict.* London, UK: Oxford University Press.

Zach, S., Raviv, S., & Inbar R. (2007). The benefits of a graduated training program for security officers on physical performance in stressful situations. *International Journal of Stress Management, 14,* 350–369.

10

Building Soldier Resilience

Michael D. Matthews

Unlike wars of the 20th century, modern warfare often involves an ill-defined enemy, asymmetric strategies and no clearly defined temporal- or even geospatial parameters. There are no lines behind which the soldier may retreat to relative safety, and lethal threat is an ever-present possibility. The recent wars in Afghanistan and Iraq are exemplars of modern conflict, but there are many other conflicts across the globe, affecting many nations, that are similar in nature.

Cost to Combat

A significant cost associated with modern war is a substantial increase in stress-related maladaptive reactions. These include post-traumatic stress disorder (PTSD), depression, conduct disorders and suicide (Cornum, Matthews, & Seligman, 2011). For example, following nine years of war, the US Army has seen its suicide rates nearly double, and are now at historic highs (Kuehn, 2009). Coupled with high rates of PTSD (Hoge et al., 2004) and other maladaptive behaviours, the cost to individual soldiers and their families is high.

The cost to the military organizations for which the affected soldiers serve is also very high. For instance, if 10 per cent of the soldiers in the US Army are sufficiently impaired to compromise their combat readiness, this would represent over 50,000 soldiers. The institutional cost is both obvious and monumental—50,000 soldiers represent approximately five infantry divisions. Moreover, the monetary cost of treating stress-related disorders is substantial. Besides direct medical costs, there are costs associated with lost productivity, training costs to replace permanently disabled soldiers and added administrative costs in dealing with affected soldiers and their families. In short, the high rates of psychological disorders associated with persistent war are of sufficient magnitude to adversely impact the combat readiness of armies.

The traditional response to dealing with combat-related stress disorders is to hire more mental health professionals and improve diagnosis and treatment. Although it is certainly necessary to diagnose and treat psychological disorders, this alone is not sufficient to reverse the costs described above. In fact, more available and aggressive psychological screening may amplify the problem by identifying even more soldiers in need of services and actually exacerbate an already overburdened healthcare system.

Cost-Cutting: An Alternative Approach

An alternative approach is to develop proactive training and development programmes to provide soldiers with the personal skills needed to display a resilient—vice a maladaptive—response to combat stress. Recently, the US army has implemented just such a programme. Comprehensive Soldier Fitness (CSF) is a programme designed to improve resilience skills in US army soldiers, their families and army civilians. Mandated by the Chief of Staff of the Army (Casey, 2011), CSF is based largely on positive psychology (e.g., Matthews, 2008). It includes ways of assessing soldier resilience, identifies different types of resilience, explores the full spectrum of responses to combat from pathology (such as post-traumatic stress disorder, PTSD) to personal growth and provides multiple strategies for improving resilience. The purpose of this chapter is to describe CSF, and to explore ways that it may be exported to other services, the militaries of other nations and to non-military settings.

History of CSF

Combat-related stress disorders date back to the very beginning of history. Stress and its pathologic consequences are thus an immutable concomitant of war. A careful reading of *The Iliad and the Odyssey* reveals various instances of combat-stress-related disorder (Shay, 1994). In the military history of the United States, these combat stress reactions have been referred to in various ways. During the Civil War, soldiers were said to suffer from 'soldier's heart'. In the 20th century, the stress reactions were referred to as 'combat shock' or 'battle fatigue'. It is also important to recognize that not all—in fact, not even most—soldiers who are exposed to combat display stress-related symptoms. These symptoms, which include dissociation, memory disturbances, hyper-reactivity, anxiety, depression and attention deficits, occur in at most about 30 per cent of those exposed to combat. Constitutional and experiential factors that predate combat exposure may be strong predisposing factors. Social and familial factors may also influence susceptibility.

What is now known as PTSD, as a formal diagnostic psychiatric disorder, emerged following the Vietnam War. The constellation of symptoms described above was codified in the 1980 edition of the *Diagnostic and Statistical Manual of the American Psychiatric Association*. Once reified with a formal diagnostic label, research into PTSD blossomed and the disorder became almost a cultural icon associated with Vietnam Veterans. Interestingly, despite widespread attention in both the science and practice of psychology as well as the popular press, evidence suggests that incidence rates of PTSD, at their very highest, did not exceed 30 percent during the Vietnam War (see Ritchie, Schneider, Bradley, & Forsten, 2008 for a review of PTSD incidence rates).

Following the events of 11 September 2001 and the subsequent involvement of the United States and many of its allies in the Global War on Terror (GWOT), soldiers are once again being repeatedly exposed to combat trauma. This has spurred another round of interest in PTSD and related research with a concomitant explosion of research and press coverage of PTSD and other maladaptive responses to combat. In the past five years, for example, nearly 2,000 scientific studies of PTSD alone have appeared in the psychological literature (Cornum et al., 2011).

With no end in sight to the GWOT, the strain of "long war" continues to take its toll on the military of the nations involved most directly in the war. The remainder of this chapter will describe the US Army's approach to dealing with the institutional and social costs associated with prolonged war.

CSF

The military response to any health problem has always been proactive and systematic. As discussed by Cornum et al. (2011), malaria was a major threat to military operations during the 19th century. The Union

Army suffered over one million cases and at least 10,000 deaths. A purely reactive approach of diagnosis and treatment was not successful in reducing the incidence of malaria. However, the Army Surgeon General initiated proactive, prevention-related initiatives that ultimately reduced the malaria threat dramatically. By the time of the building of the Panama Canal, in a region rife with malaria, this approach reduced the malaria incidence rate from 800 cases per 1000 workers to just 16 per 1000 workers (Ockenhouse, Magill, Smith, & Milhous, 2005).

General George Casey, Chief of Staff of the United States Army, in consultation with leading psychologists and health-care specialists, led the development of a parallel approach to dealing with the psychiatric casualties of war (Casey, 2011). It was felt that a traditional, disease model, reactive approach to dealing with large numbers of psychiatric casualties could never be sufficient, in and of itself, to reduce the incidence rates of PTSD and related disorders. Instead, CSF represents a double-pronged strategy of (a) enhancing diagnoses and treatment, and (b) developing proactive interventions and training strategies to imbue soldiers with the personal skills needed to bolster their resilience, and thus prevent the development of combat-related stress disorders and/or reduce the seriousness of those developed.

Elements of CSF

Defining Resilience and Developing a Metric

"What doesn't get measured doesn't get done" is axiomatic in psychology. Thus, the first step in CSF was to define the components of resilience and to develop a metric to assess soldier resilience. Because of the pressing nature of the problem, it was not possible to spend years of basic research developing Army-specific measures. Instead, a panel of psychologists from across the United States, led by former American Psychology Association President Martin E. P. Seligman, convened at the University of Pennsylvania for two days with the task of developing both a conceptual model of resilience and identifying existing psychological tests possessing good psychometric properties that could be used to assess the various dimensions of resilience. Due to operational constraints, the metric would have to be relatively short and be couched in language easily read by soldiers who might possess limited verbal and reading skills.

After considerable discussion, the team specified five domains that were thought to be critical to resilience in soldiers. These domains were physical, emotional, social, family and spiritual. Since the original inception of CSF, a fifth 'pillar' of resilience—enhanced performance—has been added. Because the Army already has well developed metrics for tracking physical fitness and programmes in place to monitor, train and enhance it, the remainder of the discussion focused on the other resilience domains. *Emotional resilience* refers to the soldier's personal cognitive and affective skills and strategies needed to respond adaptively in the face of personal challenge and adversity. *Social resilience* focuses on the soldier's network of friends and, especially in a military setting, his or her fellow soldiers, and knowing how to rely on them to help face difficult challenges. *Family resilience* is closely related to social resilience, but instead deals with the soldier's family situation. For young, unmarried soldiers, this could be parents and siblings. For married soldiers, this would expand to include spouse and children. In some instances, extended family members including grandparents, aunts and uncles and others would also fall into this category. *Spiritual resilience* was considered relevant for the soldiers because they must deal with the possibility of their own death and/or taking the lives of others. It is important to note that spiritual, while it might include religiosity, is a broad concept. Perhaps it is best thought of as "meaning and purpose" of life. Finally, *enhanced performance* involves programme aimed to improve and enhance mental and emotional skills that soldiers need to succeed in training, combat and their personal lives. Examples include self-confidence, controlling attentional processes, goal setting, energy maintenance and mental imagery skills.

Measuring Resilience

The group listed and discussed a large number of existing psychological tests that tapped into the various resilience domains. Psychometrics, availability, length and readability were all considered. In some cases, items were written exclusively for the instrument, the Global Assessment Tool (GAT). In the end, the following measures were included:

EMOTIONAL FITNESS

1. The Brief Strengths Test: A 24-item short form of the Values-in-Action–Inventory of Strengths (VIA–Is) (Peterson & Seligman, 2004).
2. Catastrophizing: 10 items adapted from existing tests or written by Professors Christopher Peterson and Nansook Park of the University of Michigan to measure emotional flexibility.
3. Good and Bad Coping: Eight items written by Peterson and Park.
4. Optimism: Four items adapted from the Life Orientation Test— Revised (Scheier, Carver, & Bridges, 1994).
5. Depression: 10 items adapted from the Patient Health Questionnaire (Kroenke, Spitzer, & Williams, 2001).
6. Positive and Negative Affect: 21 items from the Positive and Negative Affect Scale (PANAS; Watson, Clark, & Tellegen, 1988).

SOCIAL RESILIENCE

1. Friendship: 3-item scale, from the UCLA Loneliness Scale (Russell, Peplau, & Cutrona, 1980).
2. Engagement: 4-item scale derived from the Works as a Calling Scale (Wrzesniewski, McCauley, Rozin, & Schwartz, 1997) and the Orientation to Happiness Scale (Peterson, Park, & Seligman, 2005).
3. Organizational Trust: 5-item scale derived from various organization trust metrics (see Peterson, Park, & Castro, 2011 for specific sources).
4. Friendship: Seven additional items written by Peterson and Park specifically for the GAT.

FAMILY FITNESS

1. Family Fitness: Two items written by Peterson and Park specifically for the GAT.
2. Family Fitness: Three items derived from the Military Family Fitness Scale (derived from the Military Family Fitness Scale, developed by the Directorate of Basic Combat Training's Experimentation and Analysis Element, Fort Jackson, for an in-progress study).

SPIRITUAL FITNESS

Spirituality: Five items from the Brief Multidimensional Measure of Religiousness and Spirituality (Fetzer Institute, 1999).

These sum to 106 items at the time of this writing. But, it should be noted that items and scales are subject to modification depending on how well they predict important outcome criteria. Thus, the GAT represents a somewhat dynamic test and is still in the process of development.

Peterson et al. (2011) note that the GAT is "notable for several reasons". It is a comprehensive inventory of psychological and social fitness measured by a variety of measures. Importantly, it presents a "common vocabulary" that may begin to build a culture of resilience in the Army. The GAT is available to every soldier in the Army via online testing, and it provides immediate feedback on each respondent's relative standing in the domains of emotional, social, family and spiritual fitness. Since it is Army policy that all soldiers complete the GAT, the stigma associated with taking "psychological" tests (historically only taken if one has a "psychological problem") is reduced. Last, the feedback from the GAT includes suggestions for individually tailored programmes that the respondent may elect to employ to improve psychological fitness in any one or more of the resilience/fitness domains.

It is critical to point out that the scores of individual soldiers are held absolutely confidential—only the soldier taking the GAT will know his or her scores. This is necessary to preserve the integrity of the testing protocol and to give soldiers the confidence that they can answer questions honestly without concern about repercussions from their chain of command or fellow soldiers. However, it is possible to present commanders with unit-level indicators of resilience. For instance, the commander of a combat brigade preparing to deploy may look at the averaged scores of the units within his or her command and determine in what dimensions of resilience the brigade or its constituent units are "green" (i.e., fit), "yellow" (perhaps requiring some additional training) or "red" (seriously low in fitness, thus necessitating additional unit level training). Thus, summed GAT feedback can provide a useful command tool for assessing psychological combat readiness at the unit level.

A full description of the GAT, how it is being used, and its utility in predicting both resilient and pathologic responses to combat stress are being the scope of the current chapter. The reader is referred to Peterson et al. (2011) for a detailed description of the origin, use and psychometric properties of the instrument. However, preliminary results show that GAT is related to successful performance at the US Military Academy (Kelly & Matthews, 2011). Currently, unpublished results from GAT data on active duty US soldiers show that higher GAT scores are associated with desirable institutional outcomes such as retention, and low scores are associated with negative outcomes including suicide, attrition and disciplinary problems.

Developing Interventions

A hallmark characteristic of CSF is that training protocols and interventions designed to improve psychological fitness are not a 'one-size-fits-all' approach. An analogy can be made to physical fitness. The Army Physical Fitness Test (APFT) has three components: a two-mile run, sit-ups and pull-ups. A soldier must score at a minimum level—based on age-related standards—on each of the three component tests, and an overall score is also generated which must exceed a minimum standard. If a soldier fails a component of the APFT, he or she can turn to master physical fitness instructors who are found in every Army organization for training programmes and advice on how to improve performance on the failed component(s). For instance, if a soldier passes the pull-ups and sit-ups tests, but fails to complete the two-mile run fast enough to meet the minimum requirement, he or she will be placed into a programme specifically tailored to improve running speed.

The resilience training, development and intervention strategy adopted by CSF follows a parallel strategy. This includes a variety of individualized and unit training. Some of it will be institutionalized within the training component of the Army such that all soldiers receive 'schoolhouse' education about resilience and the dimensions of psychological fitness. Individualized training and interventions may be suggested from personal feedback on the GAT. Other training and interventions may be driven, as discussed above, from unit level feedback.

Specifically, the CSF involves three levels of training strategies.

UNIVERSAL RESILIENCE TRAINING

It begins at the entry of the soldier into the Army. This includes instruction on what resilience means, its importance to the individual soldier and larger units and specific skills that soldiers can employ to improve and maintain resilience. Small-unit leaders and trainers are taught specific strategies and tools for training and maintaining resilience. "This will be continuous, progressive, and sequential sustained resilience training of both enlisted soldiers and officers, given at every level of professional military development" (Cornum et al., 2011). For example, officers will receive resilience instruction during their pre-commissioning training and at all subsequent officer development programmes beginning with the officer basic course through the Army War College.

INDIVIDUALIZED TRAINING

It is the second level of training. Following the 'one-size-does-not-fit-all' approach, individualized training varies as a function of the needs of the soldier taking the GAT. It may include self-development opportunities, web-based interactive resilience modules or other protocols aimed at helping each soldier build the personal and social tools needed to be resilient in the face of adversity. Conceptual and initial approaches to individualized training have been described in a special issue of the *American Psychologist* (Seligman & Matthews, 2011). Emotional resilience is described by Algoe and Fredrickson (2011), social resilience by Cacioppo, Reis and Zautra (2011), family resilience by Gottman, Gottman and Atkins (2011) and spiritual resilience by Pargament and Sweeney (2011). These approaches are strictly evidence based and will evolve over time as it is learned which techniques are and are not successful. Collectively, training in these areas is offered to soldiers through comprehensive resilience modules (CRMs). Offered online, there are currently 32 CRMs that aim to help soldiers build and maintain strong personal relationships, learn how to use strength of character to respond effectively under stress and challenge, build mental toughness and strengthen emotional fitness.

MASTER RESILIENCE TRAINERS (MRTs)

The third approach to training resilience under CSF is (MRTs). Just as the Army for many years has had master physical fitness trainers embedded in all units, MRTs are now being developed. In the organizational structure of the Army, the senior non-commissioned officer (NCO) has the most frequent and influential contact with soldiers. Thus, a training programme currently conducted in conjunction with the Positive Psychology Center at the University of Pennsylvania is putting senior NCOs through a 10-day protocol in order to train MRTs. The programme is based on the Penn Resilience Program (PRP), originally designed to imbue school children with workable resilience skills including optimism, problem solving, self-efficacy, self-regulation, emotional awareness, flexibility, empathy and strong social relationships. The MRT course incorporates principle elements and strategies of the PRP, and adds approaches emerging from positive psychology such as identifying and employing signature strengths (Reivich, Seligman, & McBride, 2011). These strategies are grouped into four modules; resilience, building mental toughness, identifying character strengths and strengthening relationships. In order to bolster face validity, nomenclature and course content have been "greened" to be more familiar to soldiers and their experiences. After completing the course, the NCOs are assigned a formal skill identifier that certifies them as an MRT. They are then assigned back to operational units, provided with training materials and protocols, and then serve to educate and train fellow soldiers on resilience. Thus, knowledge about resilience may be infused into the Army from the "bottom up," rather than from the "top down". Feedback from certified MRTs has been exceptionally positive (Reivich et al., 2011).

Other Elements of CSF

Besides assessing and training psychological resilience, there are several other aspects to CSF that bear description. The programme is based on the principles of positive psychology and thus focuses on what is good and healthy about people and building on those capacities to improve the quality of life. It also extends to the families of soldiers as well as to Army civilian employees, who are subject to many of the stressors encountered by soldiers. The following discussion describes additional aspects of CSF that are important in understanding its impact and implications for individual and collective adjustment, as well as programme success.

Post-traumatic Growth (PTG)

Most psychological studies of the effects of combat and stress on adjustment focus on pathologic responses including PTSD, depression, anxiety, suicide and other manifestations of personal and social maladaptation. Based on these studies, extensive programmes to diagnose and treat psychological problems emerge. But, this disease-based model does not allow attention to be given to other possible consequences of stress and adversity, which, especially within the military context, remain virtually unexplored. One can imagine a normal curve that describes the full spectrum of sequels to combat stress, ranging from suicide and other extreme pathologic responses to resilience (i.e., remaining unchanged in the face of adversity) to psychological growth.

Tedeschi and McNally (2011) describe the phenomenon of PTG. It is clear from their discussion that PTG may be as likely an outcome stemming from trauma and stress as is PTSD. An extensive and growing literature exists on the after effects of cancer, accidents and other traumatic experiences and suggests that people often—after facing mortal circumstances—derive additional meaning in their lives. Tedeschi and McNally review the PTG literature, and suggest training strategies for increasing the likelihood of PTG occurring following combat. These strategies include (a) educating soldiers about trauma and that it is a possible precursor to PTG, (b) training in emotional regulation and enhancement, (c) training in constructive self-disclosure, (d) creating a trauma narrative that includes the possibility of PTG, not just PTSD and other negative consequences and (e) developing life principles that are robust to challenges.

The evidence is that PTSD is a less likely sequel to combat stress than resilience and PTG. But given the amount and frequency of press coverage of PTSD, it is very important to create an understanding in soldiers that positive effects may also occur. Because unpleasant and disturbing reactions to adversity (e.g., bad dreams, intrusive thoughts) are common and normal immediate after effects of trauma, they could easily be interpreted as "sick" by the individual affected who may then self-label as PTSD. Thus, creating new trauma narratives and expectations are critical. Yes, it is normal to have bad dreams and other unpleasant reactions following exposure to combat, but these will pass and may be replaced by a sense of accomplishment, greater appreciation of family and perhaps enhanced appreciation for and understanding of meaning and purpose in life. It is not proposed that one seek to come down with cancer or get into a firefight in order to improve their sense of well-being—that is absurd—but knowing that positive things may ultimately follow is a vital part of building a robust and resilient Army.

Extension to Army Families, Children and Civilians

The psychological fitness of soldiers cannot be viewed in isolation from their families, spouses and children. Therefore, CSF explicitly includes programmes for military families, spouses and children. The GAT has been adapted for use with these populations and training and intervention programmes that parallel those developed for the service member are being developed and institutionalized.

Gottman et al. (2011) underscore the importance of integrated support and training programmes that involve the soldier and family members. They maintain that the 'signature event' that precedes suicidal ideation or actions in soldiers is some sort of turbulence in their relationship with their domestic partner. Gottman et al. then describe the application of a pilot programme based on the 'Seven Principles of Making Marriage Work' (Gottman & Silver, 1999) wherein small groups of soldiers are given social skills training. Gottman et al. suggest that CSF develop further face-to-face and virtual training programmes to meet the needs of all Army families.

Children are also an integral part of the Army family. Even in the best of times, the children of US Army soldiers face challenges resulting from frequent moves and absences from the soldier parent or parents during deployments. In times of war, these challenges are multiplied by the possibility that their parent(s) could be killed or wounded in action. Although military children tend to be remarkably adaptable and resilient (Park, 2011), more systematic and evidence-based programmes are being developed under CSF to improve education, training and services for military children.

In contemporary war, civilians play an increasingly important role in the Army. Modern combat systems are complex and sometimes require civilian experts to deploy with the weapon system to keep it functioning. Civilian law enforcement experts are present in large numbers in Afghanistan and Iraq as they train the police and military forces of those countries how to conduct effective police and security operations. Civilians who are not deployed experience a heavy increase in workload and concern for military colleagues and friends who deploy into harm's way. Modifications are being made to CSF programmes to extend them to Army civilians in response to these concerns.

Programme Assessment

An important premise of CSF is that it will only use programmes that are of proven effectiveness. Lester, McBride, Bliese and Adler (2011) describe the programme evaluation efforts aimed to evaluate the effectiveness of various CSF programmes. A large scale research project is ongoing that tracks the impact of CSF on three broad classes of indicators. Eight combat brigade teams (CBTs) were selected for study. Employing a quasi-experimental design, the CBTs received different combinations of interventions including two with no formal resilience training. Three broad classes of outcome variables are being studied. First, GAT scores are being tracked to see if units that received training show an increase in overall resilience, or in any one or more of the fitness domains (emotional, social, family or spiritual). Second, institutional markers are being tracked. These include retention, reenlistments, promotions, suicides, PTSD, crime, divorce and other variables. All of these are conceptually linked to resilience. Finally, physiological and neurobehavioural indicators are being developed. These are intended to look at biological (e.g., stress hormones) and behavioural markers (e.g., performance on cognitive tasks) that have been linked to stress and resilience in past research.

Preliminary analyses suggest that CSF is achieving many of its objectives. Lester et al. (2011) report about units that received CSF training performed by certified MRT trainers:

1. Display significantly higher levels of resilience and psychological health than units not receiving the training.
2. Tend to show higher rates of posttraumatic growth.
3. CSF training appears to be most effective among soldiers age 18–24.
4. Is more effective when the training is conducted in formal settings, such as officially scheduled classes; when MRT trainers are perceived as confident and competent and where the command strongly supports the training.
5. Importantly, there is no evidence that the training results in poorer resilience or psychological health.
6. The statistical effect sizes compare favourably to other successful public health initiatives.

Collectively, this and related evaluations will allow CSF leaders to build and enhance programme elements that work, and to eliminate those that do not. Thus, CSF will continue to be an evolving approach to improving soldier, family and civilian resilience.

Extension to Other Military Organizations

Within the US military, there is a perceived need and interest in adopting CSF or a similar programme. The current need may be greatest for combat elements that operate on the ground, notably the Marine Corps. But, men and women who serve in the Navy and the Air Force are also exposed to many of the same combat stresses and challenges, and can benefit from institutional-wide efforts to improve resilience skills.

The CSF programme may also be of interest to the military organizations of other nations. Throughout the world, nations are faced with combating terrorism and other threats to their safety and sometimes sovereignty. Especially for nations involved in protracted conflict, CSF-like initiatives may improve combat readiness.

In the United States, for example, the Air Force is developing a programme similar to CSF (Air Force Resilience Program Overview, 2011). This programme includes foundational training for new enlistees and officers candidates, unit level training and mandatory and targeted mental-health training. Also like CSF, the Air Force programme includes family members.

Extension to Non-Military Organizations

The US health-care system is based on the disease model. It has created a huge industry aimed at diagnosing and treating diseases of all sorts, but invests relatively little—other than occasional 'lip-service'—to the prevention of disease. The CSF programme represents a complementary approach that is health-based. In addition to identifying and treating disease, greater gains can be made from giving people the life skills to prevent them. Therefore, CSF provides a model for national healthcare systems.

Historical Significance of CSF to Psychology

The history of war and the history of psychology are closely intertwined (Seligman & Fowler, 2011). The two World Wars of the 20th century stimulated the development of mass psychological testing, an explosive growth in clinical psychology and the origins of human factors engineering. Broadly speaking, Scales (2009) argues that each major war beginning with First World War (WWI) leveraged paradigm shifts in science. In his view, WWI was associated with rapid growth in chemistry, Second World War (WWII) in physics and the "Cold" War (Korea and Vietnam, notwithstanding) with information technology. Scales maintain that the GWOT will drive rapid improvements in the behavioural and social sciences because success or failure will depend more on winning "hearts and minds" than killing enemy soldiers. The American experience in Iraq is a good example. The Iraqi Army was quickly defeated, but US continues to face huge challenges in establishing a functional and modern government in place of the dictatorship that fell. The ability to do this depends more on understanding the psychology and culture of Iraq than the ability to demonstrate clear military superiority in the traditional sense.

Seligman and Fowler (2011) view CSF as an exemplar of the importance of psychology to modern militaries and their nations. The programmes developed under CSF will improve combat readiness, but also provide a model for the shaping of more general public health initiative outside the military context. The fact that the US Army would turn to psychology to respond to its needs is historically unique. The unique needs of today's Army combine with the recent developments of positive psychology to form a union that is especially well suited to respond to the challenges faced by today's Army.

Conclusion and Future Directions

Nearly two million US military personnel have served in Iraq or Afghanistan since 2001. The US military is an all-volunteer force. Unlike previous wars, it does not depend on conscripts who serve relatively short tours and who are then quickly returned to civilian society. As such, the professional US soldier continues to bear a great burden in the long war against terrorism. Many have deployed several times with no relief in sight. This provides the seedbed for significant psychosocial problems. In the absence of large-scale and aggressive programmes such as CSF, the adverse effects of the long war on soldier adjustment could be devastating.

It is also true that all soldiers eventually return to civilian life. Whether they serve a single four-year tour or retire after 30 years or more of service, they will eventually re-integrate into civilian society. To the extent they are psychologically and physically healthy and robust, they will contribute positively to the nation's economy and national well-being. For those who suffer from physical and psychological wounds, the costs to society can be very high. Thus, the payoff from CSF will continue to be felt long after soldiers leave the Army. Even a modest reduction in PTSD and related pathologies will, over the course of scores of years, result in huge dividends to the US, both in terms of "hard" economic indicators such as reduced health costs, but also in less objective indicators, such as former soldiers who live meaningful and productive lives. Therein may lie the biggest benefit of CSF.

References

Air Force Resiliency Program Overview. (n.d.). Washington, DC: Headquarters Air Force Resiliency Division.

Algoe, S. B., & Fredrickson, B. L. (2011). Emotional fitness and the movement of affective science from the lab to the field. *American Psychologist, 66,* 35–42.

Casey, G. W., Jr. (2011). Comprehensive soldier fitness: A vision for psychological resilience in the U.S. Army. *American Psychologist, 66,* 1–3.

Cacioppo, J. T., Reis, H. T., & Zautra, A. J. (2011). Social resilience: The value of social fitness with an application to the military. *American Psychologist, 66,* 43–51.

Cornum, R., Matthews, M. D., & Seligman, M. E. P. (2011). Comprehensive soldier fitness: Building resilience in a challenging institutional context. *American Psychologist, 66,* 4–9.

Fetzer Institute. (1999). *Multidimensional measurement of religiousness/spirituality for use in health research.* Kalamazoo, MI: Fetzer Institute.

Gottman, J. M., Gottman, J. S., & Atkins, C. L. (2011). The comprehensive soldier fitness program: Family skills component. *American Psychologist, 66,* 52–57.

Gottman, J. M., & Silver, N. (1999). *The seven principles for making marriage work.* New York, NY: Three Rivers Press.

Hoge, C. W., Castro, C. A., Messer, S. C., McGurk, D., Cotting, D. I., & Koffman, R. L. (2004). Combat duty in Iraq and Afghanistan, mental health problems, and barriers to care. *New England Journal of Medicine, 351,* 453–463.

Kroenke, K., Spitzer, R. L., & Williams, J. B. (2001). The PHQ-9: Validity of a brief depression severity measure. *Journal of General Internal Medicine, 16,* 606–613.

Kelly, D. R., & Matthews, M. D. (2011, May). *A measure of comprehensive soldier fitness and early attrition at West Point.* Poster presented at the annual meeting of the Association for Psychological Science, Washington, DC.

Kuehn, B. H. (2009). Soldier suicide rates continue to rise: Military, scientists work to stem the tide. *JAMA: Journal of the American Medical Association, 301,* 1111–1113.

Lester, P. B., McBride, S., Bliese, P. D., & Adler, A. B. (2011). Bringing science to bear: An empirical assessment of the comprehensive soldier fitness program. *American Psychologist, 66,* 77–81.

Matthews, M. D. (2008). Positive psychology: Adaptation, leadership, and performance in exceptional circumstances. In A. Hancock & J. L. Szalma (Eds.), *Performance under stress.* Aldershot, UK: Ashgate.

Ockenhouse, C. F., Magill, A., Smith, D., & Milhous, W. (2005). History of U.S. military contributions to the study of malaria. *Military Medicine, 170,* 12–16.

Pargament, K. I., & Sweeney, P. J. (2011). Building spiritual fitness in the Army: An innovative approach to a vital aspect of human development. *American Psychologist, 66,* 58–64.

Park, N. (2011). Military children and families: Strengths and challenges during peace and war. *American Psychologist, 66,* 65–72.

Peterson, C., & Seligman, M. E. P. (2004). *Character strengths and virtues: A handbook and classification.* New York, NY: Oxford.

Peterson, C., Park, N., & Seligman, M. E. P. (2005). Orientations to happiness and life satisfaction: The full life versus the empty life. *Journal of Happiness Studies, 6,* 25–41.

Peterson, C., Park, N., & Castro, A. (2011). Assessment for the U.S. Army Comprehensive Soldier Fitness program: The Global Assessment Tool. *American Psychologist, 66,* 10–18.

Reivich, K. J., Seligman, M. E. P., & McBride, S. (2011). Master resilience training in the U.S. Army. *American Psychologist, 66,* 25–34.

Ritchie, E. S., Schneider, B., Bradley, J., & Forsten, R. D. (2008). Resilience and military psychiatry. In B. J. Lukey & V. Tepe (Eds.), *Biobehavioral resilience to stress.* Boca Raton, FL: Taylor and Francis.

Russell, D., Peplau, L. A., & Cutrona, C. E. (1980). The revised UCLA Loneliness Scale: Concurrent and discriminant validity evidence. *Journal of Personality and Social Psychology, 39,* 472–480.

Scales, R. H. (2009). Clausewitz and World War IV. *Military Psychology, 21*(Suppl. 1), S23–S35.

Scheier, M. F., Carver, C. S., & Bridges, M. W. (1994). Distinguishing optimism from neuroticism (and trait anxiety, self-mastery, and self-esteem): A re-evaluation of the Life Orientation Test. *Journal of Personality and Social Psychology, 67,* 1063–1078.

Seligman, M. E. P., & Fowler, R. D. (2011). Comprehensive soldier fitness and the future of psychology. *American Psychologist, 66,* 82–86.

Seligman, M. E. P., & Matthews, M. D. (2011). (Eds.). Comprehensive soldier fitness. Special Issue, *American Psychologist, 66,* 1–87.

Shay, J. (1994). *Achilles in Vietnam: Combat trauma and the undoing of character.* New York, NY: Scribner.

Tedeschi, R. G., & McNally, R. J. (2011). Can we facilitate posttraumatic growth in combat veterans? *American Psychologist, 66,* 19–24.

Watson, D., Clark, L. A., & Tellegen, A. (1988). Development and validation of brief measures of positive and negative affect: The PANAS scales. *Journal of Personality and Social Psychology, 54,* 1063–1070.

Wrzesniewski, A., McCauley, C. R., Rozin, P., & Schwartz, B. (1997). Jobs, careers, and callings: People's relations to their work. *Journal of Research in Personality, 31,* 21–33.

11

Training Hardiness for Stress Resilience

Paul T. Bartone, Jarle Eid and Sigurd W. Hystad

The military occupation involves many risks and stressors to include combat exposure, potential death and injury, witnessing others being injured or killed, uncertainty, powerlessness, boredom, heavy workload and dangerous training activities (Alford & Cuomo, 2009; Bartone, Adler, & Vaitkus, 1999). In international operations, all of these stress factors can come into play, as well as the sense of isolation associated with being distant from one's own familiar culture, family and friends. In training military personnel, it thus makes sense to invest in programmes to better prepare personnel to cope positively, and be resilient in the face of such stressors. This chapter describes various ways that leaders can shape their organizations so as to generate more positive or "hardy" and resilient stress responding in soldiers, and also suggest some "self-help" steps to increase hardiness.

Resilience

The last decade has seen a rapid increase in studies that use the word "resilience" (Layne, Warren, Watson, & Shalev, 2007). But, there is also a multitude of definitions and meanings attached to the term "resilience", and the related concepts of vulnerability and stress resistance, leading to considerable terminological inconsistency and confusion. Historically, the term resilience was first applied in developmental psychopathology to describe children who developed normally and adapted well despite growing up in environments that put them at higher risk for psychopathology and other poor outcomes (Garmezy, 1971; Garmezy, 1974; Werner, Bierman, & French, 1971). Many of those working in the field still define resilience as basically the absence of pathology following stress exposure (Stroufe, 1997; see Witmer & Culver, 2001, for a review).

But, resilience is increasingly seen as involving positive processes that are distinct from those associated with heightened vulnerability (Carver, 1998; Friborg, Hjemdal, Martinussen, & Rosenvinge, 2009; Werner & Smith, 2001). Even among those definitions that focus only on positive features, there is a wide diversity of views as to what factors constitute or contribute to resilience (Luthar, 2006). These range from individual attributes such as intellectual ability, self control, flexibility, calm demeanour, optimism, self-efficacy, spirituality and confidence, to social factors such as support from family and friends, co-workers and the broader socioeconomic environment (Layne et al., 2007).

A good example of a broad, generic view of resilience is provided through the "Resilience Scale for Adults" (RSAs) by Friborg, Barlaug, Martinussen, Rosenvinge and Hjemdal (2005). The RSA stipulates five or six factors that span both individual (personal strength, social competence, structured style) and social (family cohesion and social resources) dimensions of cohesion. Efforts such as this to identify the attributes of people who adapt well (are resilient) in the face of stress are an important step. But, as Layne et al. (2007) correctly observe, this information by itself does not reveal what are the underlying processes or dynamics by which some people are more stress resilient than others. Despite a lack of scientific knowledge regarding resilience processes, recent years have seen a flood of training programmes, web sites and self-help guides all claiming to increase resilience "skills" or abilities (e.g., Reivich & Shatte, 2002; Siebert, 2005), mostly without empirical support for their efficacy (Watson, Ritchie, Demer, Bartone, & Pfefferbaum, 2006). More good theory and research are needed to shed light on resilience factors and processes—what highly resilient people are actually doing that allows them to cope so well with stress. For present purposes, resilience shall be defined simply as bouncing back quickly from adversity or setbacks. The resilient person finds ways to cope effectively with stress and is able to continue functioning despite difficulties.

Resilience in Military Context

Although stress resiliency at the individual level is important and gets the most attention, many factors outside of the individual can also influence resilient (or non-resilient) responding. Although training individuals to be more resilient may be valuable, efforts to increase resilience need to go beyond the individual level to incorporate a broad understanding of the multiple factors in life that can bear upon resilient outcomes. Taking the military as an example, factors at several levels can be seen to impact resilient responding to stress, including (a) individual level, (b) organizational policy, procedures and culture and (c) organizational structure (Bartone, Barry, & Armstrong, 2009).

At the individual level, factors that can influence resilience include social background characteristics such as early family environment, personality (including psychopathology), previous work experience, education, maturity, intelligence, physical fitness, diet, exercise and current family circumstances. Organizational policies also can affect resilience in terms of how the organization as a whole and its members respond to challenging or stressful events. Such policies include higher-level agency rules, regulations and directives, mission statements, deployment and rotation policies and rules-of-engagement. At a lower organizational level, resilience is influenced by leader directives and communications, training schedules and policies, and unit Standard Operating Procedures (SOPs). To be sure, lower-level organizational policies and procedures are influenced by policies set at higher organizational levels, as for example with rules-of-engagement or unit rotation policies. At the same time, junior leaders can establish policies within their own units that can foster greater resilience, within the bounds of broad organizational standards. For example, junior leaders can implement various strategies, such as "commander's calls" to increase open information sharing and situational awareness within their units. Junior leaders also often have the latitude to set work schedules and time-off policies.

Structural factors also have an influence on how organizations respond to challenges. In military organizations the size, type and configuration of units may be more or less appropriate for the demands of the environment at a particular time. Other structural considerations include where units are based and how they are staffed or manned, the ratio of leaders to troops, and the integration of reserve and civilian contractor security forces, as well as joint and coalition forces such as North Atlantic Treaty Organization (NATO) operations in Afghanistan (Alford & Cuomo, 2009). For example, organizational policies clearly influence (and in some cases determine) structures, whereas existing structures, force levels and types also influence the policies that are developed and implemented regarding the utilization of forces. Structures and policies have an influence on individuals in myriad ways, as for example when force structures and rotation policies determine when and for how long personnel will be deployed away from home. All of these factors—individual, organizational policies and organizational structures—are affected by what resources are available—or not.

Building Resilience

Programmes to increase resilience in high-stress occupations such as the military are surely needed, but should be well-grounded both theoretically and empirically. The discussion to this point leads to some minimum criteria that should be met by any training approach that seeks to increase stress resilience:

1. The training should be based upon a clear definition and theoretical formulation regarding the construct(s) that are presumed to influence stress resilience. As previously discussed, the construct of resilience by itself is now so general and ambiguous that an essential first step is one of definition;
2. There should be empirical support in the form of studies, showing that the resilience construct(s) under consideration actually distinguishes people who respond favourably (resiliently) to stress, from those who do not. This also means that the construct(s) should be found to moderate the effects of stress on relevant outcome variables (i.e. interaction effects);
3. There needs to be a recognition that individual and group differences may influence the training. This is to say that the assumption of uniformity (that all people are alike and can be expected to respond alike) is misplaced when it comes to resiliency training. The training may need to be adjusted and adapted to different groups, different contexts and different times in order to be effective and;
4. There should be empirical support, showing that the training is effective in increasing those qualities assumed to influence resilient responding to stress.

In what follows, emphasis in this chapter is placed on an individual level stress resilience construct that meets all of these criteria: psychological hardiness.

Hardiness as a Resilience Factor

The construct of hardiness has a clear theoretical background, and has been proven empirically as a significant stress resistance resource in a wide diversity of groups, including those involved in military and security operations. Conceptually, psychological hardiness is an individual disposition or style that develops early in life and is reasonably stable over time, although amenable to change and trainable under certain conditions (Kobasa, 1979; Maddi & Kobasa, 1984). Hardiness was first described by Kobasa (1979) as a collection of related personality qualities or traits that distinguished healthy executives under stress from unhealthy ones. More recently, Maddi and Khoshaba (2005) characterize hardiness as three related attitudes of commitment, control and challenge or the "3 Cs".

According to Bartone (2006), hardiness is more global and encompassing than a definition in terms of mere attitudes would imply. Rather, it is a broad personal style or approach to life, a *generalized mode of functioning* that incorporates commitment (conviction that life is interesting and worth living), control (belief one can control or influence outcomes) and challenge (adventurous, exploring approach to living) perspectives. In addition, the "hardy-resilient style" person has a strong future orientation, or tendency to look to the future while learning from the past. The hardy-resilient person is also courageous in the face of new experiences as well as disappointments, is action-oriented, competent and has a sense of humour (Priest & Bartone, 2001).

Since Kobasa's seminal article on hardiness in 1979, an extensive body of research has accumulated showing that psychological hardiness protects against the ill effects of stress on health and performance. Research studies with a variety of occupational groups have found that hardiness operates as a moderator or buffer of stress (e.g. Bartone, 1989; Contrada, 1989; Kobasa, Maddi, & Kahn, 1982; Roth, Wiebe, Fillingim, & Shay, 1989; Wiebe, 1991). In military samples, hardiness has been identified as a significant moderator of combat exposure stress in US Gulf War soldiers (Bartone, 1993, 1999, 2000). Hardiness operates as a stress buffer in other military groups as well, including the following: US Army casualty assistance workers (Bartone, Ursano, Wright, & Ingraham, 1989); peacekeeping soldiers (Bartone, 1996); Israeli soldiers in combat training (Florian, Mikuluncer, & Taubman, 1995); Israeli officer candidates (Westman, 1990) and Norwegian Navy cadets (Bartone, Johnsen, Eid, Brun, & Laberg, 2002). High hardy persons are not impervious to the ill-effects of stress, but they do not suffer from the same level of symptoms and performance decrements as low-hardy persons do under stressful conditions.

There is good evidence pointing to the cross-cultural validity of the hardy-resilient style. In a review of relevant studies addressing the issue of hardiness across cultures, Maddi and Harvey (2006) conclude that available evidence shows little or no cultural differences in the role of hardiness, and suggest that hardiness appears to be a factor in resilience under stress across cultures. More recently, confirmatory factor analysis in a large Norwegian sample found three facets (commitment, control and challenge) nested beneath a superordinate hardiness construct, lending support to the theoretical structure of hardiness (Hystad, Eid, Johnsen, Laberg, & Bartone, 2010). Sinclair and Tetrick (2000) found a similar structure in an American sample, using an older version of the Dispositional Resilience Scale (DRS) (Bartone, 1995).

Hardiness Training Programmes

Programmes designed to train hardiness have ranged in complexity from fairly simple self-paced learning modules, to more elaborate approaches that include additional health-related factors. One of the first efforts to train hardiness was reported by Maddi (1987). Using a small group format with multiple "hardiness induction" sessions spaced over a two- or three-month period, Maddi and colleagues were able to show significant increases in hardiness levels following the training (Maddi, Kahn, & Maddi, 1998). This training approach involves first of all teaching clients about the concept of hardiness, and how hardiness "attitudes" can influence positive coping strategies and ultimately continued good health under stress. This is followed by some guided role play and practice of effective coping strategies in response to real or imagined stressful situations. The first step involves a detailed "situational reconstruction" of a stressful situation that the person has recently experienced. After describing the situation to the group, the person is asked to imagine some ways in which the situation could be worse, and some ways it could be better. The process of "focusing" (Gendlin, 1978) is sometimes used to help clients work through any emotions that may be impeding them from a close examination of the stressful situation and their own reactions to it. This application of situational reconstruction and focusing leads to a discussion of the person's actual coping responses, as well as other responses that might be possible. The hardiness trainer guides the client to consider more proactive or "transformational" coping behaviours, directly addressing the source of the stress (Maddi, Kahn, & Maddi, 1998). A third technique used in hardiness induction training, "*compensatory self-improvement*", is based on Adler's (1956) concept that people strive to improve themselves as compensation against restrictions and constraints commonly experienced in life. In hardiness induction training, clients are encouraged to work on some self-improvement activity whenever confronted by stressful conditions that cannot be changed or readily controlled. This is believed to re-establish a sense of control, a core feature of the hardiness response pattern.

Hardiness induction training has since been further elaborated by Maddi and colleagues to include instruction regarding additional factors that can influence healthy and unhealthy reactions to stress, such as nutrition, exercise, social support and constitutional background (Khoshaba & Maddi, 2001; Maddi, 2002).

Empirical Support

Several studies have shown promising increases in hardiness and a variety of health and performance benefits following such training (Maddi, 1987; Maddi et al., 1998; Maddi, Harvey, Khoshaba, Fazel, & Resurreccion, 2009; Maddi, Khoshaba, Jensen, Carter, Lu, and Harvey, 2002). Nevertheless, there remains some question regarding how permanent are the training effects. Maddi (1987) and also Rowe (1998) suspect that periodic re-training is needed in order to maintain the increases in hardiness that result from this kind of training.

A similar classroom-type approach to hardiness training was developed by Judkins and Ingram (2002). Their training programme, aimed mainly at nurses and nurse managers, includes providing information about hardiness, while also analysing case studies with an emphasis on detecting threats and coping strategies. Results of the training showed that levels of hardiness had increased by the time managers had finished the training modules. Tierney and Lavelle (1997) also developed a classroom-based hardiness training programme for staff nurses. Theirs is a six-hour educational course that includes an introduction of key hardiness concepts, identification of significant stressors, role playing and group feedback. Results have also been positive, with significant increases in hardiness observed following course completion (Tierney & Lavelle, 1997). Similar results were reported in a study by Maddi et al. (2009), in which college students took a special class on hardiness concepts, and how to employ effective coping strategies. This training resulted in hardiness increases, and also was related to improved academic performance as measured by course grades (GPA: Grade Point Average).

Judkins, Reid and Furlow (2006) followed up on their earlier work with a more refined programme to train hardiness in nurse managers. The training includes basic information about hardiness, stress management concepts, positive communication strategies, adaptive coping, conflict management and problem-focused resolution. Results have been positive, with significant increases in hardiness that persisted up to six months post-training. These hardiness increases were also related to lower levels of turnover in this group of nurse managers, despite high levels of organizational stress. But, these results must be viewed as tentative due to the small number of research participants. Also, like other approaches, considering that the training includes such varied material and methods, it is difficult to ascertain which features of the training are effective in increasing hardiness and improving outcomes, and which are not effective.

On the whole then, direct hardiness training interventions have shown modest success in increasing levels of hardiness and bringing a range of health and performance benefits. But still some questions remain. It is uncertain how effective these interventions are in producing lasting positive effects, and related to this, what type of reinforcement may be needed to sustain increases in hardiness. Some of the training programmes have assessed changes immediately following the training, raising the possibility that the increases found are learning effects that occur because of test familiarization. For example, Tierney and Lavelle (1997) found that hardiness levels increased immediately after the intervention, but returned to baseline levels six months later. To produce lasting effects, this type of hardiness training may need to include regular follow-ups and re-training over an extended period of time.

Leader Influence on Group Hardiness

In contrast to direct training, another approach to increasing hardiness-resilient responding takes advantage of the powerful influence that leaders can have in groups. Leaders in organizations who themselves are high in hardiness can exert a positive influence on how the entire workforce responds to stress. This happens in multiple ways, but largely through the example leaders set in responding to stressful circumstances (Bartone, 2006). If leaders also have a basic understanding of the underlying dynamics of hardiness, they can be even more effective in influencing hardy responses to stress. Many authors have commented on how social processes can influence the creation of meaning by individuals. Examples include Berger and Luckmann (1967) on the "social construction of reality", Janis (1972) on "groupthink", and Weick (1995) on the process of "sensemaking in organizations". Gordon Allport (1937), the distinguished personality psychologist, viewed individual meaning as often largely the result of social influence processes.

A key aspect of the hardiness resiliency mechanism thus involves the interpretation or meaning that people attach to events around them and to their own place in the world. Recall that high hardy people tend to interpret experience as (a) interesting and worthwhile, (b) something they can exert control over and (c) challenging, providing opportunities to learn and grow. If stressful or painful experiences can be cognitively framed and made-sense-of within a broader perspective, one who sees all of existence as essentially interesting, worthwhile, fun, a matter of personal choice and providing chances to learn and grow, then the stressful experience can actually have beneficial psychological effects rather than harmful ones.

In organized work groups, including police and military organizations, this "meaning-making" process is something that leaders can influence fairly directly. For example, military units by their nature are group-oriented, hierarchical and highly interdependent. The same is true for most police and security organizations. The typical tasks and missions are group ones and the hierarchical authority structure enables leaders to exercise considerable control and influence over subordinates. Through the policies and priorities they set, the directives they give, the advice and counsel they offer, the stories they tell and especially by their own example, leaders can begin to shift the "mental models" of their subordinates (Saus, Espevik, & Eid, 2010), and thus the manner in which experience gets interpreted. The influence of hardy leaders is likely even greater under high-stress conditions, when the group tends to focus on leaders even more for guidance, and outcomes are critical. In myriad ways, these leaders encourage subordinates to interpret stressful events as interesting challenges which they are capable of meeting, and in any case which they can learn and benefit from. This process results in a shared understanding of the stressful event as something worthwhile and beneficial, while incidentally generating higher group cohesion.

Several studies support the notion that leaders may influence subordinates to think and behave in more hardy or resilient ways. The idea that hardiness is linked to meaning-making is supported by a study of US soldiers deployed to Bosnia (Britt, Adler, & Bartone, 2001), which found that hardiness levels influenced perceptions that the deployment work was meaningful, and was also associated with positive benefits. Along the same lines, McNeese-Smith (1997) found that nurse managers who actively cultivated characteristics of hardiness in their organizations had employees with significantly higher job satisfaction, productivity and organizational commitment, and also showed fewer stress-related problems.

Several leadership theories may help to understand how leaders who are high in psychological hardiness may have positive influence on their subordinates. Transformational leadership theory (Bass, 1998; Burns, 1978) emphasizes the importance of "inspirational motivation" for stimulating extra effort and performance in work groups. According to Bass and Avolio (1994, p. 3)

> Transformational leaders behave in ways that motivate and inspire those around them by providing meaning and challenge to their followers' work. Team spirit is aroused. Enthusiasm and optimism are displayed. The leader gets followers involved in envisioning attractive future states. The leader clearly communicates expectations that followers want to meet and also demonstrates commitment to goals and the shared vision.

Thus, transformational leadership is believed to work in part through some process, whereby leaders generate an increased sense of meaning, commitment and challenge amongst their subordinates. The positive meaning-making influence of high-hardy leaders corresponds closely to the "inspirational motivation" aspect of transformational leadership. In a relevant paper, Gal (1987) maintains that what makes transformational leaders effective comes down to their ability to somehow increase the overall commitment levels of their subordinates. The research on hardiness and leader performance summarized above suggests that leaders who are high in hardiness may be especially skilled at building up this sense of commitment in subordinates, and further suggests that how experiences get interpreted (interpretations shaped by leaders) is a critical part of the process.

Increasing commitment and motivation is also an important feature of "path-goal" leadership theory. Path-goal theory focuses attention on how leaders influence the motivation of subordinates by identifying significant goals, structuring situations so that subordinates experience personal rewards for goal-attainment, and clarifying the pathways for achieving these desired goals (House, 1971, 1996). Leaders are said to demonstrate supportive, directive, participative or achievement leadership depending upon their personal style and preference, as well as the contingencies of particular situations or tasks (House & Mitchell, 1974). Most relevant to the hypothesis of hardy leaders exercising positive influence is the achievement leadership orientation of path-goal theory. The achievement-oriented leader is able to tap into and even increase followers' motivation to surmount obstacles and achieve goals, and to direct this motivation toward achieving important group goals. This sounds very much like how people high in hardiness react to unexpected or highly stressful situations; they tend to interpret these situations as challenges that must be met head-on, and as opportunities to learn and grow. At the other end of the spectrum is the low-hardy, low-achievement person who sees changes more as threats or disruptions to be avoided. Path-goal leadership theory thus provides a broader framework for understanding how high-hardy leaders can influence the motivation, thinking and behaviour of subordinates. More research is needed to explore this possibility, as well as to test more directly in various groups and conditions for the influence that high-hardy leaders may have on the hardiness levels of subordinates.

Self-help Steps to Increase Hardiness

Just as leader actions and policies can lead to increases in hardy-resilient response patterns, there are also things that soldiers can do that over time may increase their own hardiness. Current hardiness theory identifies three inter-related core features or facets of hardiness: commitment, control and challenge.

HARDINESS—COMMITMENT

Commitment is all about active involvement and engagement in one's activities and the surrounding world, as well as a sense of competence and self-worth. At the opposite pole of commitment is alienation, or meaninglessness. Considering this, here are a few simple steps to help build up hardiness—commitment:

1. Take some time to think about what is important and interesting to you… your personal values and goals.
2. Work on increasing your skills and competence in some area that is important to you.
3. Take pride in your past successes and achievements.
4. Remember the good things in your life… count your blessings!
5. Spend time with family, friends, people you care about.
6. Pay attention to what's going on in the world around you, read, observe.
7. Try out new things.

HARDINESS—CONTROL

Hardiness—control is the belief that you can control or influence what is happening and what is going to happen. The opposite of control is a sense of powerlessness or helplessness to do anything that will make a difference. With this in mind, here are some steps to increase your sense of control:

1. Focus your time and energy on things you can control or influence.
2. Work on tasks that are within your capabilities, moderately difficult but not overwhelming.
3. For difficult jobs, break them up into manageable pieces so you can see the progress.
4. Plan ahead, and gather up the right tools and resources for the task
5. Ask for help when you need it.
6. Recognize your successes.
7. When you just cannot solve a problem…. Turn your attention to other things you can control!

HARDINESS—CHALLENGE

Hardiness—challenge is a positive perspective on change and variety in life. People high in challenge tend to take changes in stride, see variety as part of the richness of life, and are optimistic about the future. On the opposite pole, people low in challenge are always seeking security, want everything to be simple and predictable and are fearful of the future. Considering this, here are some steps to build challenge:

1. Remind yourself that no matter what happens, change is always an opportunity to learn and get better.
2. Do not live every day by a rigid schedule… allow for variation and surprises.
3. Be willing to change your plans to meet changing conditions.
4. Whenever you do fail at something, ask: what can I learn from this?
5. Try out new things, take reasonable risks.
6. Use your imagination to think about future positive outcomes.
7. Do not dwell on past disappointments…. learn, forgive and look ahead.

These self-help steps for increasing one's hardiness sound simple, and in some ways they are. There is nothing complicated about it. The difficulty really comes in breaking down old habits and ways of looking at the world and oneself, mind sets that leave a person more vulnerable to stress. When the attitudes of commitment, control and challenge are strong, individuals tend to see stressful situations as manageable, and so less stressful. They look for solutions, take actions where they can to cope positively with problems, and accept setbacks as a normal part of living. In other words, they are more resilient. For the soldier who wants to be more resilient, it is believed that the above self-help steps for increased hardiness can help.

Future Directions in Hardiness Training

Based on this brief review, training soldiers to be more resilient under stress would seem to be possible, but is not likely quick and easy. For most soldiers, by the time they join services they have well-established patterns for responding to stressful or changing conditions in life. Learning new and different response patterns is not just a matter of acquiring new knowledge about how to respond in more healthy ways. Instead, developing resilience involves the more difficult task of replacing entrenched, characteristic response patterns or habits with new ones. So any training for resilience that focuses just on content— information is not likely to have much impact.

There is now reasonably good evidence that direct training and edu- cation programmes work to increase individual hardiness. A key feature of the hardiness resiliency mechanism involves the interpretation and the meaning that gets attached to events. High-hardy people typically interpret experience as follows: overall interesting and worthwhile (commitment), something they can exert control over (control) and challenging opportunities to learn and grow (challenge). Thus, the key to creating and sustaining increases in hardiness is to help soldiers change their "mental models" or lenses through which they see the world, to develop a broader perspective of stressful circumstances and find other ways of understanding themselves and their experience. So training programmes need to find effective ways to accomplish this.

Another approach to increasing hardiness involves shaping or structuring the organizational environment in ways that encourage hardy responses in the workforce. As previously discussed, leaders who themselves are high in hardiness can have positive influence on their followers. Leaders can have a dual role in that they lead by example and act as mentors, and at the same time implement policies that will enhance and maintain hardiness in their subordinates.

Finally, a promising avenue for future development involves incor- porating knowledge and ideas from the hardiness framework into exist- ing selection and training programmes for security and police forces. In one relevant study, Zach, Raviv and Inbar (2007) examined military officers undergoing a rigorous selection and training programme for Special Forces security officers. The training course lasted nine weeks and included training and simulation of real-life events under gradu- ally increasing stressful conditions (e.g., running an obstacle course, overpowering terrorists and hostage-taking incidents). The results showed that individual hardiness levels increased as a result of the training course. Similarly, Norwegian officer candidates were seen to increase in hardiness over the time period of the stressful selection and training course, even though increasing hardiness was not a goal of the course (Eid, personal communication). Later interviews with trainers in the officer candidate course confirmed that on an implicit level, the qualities of hardiness are valued in candidates and there is some effort to boost these during training. These studies suggest that a potentially fruitful approach to hardiness training would involve adjusting existing training programmes in ways that emphasize more explicitly the qualities associated with hardiness.

Conclusion

The military is a high-stress, high-risk occupation. As such, it is important to select and train resilient people who can continue to function effectively and remain healthy under stress. Psychological hardiness presents one promising pathway to resilience, and there is evidence that it can be trained and also increased through leadership and organizational policies. It remains now for researchers, practitioners and leaders to apply what is known about resilient, hardy responding to stress; to the design of enhanced training programmes and organizational policies for various groups and environments.

References

Adler, A., (1956). *The individual psychology of Alfred Adler*, H. L. Ansbacher & R. R. Ansbacher (Eds.). New York, NY: Harper Torchbooks.

Alford, J. D., & Cuomo, S. A. (2009). Operational design for ISAF in Afghanistan: A primer. *Joint Force Quarterly, 53*, 92–98.

Allport, G.W. (1937). *Personality: A psychological interpretation.* New York, NY: Holt.

Bartone, P. T. (1989). Predictors of stress-related illness in city bus drivers. *Journal of Occupational Medicine, 31*, 657–663.

———. (1993, June). *Psychosocial predictors of soldier adjustment to combat stress.* Paper presented at the Third European Conference on Traumatic Stress, Bergen, Norway.

———. (1995, July). *A short hardiness scale.* Paper presented at the Annual Convention of the American Psychological Society, New York, NY.

———. (1996, August). *Stress and hardiness in US peacekeeping soldiers.* Paper presented at the convention of the American Psychological Association, Toronto, ON, Canada.

———. (1999). Hardiness protects against war-related stress in Army Reserve forces. *Consulting Psychology Journal, 51*, 72–82.

———. (2000). Hardiness as a resiliency factor for United States forces in the Gulf War. In J. M. Violanti, D. Paton, & C. Dunning (Eds.), *Posttraumatic stress intervention: Challenges, issues, and perspectives* (pp. 115–133). Springfield, IL: C. Thomas.

Bartone, P. T. (2006). Resilience under military operational stress: Can leaders influence hardiness? *Military Psychology, 18,* s131–s148.

Bartone, P. T., Adler, A. B., & Vaitkus, M. A. (1998). Dimensions of psychological stress in peacekeeping operations. *Military Medicine, 163,* 587–593.

Bartone, P. T., & Hystad, S. W. (2010). Increasing mental hardiness for stress resilience in operational settings. In P. T. Bartone, B. H. Johnsen, J. Eid, J. M. Violanti, & J. C. Laberg (Eds.), *Enhancing human performance in security operations* (pp. 257–272). Springfield, IL: Charles C. Thomas.

Bartone, P. T., Johnsen, B. H., Eid, J., Brun, W., & Laberg, J. C. (2002). Factors influencing small unit cohesion in Norwegian Navy officer cadets. *Military Psychology, 14,* 1–22.

Bartone, P. T., Ursano, R. J., Wright, K. M., & Ingraham, L. H. (1989). The impact of a military air disaster on the health of assistance workers: A prospective study. *Journal of Nervous and Mental Disease, 177,* 317–328.

Bass, B. M. (1998). *Transformational leadership: Industry, military, and educational impact.* Mahwah, NJ: Erlbaum Associates.

Bass, B. M., & Avolio, B. J. (1994). Introduction. In B. M. Bass & B. J. Avolio (Eds.), *Improving organizational effectiveness through transformational leadership* (pp. 1–9). Thousand Oaks, CA: SAGE Publications.

Berger, P. L., & Luckmann, T. (1967). *The social construction of reality.* London, UK: Penguin.

Britt, T. W., Adler, A. B., & Bartone, P. T. (2001). Deriving benefits from stressful events: The role of engagement in meaningful work and hardiness. *Journal of Occupational Health Psychology, 6,* 53–63.

Burns, J. M. (1978). *Leadership.* New York, NY: Harper & Row.

Carver, C. S. (1998). Resilience and thriving: Issues, models, and linkages. *Journal of Social Issues, 54,* 245–266.

Contrada, R. J. (1989). Type A behavior, personality hardiness, and cardiovascular responses to stress. *Journal of Personality and Social Psychology, 57,* 895–903.

Florian, V., Mikulincer, M., & Taubman, O. (1995). Does hardiness contribute to mental health during a stressful real life situation? The role of appraisal and coping. *Journal of Personality and Social Psychology, 68,* 687–695.

Friborg, O., Barlaug, D., Martinussen, M., Rosenvinge, J., & Hjemdal, O. (2005). Resilience in relation to personality and intelligence. *International Journal of Methods in Psychiatric Research, 14,* 29–42.

Friborg, O., Hjemdal, O., Martinussen, M., & Rosenvinge, J. (2009). Empirical support for resilience as more than the counterpart and absence of vulnerability and symptoms of mental disorder. *Journal of Individual Differences, 30,* 138–151.

Gal, R. (1987). Military leadership for the 1990s: Commitment-derived leadership. In L. Atwater & R. Penn (Eds.), *Military leadership: Traditions and future trends* (pp. 53–59). Annapolis, MD: US Naval Academy.

Garmezy, N. (1971). Vulnerability research and the issue of primary prevention. *American Journal of Orthopsychiatry, 41,* 101–116.

———. (1974). The study of competence in children at risk for severe psychopathology. In E. J. Anthony & C. Koupernik (Eds.), *The child in his family: Children at psychiatric risk* (Vol. 3, pp. 77–97). New York, NY: Wiley.

Gendlin, E. T. (1978). *Focusing.* New York, NY: Bantam Dell.

House, R. J. (1971). A path-goal theory of leader effectiveness. *Administrative Science Quaterly, 16,* 321–338.

———. (1996). Path-goal theory of leadership: Lessons, legacy, and a reformulated theory. *Leadership Quarterly, 7,* 323–352.

House, R. J., & Mitchell, T. R. (1974). Path-goal theory of leadership. *Contemporary Business, 3*, 81–98.

Hystad, S. W., Eid, J., Johnsen, B. H., Laberg, J. C., & Bartone, P. (2010). Psychometric properties of the revised Norwegian Dispositional Resilience (hardiness) Scale. *Scandinavian Journal of Psychology, 51,* 237–245.

Janis, I. L. (1972). *Victims of groupthink: A psychological study of foreign policy decisions and fiascoes.* Boston, MA: Houghton Mifflin.

Judkins, S. K., & Ingram, M. (2002). Decreasing stress among nurse managers: A long term solution. *The Journal of Continuing Education in Nursing, 33,* 259–264.

Judkins, S. K, Reid, B., & Furlow, L. (2006). Hardiness training among nurse managers: Building a healthy workplace. *The Journal of Continuing Education in Nursing, 37,* 202–207.

Khoshaba, D. M., & Maddi, S. R. (2001). *HardiTraining* (4th ed.). Newport Beach, CA: Hardiness Institute.

Kobasa, S. C. (1979). Stressful life events, personality and health: An inquiry into hardiness. *Journal of Personality and Social Psychology, 37,* 1–11.

Kobasa, S. C., Maddi, S. R., & Kahn, S. (1982). Hardiness and health: A prospective study. *Journal of Personality and Social Psychology, 42,* 168–177.

Layne, C. M., Warren, J. S., Watson, P. J., & Shalev, A. Y. (2007). Risk, vulnerability, resistance, and resilience: Toward an integrative conceptualization of posttraumatic adaptation. In M. J. Friedman, T. M. Keane, & P. A. Resick (Eds.), *Handbook of PTSD: Science and practice* (pp. 497–520). New York, NY: Guilford.

Luthar, S. S. (2006). Resilience in development: A synthesis of research across five decades. In D. J. Cohen & D. Cicchetti (Eds.), *Developmental psychopathology: Risk, disorder, and adaptation* (pp. 739–795). Hoboken, NJ: Wiley.

Maddi, S. R. (1987). Hardiness training at Illinois Bell Telephone. In J. P. Opatz (Ed.), *Health promotion evaluation.* Stevens Point, WI: National Wellness Institute.

———. (1999). The personality construct of hardiness: Effects on experiencing, coping, and strain. *Consulting Psychology Journal, 51,* 83–94.

———. (2002). The story of hardiness: Twenty years of theorizing, research and practice. *Consulting Psychology Journal, 54,* 173–185.

Maddi, S. R., & Harvey, R. H. (2006). Hardiness considered across cultures. In P. T. P. Wong & L. C. J. Wong (Eds.), *Handbook of multicultural perspectives on stress and coping* (pp. 409–426). New York, NY: Springer.

Maddi, S. R., Harvey, R. H., Khoshaba, D. M., Fazel, M., & Resurreccion, N. (2009). Hardiness training facilitates performance in college. *The Journal of Positive Psychology, 4,* 566–577

Maddi, S. R., Kahn, S., & Maddi, K. L. (1998). The effectiveness of hardiness training. *Consulting Psychology Journal, 50,* 78–86.

Maddi, S. R., & Khoshaba, D. M. (2005). *Resilience at work.* New York, NY: Amacom.

Maddi, S. R., Khoshaba, D. M., Jensen, K., Carter, E., Lu, J. L., & Harvey, R. H. (2002). Hardiness training for high-risk undergraduates. *The Journal of the National Academic Advising Association, 22,* 45–55.

Maddi, S. R., & Kobasa, S. C. (1984). *The hardy executive.* Homewood, IL: Jones-Irwin.

McNeese-Smith, D. K. (1997). The influence of manager behavior on nurses' job satisfaction, productivity, and commitment. *Journal of Nursing Administration, 27,* 47–55.

Priest, R., & Bartone, P. T. (2001, July). *Humor, hardiness and health.* Paper presented at the Annual Conference of the International Society for Humor Studies, College Park, MD.

Reivich, K., & Shatte, A. (2002). *The resilience factor.* New York, NY: Broadway Books.

Roth, D. L., Wiebe, D. J., Fillingim, R. B., & Shay, K. A. (1989). Life events, fitness, hardiness, and health: A simultaneous analysis of proposed stress-resistance effects. *Journal of Personality and Social Psychology, 57,* 136–142.

Rowe, M. (1998). Hardiness as a stress mediating factor of burnout among healthcare providers. *American Journal of Health Studies, 14,* 16–20.

Saus, E. R., Espevik, R., & Eid, J. (2010). Situational awareness and shared mental models: Implications for training in security operations. In P. T. Bartone, B. H. Johnsen, J. Eid, J. M. Violanti, & J. C. Laberg (Eds.), *Enhancing human performance in security operations* (pp. 161–178). Springfield, IL: Charles C. Thomas.

Siebert, A. (2005). *The resiliency advantage.* San Francisco, CA: Berret-Koehler.

Sinclair, R. R., & Tetrick, L. E. (2000). Implications of item wording for hardiness structure, relation with neuroticism, and stress buffering. *Journal of Research in Personality, 34,* 1–25.

Sroufe, L. A. (1997). Psychopathology as an outcome of development. *Development and Psychopathology, 9,* 251–268.

Tierney, M. J., & Lavelle, M. (1997). An investigation into modification of personality hardiness in staff nurses. *Journal of Nursing and Staff Development, 13,* 212–217.

Violanti, J. M., & Paton, D. (1999). *Police trauma: Psychological aftermath of civilian combat.* Springfield, IL: Charles C Thomas.

Watson, P. J., Ritchie, E. C., Demer, J., Bartone, P., & Pfefferbaum, B. J. (2006). Improving resilience trajectories following mass violence and disaster. In E. C. Ritchie, P. J. Watson, & M. J. Friedman (Eds.), *Interventions following mass violence and disasters: Strategies for mental health practice* (pp. 37–53). New York, NY: Guilford.

Weick, K. E. (1995). *Sensemaking in organizations.* Thousand Oaks, CA: SAGE Publications.

Werner, E. E., Bierman, J. M., & French, F. E. (1971). *The children of Kauai.* Honolulu: University of Hawaii Press.

Werner, E. E., & Smith, R. S. (2001). *Journeys from childhood to midlife: Risk, resilience, and recovery.* Ithaca, NY: Cornell University Press.

Westman, M. (1990). The relationship between stress and performance: The moderating effect of hardiness. *Human Performance, 3,* 141–155.

Wiebe, D. J. (1991). Hardiness and stress moderation: A test of proposed mechanisms. *Journal of Personality and Social Psychology, 60,* 89–99.

Witmer, T. A. P., & Culver, S. M. (2001). Trauma and resilience among Bosnian refugee families: A critical review of the literature. *Journal of Social Work Research, 2,* 173–187.

Zach, S., Raviv, S., & Inbar, R. (2007). The benefits of a graduated training program for security officers on physical performance in stressful situations. *International Journal of Stress Management, 14,* 350–369.

12

Promoting Psychosocial Health of Disaster First Responders

Sujata Satapathy

Disaster response is amongst the prime functions conducted by military and paramilitary forces under the category of Military Operations Other Than War (MOOTW). One of the crucial yardsticks for assessing the success of disaster management is the quick and effective rescue and response operations immediately after any disaster. Trained, skilled and experienced professionals during this period play a critical role in disaster preparedness and management, especially in limiting the death toll and injury profile. The broad and diverse category of disaster workers (DWs) work from an early phase till the rehabilitation phase of the disaster management. And within this broad category, a specific group referred to as disaster first responders (DFRs) render their services during the immediate or early phase of the disaster management. DFRs are the individuals who provide services in the immediate aftermath of a disaster and may include primarily firefighters, search and rescue teams, ambulance drivers, medical personnel, local community and the local administration. Both represent a variety of disciplines and are generally used interchangeably in the literature. Both the groups face the possibility of physical harm from environmental and other exposures during performing various job tasks, but the DFRs are generally at higher risks due to rescue and evacuation operations in dangerous circumstances.

This chapter focuses on promoting the psychosocial health of professionals, especially service personnel from military and paramilitary forces who provide services during various disasters and developmental activities.

Indian Scenario

In India, the Disaster Management Act (DM Act, 2005) has made the statutory provisions for constitution of National Disaster Response Force (NDRF) for the purpose of specialized response to natural and man-made disasters. Accordingly, in 2006, NDRF was constituted with 08 battalions and currently, is having strength of 10 battalions and each consists of 1,149 personnel. Along with NDRF, the group of DFR also comprises people from Civil Defence (the Civil Defence Act was suitably amended to include the disaster management as an additional role for the Corps), Home Guards (the key role of Home Guards is to serve as an auxiliary to the police in maintenance of internal security, help the community in any kind of emergency such as an air-raid, fire, cyclone, earthquake and epidemic), fire safety officers, emergency health professionals, mental health workers, local communities and government. The NDRF battalions have been involved in many crucial disaster events such as cyclone, floods/flash floods, cloud burst, earthquakes, train and bus accidents, boat capsize and collapsed structures in the last few years. The media and various organizations repeatedly reported the successful rescue and response operation carried out by the deployed teams, which reflected their physical fitness to withstand the climatic variation and task difficulty, skills in task operation, commitment and dedication to the people and skills to use technological support in rescue and response.

A glimpse about the work profile of the DFRs in India can be obtained from the brief case studies in Box 12.1.

Work Profile of DFRs

The work profile of the DFRs generally includes undertaking rescue and evacuation operation, retrieving dead bodies from debris, cutting trees and removing electric poles, managing dangerous chemicals, constructing make-shift roads, providing medical care and essential services in the immediate aftermath of disasters. They generally remain in the disaster-affected area for weeks or months, often have to work long hours under stressful conditions and witness death, serious physical injuries, physical destruction and psychological devastation

Box 12.1
Brief case studies of DFRs in India

Uttarakhand Flash Flood, 2013

Due to heavy rain in Kedarnath valley and resulting severe flash floods on 16 June 2013, there was massive flash flood in the state of Uttarakhand in the northern part of India. Immediately 14 SAR teams (10 SAR teams of 08 NDRF and 04 SAR teams of 07 NDRF battalion) were rushed to various affected areas, viz., for rescue operation. NDRF teams did commendable job in adverse weather conditions and rescued about 9,657 people and retrieved 306 dead bodies, apart from providing medical aid to 920 needy persons and other essential assistance to the thousands of stranded people. They also recovered INR.1,16,37,683.50, and valuable objects from plunderers and handed over to state authorities.

Building Collapse in 2013

After the collapse of a seven-storey building at Shil Mumbra, District, Thane, Maharashtra, a joint operation was conducted by NDRF with other agencies with effect from 4 April 2013 to 6 April 2013 under the supervision of the Commandant, 05 Bn NDRF and rescued 62 live victims and retrieved 72 dead bodies from the debris.

During the year 2013, many incidents of building collapse took place in Andhra Pradesh, Maharashtra, Karnataka and Gujarat. NDRF rescued total 154 precious lives and retrieved 181 dead bodies during rescue operations.

Chemical Disaster, 2013

Team of NDRF was deployed with effect from 11 February 2013 to 12 February 2013 to respond to Leakage of Gas at Kolar, Karnataka where explosion in a load carrier carrying liquid carbon dioxide gas cylinders took place on Kolar-Chennai Highway. Further, another team was deployed with effect from 18 February 2013 to 19 February 2013 at Indira colony, Barakpore (WB) in c/w gas leakage incident where during operation team detected hot water vapour coming out from underground water supply pipe due to electric short circuit. Also 01 NDRF Team was deployed with effect from 21 April 2013 to 22 April 2013 to Eluru, district West Godavari, Andhra Pradesh, in c/w accident of a tanker truck containing Amonia liquid gas. Team neutralized Ammonia Gas.

Source: ndrfandcd.gov.in, accessed on 3 March 2014.

that can follow the disaster (DeWolfe, 2000). In addition, they may experience physical injuries or fatigue or psychological trauma due to the nature of their work. Pre-existing physical and mental health conditions may be exacerbated and new health conditions may arise due to extremely stressful working conditions that are physically and emotionally taxing. Research has highlighted on the possible effect of environmental exposures and other risk factors, such as structural instabilities within the built environment, on physical health of the first responders (Wheeler et al., 2007).

Psychosocial Health of DFRs

Disasters always do not have negative impacts on mental health of the DFRs. Some studies on the DFRs revealed that they find their work rewarding (Shih et al., 2002; Thoresen et al., 2009). Another study revealed that the Norwegian aid workers found their experiences after the Tsunami in 2004 a meaningful, successful and valuable personal experience (Thoresen et al., 2009).

However, disasters have also been defined as situations of massive, collective stress (Burkle, 1996). Case Study 12.2 enlists major psychosocial stresses related to disaster at operational, personal, organizational and sensation levels. The nature of job can expose them to the most gruesome sights and smells in the disaster-affected area. Even though they are trained as policemen, fire fighters, ambulance drivers, NDRF, home guards, health professionals, district administrators, etc., the impact of coming in contact with painful experiences (when this is multiplied by 100s or 1000s of bodies that have to be disposed of) could be severe. In addition, factors such as fatigue, intense dedication to the task, reluctance to be relieved from duty or taking a short break (Cohen, 2002) should be considered. Some specific situations such as overwhelming volume of task and demands, suicide/nervous breakdown/death of/injury to a co-worker (e.g., death of co-workers following a chopper crash during rescue operation during Uttarakhand floods in 2013), witnessing death, distress and injury all around, lack of logistic support, etc., may increase the vulnerability of first responders to traumatic stress-related mental health conditions. The presence of mental health conditions may significantly affect first responders' ability to function (Benedek et al., 2007) either during the disaster operation or subsequently after completion of disaster task operation. Therefore, highlighting the sources of stress for the DWs would help in planning appropriate capacity building and preventive mental health programmes (Box 12.2).

Box 12.2
Sources of stress for disaster and emergency first responders

Operational	*Personal*
• Long working hours, over worked/ overwhelming responsibilities • Lack of a clear job description/lack of understanding about the work • Lack of prior experience or inadequate skills to perform optimally • Poor communication with management • Working in unsafe and insecure areas • Environment/weather is rough and demanding	• Personal safety, family concerns • Lack of adequate sleep/rest or time to have food • Impending or existing health issues • Compassion fatigue • Not being able to share the over-whelming experiences/emotions with others • Lack of relaxation activities
Organizational	*Sensation*
• At times moral and ethical dilemmas • Feeling frustrated with the ongoing situations or decisions by others • Team conflicts • Feeling helpless looking at the huge needs • Being scolded by seniors or ridiculed by the colleagues	• Witnessing or even directly experiencing terrible things, such as destruction, injury, death or violence. They may also hear stories of other people's/ co-worker pain and suffering • Horrific and ghastly images and smell • Trapped in the disaster itself

Source: Adapted from Centre for Mental Health Services (CMHS), 2005.

Trauma and Traumatic Stress

As per the American Psychiatric Association (APA, 2000), DSM-IV, a traumatic event, is defined as experiencing *a threat* (actual or perceived) *of death or serious injury to self or others*, with a response of "intense fear, helplessness or horror". However, it is not the event itself, but the perception and meaning attached to it that makes it traumatic. In traumatic situations, individual experiences an immediate threat to oneself or to others, often followed by serious injury or harm. A range of reactions, such as anxiety, fatigue, irritability, hyper-alertness, increased emotionality, disturbed sleep, change in appetite, feeling overwhelmed, sadness, reduced social interaction, may be experienced after a disaster or any other traumatic event. These powerful, distressing emotions go along with strong even frightening physical reactions, such as rapid heartbeat, trembling and stomach dipping.

For reasons that are basic to survival, traumatic experiences, long after they are over, continue to take priority in our thoughts, emotions and behaviour. Intense fears and other strong emotions, intense physical reactions and the new way of looking at dangers in the world may recede into the background, but events and reminders may bring them to mind again. Thus, feeling intensely threatened by the traumatic event a person experienced or witnessed, then that event is called a trauma and the reactions described above are referred to as traumatic stress reactions.

During and up to four weeks of the disaster, they may exhibit the symptoms of acute stress reactions. Within two to six months of the disaster operation, they may suffer from depression, anxiety and adjustment disorders and after six months, there could be delayed manifestations of post-traumatic stress disorder (PTSD) (symptoms of acute stress reaction lasting for more than one month). As an effect of psychological trauma, PTSD is less frequent and more enduring than the more commonly seen acute stress disorders. Studies have demonstrated that, after participating in disaster responses, first responders experience elevated rates of depression, stress disorders and PTSD for months and sometimes years (Stellman et al., 2008). Those without disaster response training face a greater risk of receiving a PTSD diagnosis after the response concludes (Perrin et al., 2007).

The study on the effect of an explosion of fireworks depot in the Netherlands on the rescue workers revealed that sick leave among the workers increased substantially during 18 months after the explosion (Morren et al., 2007). Another study revealed 27 per cent of professionals who work with traumatized victims experienced extreme distress (Meldrum et al., 2002). The prevalence of PTSD among fire fighters exposed to the 11 Sepember disaster was 9.8 per cent in the first year after the disaster and this prevalence increased to 10.6 per cent in the fourth year after the disaster (Berninger et al., 2010).

Theoretical Models: Care-Givers' Trauma

Intensely traumatic events can lead to acute and long-term mental-health consequences among the DFRs, if not timely addressed.

This can range from secondary traumatic stress or compassion fatigue to VT to burnout. Many researchers in the past two decades accentuated on the linkages of exposure to pain, suffering and trauma with the physical and mental health of professionals providing care, especially nurses (Abendroth & Flannery, 2006; Adams et al., 2006; Sabo, 2006).

Thus, research works revealed that due to the impact of overtaxing work done especially in the first few days or even in the first phase of a mega disaster, the FRs and the survivors exhibit similar symptoms. The list of signs and symptoms enumerated here is not exhaustive but an indicative one:

1. Difficulty in concentrating
2. Intrusive imagery
3. Lack of interest, hopelessness
4. Exhaustion
5. Irritability and anger with trivial issues
6. Diminished sense of personal accomplishment
7. Overly high expectations of self or others
8. Inability to maintain balance of empathy and objectivity
9. Decreased ability to feel happy
10. Never taking time off for own self
11. Blaming behaviour
12. Psychosomatic symptoms of stress
13. Using/abusing drugs, alcohol and other substances
14. High attrition (changing the profession/ the field).

However, the adaptive or maladaptive bio-psycho-social reactions exhibited by DFRs vary in number, intensity, severity and duration, essentially because of three key factors, viz., disaster (type, magnitude, severity, impact), personal (age, sex, physical and mental-health condition, personality, coping skills) and external (organizational support and support received from the local community). Research works in this field revealed proximity, duration and intensity of exposure (Benedek et al., 2007), and negative outcomes of such events such as traumatic stress symptoms (Marmar et al., 1999), secondary traumatic stress or compassion fatigue (Figley, 1995; 1999) and burnout (Alexander & Klein, 2001, 2009) as the most significant predictors of mental health.

Secondary Traumatic Stress

Secondary traumatic stress is the experience of trauma symptoms (acute or post-traumatic) in mental-health workers as a result of and in relationship to their exposure to the trauma material of survivors. Raised arousal level, frequent nightmares, intrusive thoughts, distressing images and feelings, avoiding talking about the incident or experience are few of the key signs of secondary traumatic stress.

Compassion Fatigue (CF)

CF is described as the profound effect of intense care giving professions on the mental-health conditions of the caregivers (Figley, 1995). This dominant theoretical model postulating the emergence of compassion fatigue is based on a stress-process framework (Adams et al., 2006). CF refers to the profound emotional and physical erosion that takes place when helpers are unable to refuel and regenerate. They can be affected either directly through exposure to traumatic events (e.g., working as an ambulance driver, police officer, a doctor, a mental-health worker); or indirectly through secondary exposure (hearing survivors talk about trauma they have experienced, helping people who have just been victimized, working with child survivors, working with survivors who are chronically in despair, witnessing people's inability to improve their difficult life circumstances/feeling helpless in the face of poverty and encountering emotional anguish in the survivors). A DFR rescuing people from the severe flash floods may feel incredibly drained out, fatigued, exhausted, unable to serve with compassion. Despite very high job satisfaction, the direct encounter with tough and dreadful circumstance may result in reliving the scenes again and again after the task completion.

Vicarious Trauma (VT)

VT refers to the profound shift that workers experience in their world view when they work with disaster survivors who have experienced trauma. Helpers notice that their fundamental beliefs about the world are altered and possibly damaged by being repeatedly exposed to traumatic material. For example, a social worker working with child trafficking or child sexual abuse may be secondarily traumatized and deeply disturbed by the pathetic background and emotional turmoil of the children. This may, in turn, affect his/her sex life, or feelings of safety for own children or interests in pleasurable activities at home. The worker may not, however, feel too exhausted to interact with people at home or office, but repeated thoughts may disturb the worker's efficiency in performing various tasks.

DFRs can simultaneously experience compassion fatigue and VT. The concept of VT has precisely outlined the key issues of identifying the experience and manifestation of VT. According to this theory, the five components of self, namely, frame of reference, ego resources, psychological needs and cognitive schemas, memory and perception, can be potentially affected by exposure to traumatic experiences or materials. These experiences and reactions, if not addressed timely, adequately and appropriately can lead to serious mental-health problems.

Burnout

The term burnout refers to the physical and emotional exhaustion that workers can experience when they have low job satisfaction and feel powerless and overwhelmed at work. Burnout does not necessarily mean that our view of the world has been damaged, or that we have lost the ability to feel compassion for others. Most importantly, unlike compassion fatigue and VT, burnout can be resolved with reasonably less difficulty, for example, changing jobs, taking long leave, pursuing with the authority in confidence, etc.

Resilience

Several protective factors reported against the development of traumatic stress-related mental-health conditions are receiving training to deal with situations first responders may encounter (Thoresen et al., 2009), being married (Fullerton Ursano, & Wang, 2004), self-efficacy, collective efficacy, a sense of community

(Pietrantoni & Prati, 2008), social support (Marmar et al., 2006) and resilience (Avey et al., 2009; Gupta et al., 2012; Mansfield et al., 2012; Pipe et al., 2012; Van Breda, 2011).

Planning of Preventive and Promotive Psychosocial and Mental Health Services

It is mostly conclusive that the nature of duties and tasks first responders carry out in many disasters can make them susceptible to the development or exacerbation of various mental-health conditions, especially trauma-related conditions.

This necessitates the need for the planning for any protective, preventive and promotive mental health package, which has two objectives, first, to prevent long-term mental-health impacts of disasters on the responders, and second, to promote positive mental health among them. This planning is a part of disaster mental health preparedness and hence requires intentional and comprehensive efforts.

Although psychosocial health promotion programmes are more generic in nature with the objective to emphasize on the effective management of stress and enhancing positive coping; the preventive mental health is more of early diagnosis and management of mental disorders. Both types of services aim to enhance person's bio-psycho-social, occupational and daily life functioning so as to bring it back to the pre-disaster time period or even making it better than that. These types of services are highly linked to each other, and hence should be planned as a continuum of service provision rather than separate entities. The psychosocial health promotion activities can be conducted mainly before the team deputation to the disaster-affected area or even during the operation; the mental-health programmes can be conducted after the response and relief phase, upon the task completion for the purpose of early diagnosis and treatment.

Prevention and intervention efforts should occur across three dimensions: well-being (personal process of self and task management), organizational (organizational process to facilitate task performance), psycho-education (ongoing supervision of work stress and quality of work, educating them for early identification of mental-health conditions easier and sooner).

Preventive and Curative Mental Health

The purpose of preventive and curative mental health is primarily to provide medical care for people with mental illnesses (PWML existing prior to disaster and those developed mental illness after the disaster). The programmes could begin in a group or individual mode depending upon the needs.

1. Assessment of mental health status and psychopathology.
2. Diagnostic assessment of personality, etc.
3. Assessment of compassion fatigue and VT.
4. Checking possible alcohol and substance abuse indicators.
5. Curative medical treatment by psychiatrists with regular follow-up, if needed.
6. Curative psychological therapies with regular follow-up and linkages with the psychiatrists for identified people.

Cognitive-behavioural therapies have proven effective in helping children suffering from traumatic stress. These therapies generally include the features such as teaching stress management and relaxation skills, correcting untrue or distorted cognitions about what happened and why changing unhealthy and wrong views that have resulted from the trauma, involving employer in creating optimal recovery, assessing and enhancing coping skills, etc.

1. Accessible referral mechanism.
2. Follow-up mechanism.

This should be done by a professional team comprising clinical psychologists and psychiatrists. This can be carried out even long after the disaster happened for early diagnosis and management.

Psychosocial Health Promotion

The psychosocial health promotion should be done at two levels: (a) for the first responders, and (b) for their employers.

FOR THE FIRST RESPONDERS

Disaster-oriented education and training programmes on stress management specially tailored to cater to the needs of the first responders and DWs should be developed. The various components of the training/education programme meant for the first responders and DWs are as follows:

1. Assessing stressors associated with disasters.
2. Assessing know-how of self-care (personal, professional and coping).

   i. *Self-care:* Self-recognition (when do I get stressed?), self-awareness (what works for me to reduce stress?), self-regulation (how do I manage it in between the disaster operation?), self-realization (do I know when I become okay?)
   ii. *Professional care:* Peer support, organizational support and professional support.
   iii. *Coping skills:* (Positive vs. negative coping, problem focused vs. emotion focused coping, predominant coping strategies, etc.)

3. Educating on disaster trauma and stress manifestations:

   i. Understanding body and mind coordination
   ii. Signs and symptoms
   iii. Stress buster factors.

4. Self-care (reworking on previous ones or development of new basic care techniques) and care of co-workers.

Developing a "buddy system" among group members to provide constant support (Myers & Wee, 2005), increasing exercise/activity levels and taking time for yourself (Dutton & Rubenstein, 1995), doing meditation/yoga (Politsky, 2007) and appropriate use of humour (Moran, 2002) are highlighted for better self-care among the DFRs.

For Employers of First Responders

Promotional activities should also be conducted for the employers or the deploying authority for such a group of workers; however, the duration, content and mode of information dissemination could be different for different target groups. The duration of such programmes could vary between two and four days depending upon the resources available and time in hand. Programmes or activities meant for the employers or the deploying authority should be more specific in terms of:

1. Providing first-hand information on team selection, date and mode of departure, enquiring unwillingness for deployment (establishing genuineness of problems, if raised to cancel a deputation/duty posting, is also a core aspect of this).
2. Debriefing on the current situation in the affected area, quantity of damage and loss, possible stressors to be encountered by them, etc., (creation of a buddy system, nurturing team spirit, clear chain of command, clear instruction for clarity of roles and functions, disseminating updated information, establishing linkages of workers with family members, etc.).
3. Cautioning against sending same persons again and again if there are sufficient number of rescue and response workers.
4. Emphasizing on learning from past experiences (Do's and don'ts).
5. Instructions on seeking professional help and self-care, if emotional problem become overwhelming and cannot be controlled by them.

Reissman and Howard (2008) highlighted worker safety and health preparedness and leadership as essential for protecting DWs and promoting resiliency. Bilal et al. (2007) suggested the followings points to be considered while planning for mental health of DFRs:

1. Timely rotation of care providers at regular intervals;
2. Proper logistic support must be ensured;
3. A debriefing of the individual care provider and the whole team;
4. A break or time off for the care provider;
5. Expressions of appreciation from the management and judicious granting of rewards.

Model IEC Materials for Protection and Promotion of Mental Health

Information, Education and Communication (IEC) materials can be specifically developed, for the workers and employers separately, in the form of key checklists before and after deployment. Model IEC materials (Boxes 12.3, 12.4 and 12.5) could be prepared with colourful

Box 12.3

IEC material on self-care and coping

Positive Coping Abilities and Life Style
1. Sharing the problems/experiences
2. Try to take time to eat, rest and relax, even for short periods
3. Reading books, visiting temple, watching TV, etc.
4. Relaxation
5. Talking to family and friends
6. Seeking and providing support in difficulties
7. If needed, seeks professional help

Negative Coping and Life Style
1. Take alcohol or starts excessive smoking
2. Pick up fights
3. Less interactive become lonely
4. Over eating and sleeping
5. Disobeying others and supervisors

Source: Adapted from SAMHSA, 2005.

Box 12.4

IEC material on stress management during a disaster management operation

How to Manage a Stress during a Disaster Operation?
1. Provide and seek guidance and support
2. Encourage supportive peer relationship
3. Respect confidentiality as people can feel safe and seek responsibilities
4. Take break when you feel your tolerance level is diminishing
5. Have de-brief session regularly
6. Encourage the community volunteers to participate in disaster response
7. Involve the local communities
8. Effective use of the Government and Non-Government structure
9. Minimize intake of alcohol

Source: Adapted from SAMHSA, 2005.

Box 12.5

IEC material on seeking support

Seek support from someone you trust when you....
1. Have upsetting thoughts or memories about the crisis event
2. Feel very nervous or extremely sad
3. Have trouble sleeping
4. Frequent nightmares
5. Regular daily schedule is upset
6. Interest, attention and concentration are decreased
7. Drink a lot of alcohol or take drugs to cope with your experience
8. Consult a professional if these difficulties persist more than one month

Prevention of illness is definitely the best for you, your family and your employer: Timely psychosocial evaluation by professionals and appropriate referral for **treatment is definitely effective** and receiving treatment is your fundamental right **Recovery and leading a normal qualitative life is assured** with timely professional help

Source: Adapted from SAMHSA, 2005.

illustrations on a single page so that they become user friendly and handy to carry along whenever DFRs are deputed to the disaster affected area.

Conclusion

Working in MOOTW like disaster rescue and evacuation operation can be inevitably stressful for service personnel. Long working hours, volume of emotionally taxing survivors' needs and demands, ambiguous roles and exposure to human suffering can affect even the most experienced personnel. Although such operations can be personally rewarding and challenging, they also have the potential to affect the personnel in adverse ways. Thereby, both preventive and promotive mental health care is required for optimal stress management and effective functioning of personnel involved as DFRs. Preventive stress management focuses on two critical contexts: the organizational and the individual during pre- and post-deployment to disaster-affected areas. The main objective is to increase resilience among them and reduce the risk of any long-term psychological disorder. Development of IEC materials and disseminating these among the teams before deployment to the field could be a powerful mental-health preparedness tool. Conclusively, adopting a protective, preventive and promotive psychosocial health perspective allows both personnel and policy makers to anticipate stressors and to shape crises rather than simply reacting to them.

References

Abendroth, M., & Flannery, J. (2006). Predicting the risk of compassion fatigue: A study of hospice nurses. *Journal of Hospice and Palliative Nursing, 8*, 346–356.

Adams, R., Boscarino, J., & Figley, C. (2006). Compassion fatigue and psychological distress among social workers: A validation study. *American Journal of Orthopsychiatry, 76*, 103–108.

Alexander, D. A., & Klein, S. (2001). Ambulance personnel and critical incidents: Impact of accident and emergency work on mental health and emotional well-being. *British Journal of Psychiatry, 178*, 76–81.

———. (2009). First responders after disasters: A review of stress reactions, at-risk, vulnerability, and resilience factors. *Prehospital and Disaster Medicine, 24*, 87–94.

American Psychiatric Association. (2000). *Diagnostic and statistical manual of mental disorders* (4th ed., test rev.). Washington, D.C.: Author.

Avey, J. B., Luthans, F. & Jensen, S. M. (2009). Psychological capital: A positive resource for combating employee stress and turnover. *Human Resource Management, 48*, 677–693.

Benedek, D. M., Fullerton, C., & Ursano, R. J. (2007). First responders: Mental health consequences of natural and human-made disasters for public health and public safety workers. *Annual Review of Public Health, 28*, 55–68.

Berninger, A., Webber, M. P., Cohen, H. W., Gustave, J., Lee, R., Niles, J. K., et al. (2010). Trends of elevated PTSD risk in firefighters exposed to the world trade center disaster: 2001-2005. *Public Health Reports, 125*, 556–566.

Bilal, M. S., Rana, M. H., Rahim, S., & Ali, S. (2007). Psychological trauma in a relief worker: A case report from earthquake-struck areas of north Pakistan, *Prehospital and Disaster Medicine, 22*(5), 458–461. Retrieved from http://pdm.medicine.wisc.edu

Burkle, F. M. (1996). Acute-phase mental health consequences of disasters: implications for triage and emergency medical services. *Annals of Emergency Medicine, 28*, 119–128.

Centre for Mental Health Services. (2005). *Stress prevention and management approaches for rescue workers in the aftermath of terrorist acts*. Retrieved from http://www.mental health. samhsa.gov/cmhs/EmergencyServices/stress.asp.

Cohen, R. E. (2000). Mental health services for victims of disasters. *World Psychiatry, 1*(3), 149–152. Retrieved from http://www.ncbi.nlm.nih.gov/ pmc/articles/ PMC 148 9 840/

DeWolfe, S. (2000). *Field manual for mental health and human service workers* (Publication No. ADM 90-537). Rockville, MD: US Department of Health and Human Services.

DM Act. (2005). *Disaster management act*. New Delhi, India: Government of India.

Dutton, M. A., & Rubenstein, F. L. (1995). Working with people with PTSD: Research implications. In C. R. Figley (Ed.), *Compassion fatigue: Coping with secondary traumatic stress disorder in those who treat the traumatised* (pp. 51–81). New York, NY: Brunner/ Mazel.

Figley, C. R. (1995). Compassion fatigue as secondary traumatic stress disorder: An overview. In C. R. Figley (Ed.), *Compassion fatigue: Coping with secondary traumatic stress disorder in those who treat the traumatized* (pp. 1–20). New York, NY: Brunner/Mazel.

———. (1999). Compassion fatigue: Toward a new understanding of the costs of caring. In B. H. Stamm (Ed.), *Secondary traumatic stress: Self care issues for clinicians, researchers and educators* (pp. 3–28). Lutherville: Sidran.

Fullerton, C.S., Ursano, R. J., & Wang, L. (2004). Acute stress disorder, posttraumatic stress disorder, and depression in disaster and rescue workers. *American Journal of Psychiatry, 161*, 1370–1376.

Gupta, R., Sood, S., & Bakhshi, A. (2012). Relationship between resilience, personality and burnout in police personnel. *International Journal of Management Sciences, 1*(4), 1–5.

Mansfield, C. F., Beltman, S., Price, A., & McConney, A. (2012). "Don't sweat the small stuff:" Understanding teacher resilience at the chalkface. *Teaching and Teacher Education, 28*, 357–367.

Marmar, C. R., McCaslin, N., Metzler, T. J., Best, S., Weiss, D. S., Fagan, J.,…Neylan, T. (2006). Predictors of posttraumatic stress in police and other first responders. *Annals of N.Y. Academy of Science, 1071*, 1–18

Marmar, C. R., Weiss, D. S., Metzler, T. J., Delucchi, K. L., Best, S. R., & Wentworth, K. A. (1999). Longitudinal course and predictors of continuing distress following critical incident exposure in emergency services personnel. *The Journal of Nervous and Mental Disease, 187*, 15–22.

Meldrum, L., King, R., & Spooner, D. (2002). Secondary traumatic stress in case managers working in community mental health services. In C. R. Figley (Ed.), *Treating compassion fatigue* (pp. 85–106). New York, NY: Brunner-Routledge.

Moran, C. C. (2002). Humor as a moderator of compassion fatigue. In C. R. Figley (Ed.), *Treating compassion fatigue* (pp. 139–154). New York, NY: Brunner-Routledge.

Morren, M., Dirkzwager, A. J. E., Kessels, F. J. M., & Yzermans, C. J. (2007). The influence of a disaster on the health of rescue workers: A longitudinal study. *Canadian Medical Association Journal, 176*, 1279–1283.

Myers, D., & Wee F. (2005). *Stress management and prevention of compassion fatigue for psychotraumatologists. Disaster mental health services: A primer for practice.* New York, NY: Brunner-Routledge.

Perrin M. A., DiGrande, L., Wheeler, K., Thorpe, L., Farfel, M., & Brackbill R. (2007). Differences in PTSD prevalence and associated risk factors among world trade center disaster rescue and recovery workers, *American Journal of Psychiatry, 164*, 1385–1394.

Pietrantoni, L., & Prati, G. (2008). Resilience among first responders. *African Health Sciences, 8*, 14–20.

Pipe, T. B., Buchda, V. L., Launder, S., Hudak, B., Hulvey, L., Karns, K. E., et al. (2012). Building personal and professional resources of resilience and agility in the healthcare workplace. *Stress and Health, 28*, 11–22.

Politsky, S. (2007). Revitalizing yourself: Making time 4u. *Oncology Nursing Forum, 34*, 494.

Reissman, D. B., & Howard, J. (2008). Responder safety and Health: Preparing for the future disaster. *Journal of American Mental Health, 75*, 135–141.

Sabo, B. M. (2006). Compassion fatigue and nursing work: Can we accurately capture the consequences of caring work? *International Journal of Nursing Practice, 12*, 136–142.

Shih, F. J., Liao, Y. C., Chan, S. M., & Gau, M. L. (2002). Taiwanese nurses' most unforgettable rescue experiences in the disaster area after the 9-21 earthquake in Taiwan. *International Journal of Nurses Studies, 39,* 195–206.

Stellman, J. M., Smith, R. P., Katz, C. L., Sharma, V., Charney, D. S., Herbert, R., et al., (2008). Enduring Mental Health Morbidity and Social Function Impairment in World Trade Center Rescue, Recovery, and Cleanup Workers: The Psychological Dimension of an Environmental Health Disaster. *Environmental Health Perspectives, 116,* 1248–1253.

Substance Abuse & Mental Health Services Administration (SAMHSA). (2000). *Field Manual for Mental Health and Human Service Workers in Major Disasters* (Publication ID: ADM90-0537). Rockville, MD: U.S. Department of Health and Human Services.

Thoresen, S., Tonnessen, A., Lindgaard, C. V., Andreassen, A. L., & Weisaeth, L. (2009). Stressful but rewarding: Norwegian personnel mobilized for the 2004 tsunami disaster. *Disasters, 33,* 353–368.

SAMHSA-U.S. Department of Health and Human Services (2005). *A guide to managing stress in crisis response professions* (Pub. No. SMA 4113). Rockville, MD: Center for Mental Health Services, Substance Abuse and Mental Health Services Administration.

Van Breda, A. D. (2011). Resilient workplace: An initial conceptualization. *Families in Society, 92,* 33–40.

Wheeler, K., McKelvey, W., Thorpe, L., Perrin, M., Cones, J. Kass, D., et al. (2007). Asthma Diagnosed after 11 September 2001 among Rescue and Recovery Workers: Findings from the World Trade Center Health Registry, *Environmental Health Perspectives, 115,* 1584–1590.

13

Value-based Leadership

Vidushi Pathak, Anju Rani and Sneha Goswami

Breach of ethics and values in military is a rampant issue now-a-days. Problems such as these, however, are essentially little more than 'abuse of power'. It has been well-documented throughout the history of the world that power corrupts, and that absolute power corrupts absolutely (Lloyd, 2010). In the Armed Forces, individuals exercise immense powers and responsibilities. Effective balance between power and responsibilities is at the core of what is meant by values in practice. Therefore, a clear understanding of values is required. Many feel that this dimming of value perspective is the primary reason why cases of misconduct are still occurring in the Armed Forces. A job well done, backed by values, is superior to the one done just to follow an order.

Changing Faces of Military as an Organization

Military is an organization authorized by its greater society to use lethal force, including weapons, in defending its country by combating actual or perceived threats. The primary reason for the existence of the military is to engage in combat, should it be required to do so to protect the country and to win. This represents the organizational goal of any army, and the primary focus for army thought. Profession of soldiering in the Armed Forces is older than recorded history itself. The Battle of Kadesh in 1274 BC was one of the defining points of Pharaoh Ramesses II's reign and is celebrated on his monuments. A thousand years later the first emperor of unified China, Qin Shi Huang, was so determined to impress the Gods with his military might that he was buried with an army of terracotta soldiers. The Romans were dedicated to military matters, leaving behind many treatises and writings as well as a large number of lavishly carved triumphal arches and victory columns. All these accounts reflect the importance given to war and armies.

What distinguishes modern military organizations from the previous ones is not their willingness to prevail in conflict by any method, but rather the technological variety of tools and methods available to modern battlefield commanders, from submarines to satellites, from knives to nuclear warheads. Also, the nature of war and the enemy has changed drastically. Today, soldiers have to be prepared to be a global soldier, fight wars in foreign lands and participate in peace-keeping missions. With globalization of military operations, more integrative and cooperative defense efforts are required. Armed Forces across nations today appear to share common military codes, ethics and values; culturally, however, they differ. A fundamental understanding of values in relation to culture which guide people's lives in different parts of the world has become particularly relevant to military life today.

Core Military Values

Core values are the precisely stated values preferred by an organization. These are the organization's most essential and enduring tenets (Collins & Porras, 1996). They describe what the organization is about and give meaning to all its members. A small set of timeless guiding principles that require no external justification; they have intrinsic value and importance to those within the organization (ibid.). Values must be well differentiated from the objectives of the organization in order to be better understood as desirable behaviours. They define objectives and point to the actions necessary to achieve them. Objectives can be flexible at a given point in time, but values are immutable. They are like the columns supporting a building. Inside, one can make all the changes required, but we never move its foundations (Jackson, 2001). Certain values have been the central focus of military leadership also. Military tradition of duty, honor and selfless service to the country has been the core values followed by most armies. Sanwal (2011) has stated loyalty, integrity, respect and courage as core Indian Army values. The Soviet soldier is influenced by the communist values held by his countrymen. From their earliest days, Soviet children in school itself were taught about the glories of Russian and Soviet feats of arms. Although in most other societies the Army is separate from the civilian society, in Israel no such division exists because the Army is a civilian army. An established tradition of compulsory military service to army makes it both a duty and privilege for the people of Israel to serve in the forces. In case of the British Armed Forces, it is seen that besides the common factors of command and management skills, there is a large extent of personal responsibility and relationship among officers.

US Army doctrine, as expressed in Field Manual 22-100, explicitly names seven core values of overriding importance in leadership: *Loyalty, Duty, Respect, Selfless Service, Honor, Integrity and Personal Courage* (LDRSHIP; Department of the US Army, 1999, p. B-2).

Loyalty

Abide true faith and adherence to the National Constitution, the Army, unit and other soldiers. Loyalty is first to the organization, its values and principles and not to the individual. The other values closely associated with loyalty are unity, justice, altruism, respect for authority and courage of conviction.

Duty

Fulfill one's obligations.

Respect

Treat people as they should be treated. A good leader must always respect individuals, whether senior or junior. He must honor their status, value their opinion and accept inputs humbly. Individuality and self-esteem must be respected, as they will foster mutual respect, something that is imperative for value-based teamwork.

Selfless Service

Put the welfare of the nation, the Army and subordinates before one's own.

Honor

Live up to all the Army values.

Integrity

This implies honourable intentions and principles in thought, deed and actions, and demands the highest level of commitment. Broadly speaking, it means adherence to moral and ethical standards. Integrity is all encompassing and includes straightforwardness, selflessness, self-discipline, self-denial, honesty and propriety.

Personal Courage

Face fear, danger or adversity (physical or moral). Moral and physical courage are products of character-forming process for development of self-control, self-discipline, physical and mental robustness, knowledge of one's job and therefore, building up of self-confidence. Physical courage is a virtue that makes a man intrepid in the face of danger. Moral courage is the ability to discriminate between right and wrong as also stating it unequivocally. It also involves owning up one's mistake and standing by the subordinate.

Soldier's Dilemma: Personal vs Military Values

Soldiers hold two sets of values: personal values and military or organizational values. Are organizational values fundamentally different from personal values? Ideally, these are not! Often both sets are the same, but not always. The Armed Forces value selfless service to the nation. Youth entering the forces are assessed for integrity, not values. This can lead to disappointment as a gap emerges between personal and organizational values leading to job dissatisfaction and lesser effort input by the individual.

As the value system of the society changes, the Armed Forces too have been caught in this change process. Increased material aspirations, need for fame and recognition, intense competition for promotion in a pyramid structured organization, various external pressures and increased tendency to opt for short cut are causing decay of age-old traditions and values. Coupled with these, the changing nature of warfare and the enemy is also challenging the traditional values held. Similar views have been expressed by Sanwal (2010), wherein he believes that recent media reports of scams in various arenas by the Armed Forces officials, staged encounters, confidential reports being leaked, spying, misappropriation of funds, etc., are indications of the broader decline in the value system. In a study conducted by Heinecken (2009), it was found that over 64 per cent of UK, Canadian and South African and 55 per cent of German officers agree that there has been a breakdown in the traditional value system based on selfless service in the forces. Over half of the respondents felt that there was no longer a reciprocal sense of loyalty, thereby upholding the views that military leadership's responsibility towards subordinates has declined. The study confirms the general perceptions held by the public regarding the decline in the value aspect of the forces, globally.

There may be several external as well as internal factors responsible for the decline in values in the forces. Rapidly changing economic scenario resulting in changed societal dynamics, changing nature of war leading to more and more interaction between civilians and the Armed Forces, various forms of corruption, greater media infiltration in all field of society, more demanding professional commitments in comparison to civilian jobs, changing definitions of relationship between men and women coupled with more and more women entering the forces, changed priorities of the younger generation, paucity of enough role models at the top level might be some of the external factors that are responsible for the decline. Sanwal (2010) in his paper notes that other factors such as changing battlefield environment, failure on the part of the senior leadership to devote enough attention in building focused leadership, lack of transparency in many areas of working, an apparent gap between ideal work environment and reality are some of the internal factors that are contributing to the decay as well.

When personal and organizational values conflict, ethical dilemmas arise and the culture of the organization starts to wane and its members begin to disperse. Subsequently, a strong need of alignment of personal and organizational values has been felt. Alignment refers to the 'shared' aspect of definition. In other words, this means that the employees value the same things that the organization does. Shared values underpin common assumptions about what is desired or good and help soldiers learn how things are done in Armed Forces. Shared values also translate into the behaviours the military requires in order to achieve its objectives and to sustain high ethical standards. Jimenez (2009) also asserted that values must serve as more than just window dressing. Therefore, the credibility of Armed Forces depends on congruency between personal values and organizational values.

Cultivating Shared Values: Personal and Military

It is an important requisite for the organization to achieve the desired fit between the person and the job. Therefore, the organization must pay attention to individual differences in soldiers during planning and implementing strategies. Same ideas have been submitted by different researchers (Balachandran, 2012; Green, 2012; Hill & Jones, 2002; Jackson, 2001; Jimenez, 2009; Kerns, 2003) on different platforms regarding alignment of personal values with organizational values. Jimenez (2010) reported that personal values are so diverse and omnipresent in people that organizations have difficulty in bringing a large number of people from different strata of society into one common umbrella of organizational values.

If we look at core values of Armed Forces, that is, *Loyalty, Duty, Selfless Service, Personal Courage*, etc., which although sound great but means nothing if not captured clearly. This is where most organizations fail. After listing down the core values in military doctrine as it has already been done, it is also important to articulate them further into measurable actions and behaviours which can be used by leaders and managers to drive employee performance. Leaders also need to articulate how these values present the organization to the external world (Balachandran, 2012). Thus, military leadership and organization need to take the following measures at various levels to ensure value-based person-job fit.

Selection Process

Alignment between personal and organizational values has to start right from the time of selection of a soldier. Other than the qualifications and skill sets needed for the job profile, the prospective employees' fit with the organization's core values needs to be given due importance in the overall selection decision itself (Balachandran, 2012). If the organization selects someone whose values are mismatched to those of the organization, it will be hard to shift the course of their thinking and it will ultimately prove to be difficult to retain them. The smartest organizations tend to acknowledge the importance of both personality and value characteristics, and consider this duality in every action.

Values are the unseen magnets that steer the course of one's action through the choices made. If the two values sets, individual and organizational, are not in the same direction, it may lead to a lot of dissonance at all levels and will create problem in retaining such an individual for long in the organization. Armed Forces need to be selective in whom it selects and should adopt a duality approach. Currently, some of the core values are assessed by selectors but not all. Thus, the need of the hour is to assess the whole value system of individuals.

Induction Process

Merely having core values alone is not enough for Armed Forces to reap the benefits of a strong set of core values. These values must truly resonate with each individual, which requires the Armed Forces to be expert at articulating and communicating them efficiently. Armed Forces must state and clarify the values respected by its organization. Kerns (2003) also laid emphasis on the induction process as one step towards alignment of personal and organizational values. As suggested by Jimenez (2009), this can be accomplished through formal orientation programmes. Every attempt should be made to expose new soldiers to the correct values and expectations and to teach them the way things are done in the organization. Leaders must ensure their soldiers understand those values, so they can always do the right things in difficult or stressful peacetime or wartime situations. Hence, new soldiers need to be oriented and immersed in values and traditions of the forces so that they become active organizational disciples. Leaders must also be engaged in dialogues with soldiers informally about what is important to them and explain what is important to the organization and work to bring about a balance between the two.

Rewards and Reinforcement

The most efficient way to foster values is to reinforce good practices and behaviours that better reflect the desired organizational values (Jimenez, 2009). Soldiers can be motivated to assume principles with conviction through this established and effective way. Intimidation and punishment in the best of cases generate only fear, not conviction. Thus, rewards can strengthen individual values and maintain enthusiasm in support of key organizational values that are considered vital to organizational success. Soldiers cannot be forced to do well what they do not want to do. It does not mean that mistakes should go unnoticed or that leaders must be lenient. But positive reinforcement is much more than a pat in the back (ibid.). This technique can work best when soldiers receive commendation immediately for a specific behaviour, and leaders express the positive feeling that implementing the value entail. If this method is practiced methodically, the organizational environment works as a virtuous cycle of value reproduction.

Training and Development

When ethical dilemma arises, military members should be prepared to handle it. Hence, value development should be a primary focus for military education and training, not merely a strategic goal. Value inculcation must be the cognizant objective in all trainings. Armed Forces may train values intentionally and non-intentionally or incidentally. In intentional values training, leaders consciously and deliberately train the new cadets the desired values formally or informally, but it is never accidental. Unintentional or incidental values training occurs when military members form a value conclusion based on their observation of their leader's unintentional example. Most often this example is incidental to some other situational priority (Jackson, 2001).

Leaders must always exhibit military values and must ensure their soldiers understand those values so that they can always do the right things in difficult or stressful peace time or wartime situations. The existing system of training and development of future military leaders revolves around operational assignments, that is, both command and staff, and self structured at various levels. This type of training needs to be complimented through structured training; more importantly, through self-development in terms of lateral and broad-based competencies (Sanwal, 2011).

Promoting an Ethical Culture

Ethical or unethical practices usually reflect the values, attitudes, beliefs and behaviour patterns of the organization's culture (Rao, 2010). So, time and effort must be devoted to promote a culture based on shared values that are expressed through the everyday behaviour of its members. The coherence and consistency between what members of organization say and do strengthens its culture. This leads to high level of harmony and performance of the organization. An organization's values should be reflected in the specific behaviours of its members, and not just in its mission statement. To convey its core value to new members, one must possess it and implement it through; its credibility depends on it (Jimenez, 2009). The individuals internalize and learn the norms and values of the culture through socialization and adopting organizational membership. Culture is a powerful tool because once these values have been internalized the culture becomes a part of the individual's values and the individual follows organizational values without thinking about them (Hill & Jones, 2002).

Authentic Leadership

General Donald Campbell emphasized:

> You can't hit every decision that they'll face on the battlefield. But you try to instill in them values, standards that are common to the military and our profession, which is about leadership, duty, honor and integrity. And if

you do that, 99.9 percent of our soldiers will go to 100 percent, and they'll all do the right thing. (as cited in Knickmeyer, 2006, p. A10)

In military leadership, leaders cannot just lead through their position of authority; they will instead have to lead by example. When people operate from this highest level of development they focus on the needs of followers and on encouraging others to think for themselves and to engage in higher levels of moral reasoning and will make ethical decisions whatever the organization consequences are for them (Rao, 2010). This is the crux of value-based leadership that is called the *Authentic Leadership*. Authentic leaders are deeply aware of their values and beliefs, they are self-confident, genuine, reliable and trustworthy and they focus on building followers' strengths, broadening their thinking and creating a positive and engaging organizational context (Avolio & Gardner, 2005; Gardner, Avolio, Luthans, May, & Walumbwa, 2005). The four important components of Authentic Leadership, which by and large have been accepted, are *Self Awareness, Balanced Processing, Internalised Moral Perspective* and *Relational Transparency*.

SELF AWARENESS

Self awareness refers to one's awareness of, and trust in, one's own personal characteristics, values, motives, feelings and cognitions. It includes knowledge of one's inherent contradictory self-aspects and the role of these contradictions in influencing one's thoughts, feelings, actions and behaviours (May, Chan, Hodges, & Avolio, 2003). Being self aware means knowing who you are, what you believe in and what you stand for. For a leader being aware of who he is, is important as it anchors him and help him in understanding and making decisions.

BALANCED PROCESSING

It refers to taking everyone's viewpoints in consideration and objectively analysing these views before coming to a decision. Other viewpoints may not match with a person's viewpoint, but by asking for others' views a different perspective might emerge which can help in decision making (Walumbwa, Avolio, Gardner, Wernsing, & Peterson, 2008). Another important aspect of balanced processing focuses on processing the information without having any bias. Unbiased processing has been described as 'the heart of personal integrity and character', thereby significantly influencing a leader's decision making and strategic actions (Ilies, Morgeson, & Nahrgang, 2005). Thus, for a leader to be authentic he needs to have a high level of integrity and character.

INTERNALIZED MORAL PERSPECTIVE

Internalized moral perspective means to regulate behaviour so that the behaviour aligns with the individual's moral standards. Having an internalized moral compass helps an individual to guide his behaviour in times when it becomes difficult to do so. Many researchers believe this aspect to be central to authentic leadership (Gardner et al., 2005; Ilies et al., 2005; Luthans & Avolio, 2003; May et al., 2003; Walumbwa et al., 2008). In case of military leadership, it is all about following the chain of command. What the leader orders, the follower follows. This attitude of deference is something which is an integral part of the military.

RELATIONAL TRANSPARENCY

Relational transparency refers to presenting one's true self to others. This means being honest, transparent and true when dealing with others. Trusting relationships are built on the basis of the authentic leader's expressing true feelings to their subordinates. Disclosing one's true self to one's followers builds trust and intimacy, fostering teamwork and cooperation (Gardner et al., 2005) and feelings of stability and predictability (Chan, Hannah, & Gardner, 2005). It is important that leaders share their vision with their subordinates and let them know what they expect from them. This will help to win the loyalty and trust of their subordinates.

Indian Military Values

Globally, Armed Forces have played crucial role in building the history of civilizations. In India also the value and importance of army was realized very early. This led to the maintenance of a permanent militia to fight opponents. War or no war, the army was to be maintained to meet any unexpected contingency. This gave rise to the *Kshatriya* or warrior caste, and the *kshatramdharman* came to mean the primary duty of war. To serve the country by participating in war became the *svadharma* of this warrior community. In case of the Indian Armed force's core value system, it is seen that a lot of it is based on its cultural history. Indian forces place a lot of emphasis on the conduct of its men during any conflict situation. It treats its opponent with due respect and does not violate basic human rights. The ancient scriptures have also laid a lot of emphasis on the science of warfare and even today the Force tries to emulate those values as much as possible. The ancient law-givers, the reputed authors of the Dharmasutras and the Dharmasastras, codified the then-existing customs and usages for the betterment of mankind. Thus, the law books and the epics contain special sections on royal duties and the duties of common warriors. Some of them are mentioned in the succeeding text.

sva-dharmam api cāvekṣya
na vikampitum arhasi
dharmyād dhi yuddhācchreyo 'nyat
kṣatriyasya na vidyate

(Srimad Bhagvad Gita 2.31)

Translation: Considering your specific duty as a warrior (kṣhatriya), you should know that there is no better engagement for you than fighting on religious principles; and so there is no need for hesitation.

kṣatriyo hi prajā rakṣan
śastra-pāṇiḥ pradaṇḍayan
nirjitya para-sainyādi
kṣitiṁ dharmeṇa pālayet

(Srimad Bhagvad Gita 2.32)

Translation: The warrior's duty is to protect the citizens from all kinds of difficulties, and for that reason he has to apply violence in suitable cases for law and order. Therefore, he has to conquer the soldiers of inimical kings, and thus, with religious principles, he should rule over the world.

atha cet tvam imaṁ dharmyaṁ
saṅgrāmaṁ na kariṣyasi
tataḥ sva-dharmaṁ kīrtiṁ ca
hitvā pāpam avāpsyasi

(Srimad Bhagvad Gita 2.33)

Translation: If, however, you do not perform your religious duty of fighting, then you will certainly incur sins for neglecting your duties and thus lose your reputation as a fighter. Conclusively, The Bhagavad Gita's great message that:

> *Violence is sometimes necessary,*
> *if it flows from Dharma*
> *For a warrior, nothing is higher than a war against evil.*
> *The warrior confronted with such a war should be pleased,*
> *Arjuna, for it comes as an*
> *Open gate to heaven.*
> *But if you do not participate in this battle against evil, you will incur sin,*
> *violating your dharma and your honor....*

(Srimad Bhagvad Gita 2.31-33)

Indian military science recognizes two kinds of warfare– the *Dharmayuddha* and the *Kutayuddha*. *Dharmayuddha* was carried on the principles of dharma, meaning that the *Kshatriyadharma* or the law of Kings and Warriors was a just and righteous war which had the approval of society. *Kutayuddha*, on the other hand, was unrighteous war, a crafty fight carried on in secret. The Hindu science of warfare values both '*niti*' and '*shaurya*', that is, ethical principles and valor. The principles regulating the two kinds of warfare are elaborately described in the *Dharmasutras* and *Dharmashastras*, the epics (Ramayana and Mahabharata), the Arthashastra treatises of Kautilya, Kamandaka and Shukra. It was therefore realized that the waging of war without regard to moral standards degraded the institution into mere animal ferocity.

Chivalry, individual heroism, qualities of mercy and nobility of outlook even in the grimmest of struggles were not unknown to the soldiers of ancient India. Thus among the laws of war, it can be found that (a) a warrior (Kshatriya) in armor must not fight with one not so clad (b) one should fight only one enemy and cease fighting if the opponent is disabled, (c) aged men, women and children, the retreating, or one who held a straw in his lips as a sign of unconditional surrender should not be killed. As early as the 4th century B.C., Megasthenes noticed a peculiar trait of Indian warfare which he later mentioned in his book *Indika*.

Whereas among other nations it is usual, in the contests of war, to ravage the soil and thus to reduce it to an uncultivated waste, among the Indians, on the contrary, by whom husbandmen are regarded as a class that is sacred and inviolable, the tillers of the soil, even when battle is raging in their neighborhood, are undisturbed by any sense of danger, for the combatants on either side in waging the conflict make carnage of each other, but allow those engaged in husbandry to remain quite unmolested. Besides, they never ravage an enemy's land with fire, nor cut down its trees.

When a conqueror felt that he was in a position to invade the foreigner's country, he sent an ambassador with the message: 'Fight or submit.' More than 5,000 years ago, India recognized that the person or the ambassador was inviolable. This was a great service that ancient Hinduism rendered to the cause of international law. It was the religious force that invested the person of the herald or ambassador with an inviolable sanctity in the ancient world. The Mahabharata rules that the king who killed an envoy would sink into hell with all his ministers.

Although each ruler brought with him a distinct set of values, in general, it can be observed that the basic values formed during the early Vedic period constituted the crux of armies in India across centuries till foreign invasions started becoming common. Foreign invaders generally brought their own armies to conquer but once the conquest was won and they settled, they started recruiting from the natives. As a result, the natives had to abide by their values which was not without clashes and uprisings. With the advent of Britishers, there was a shift. The employed Indian soldiers held on to their Indian value systems and did not embrace the values of the British Armed Forces. Although there were simmering discomforts because of basic clashes in values, the high point reached with the First War of Independence. The Mutiny was a result of various grievances. However, the flashpoint was reached when the soldiers were asked to bite off the paper cartridges for their rifles which they believed were greased with animal fat, namely beef and pork. This was, and is, against the religious beliefs of Hindus and Muslims, respectively. So the soldiers fought for their values. Thus, it can be said that the Indian Armed Forces have followed the traditional values inspired by its larger society and have always tried to maintain its values down the ages.

Conclusion

Armed Forces have benchmarked very high moral and ethical standards for its men to follow. However, in saying this, it cannot be denied that though the forces strive to achieve such high standards from its men in reality this often fails. The changing sociocultural and economic environment is having an impact on the very foundations of value held deeply by the Forces. Armed Forces need leaders who apart from having skills of high level and of new kinds, ought to be necessarily value based. Value-based soldiering and leadership is mandatory for ethical conduct in warfare as use of more brute force cannot breed a peaceful internal as well as external environment. Emphasis must be on fostering a culture of shared values.

References

Avolio, B. J., & Gardner, W. L. (2005). Authentic leadership development: Getting to the root of positive forms of leadership. *The Leadership Quarterly, 16*, 315–338.

Balachandran, R. (2011). *Power of shared values.* Retrieved from http://magicofteams.word-press.com/2011/11/20/power-of-values on 28th October 2012.

Bhagwad Gita. (n.d.). *Bhagwad Gita sloka.* Retrieved from http://vedabase.net/bg/2/31/en

Chan, A., Hannah, S., & Gardner, W. (2005) Veritable authentic leadership: Emergence, functioning, and impacts. In W. Gardner, B. Avolio & F. Walumbwa (Eds.), *Authentic leadership theory and practice: Origins, effects and development* (Vol. 3, pp 3–42). Boston, MA: Elsevier Ltd.

Collins, J. C., & Porras, J. I. (1996). Building your company's vision. *Harvard Business Review, 74*, 65–77.

Department of the US Army (1999). *Military leadership.* (Field Manual 22-100). Washington, D.C.: Headquarters, Department of the Army.

Gardner, W. L., Avolio, B. J., Luthans, F., May, D. R., & Walumbwa, F. (2005). "Can you see the real me?" A self-based model of authentic leader and follower development. *The Leadership Quarterly, 16*, 343–372.

Green, D. D. (2012). *Evolution of value formation – Nu leadership series.* Retrieved from http://ezinearticles.com/Evolution-of-Value-Formation-Nu-Leadership

Heinecken, L. (2009). Discontent within ranks? Officer's attitudes toward military employment and representation – A four-country comparative study. *Armed Forces & Society, 35*, 477–491.

Hill, C. W. L., & Jones, G. R. (2002). *Strategic Management: An Integrated Approach.* Boston, MA: Houghton Mifflin Company.

Ilies, R., Morgeson, F. P., & Nahrgang, J. D. (2005). Authentic leadership and eudaemonic wellbeing: Understanding leader-follower outcomes. *The Leadership Quarterly, 16*, 373–394.

Jackson, K. A. (2001). *The army's institutional values: Current doctrine and the army's values training strategy* (Masters dissertation, Faculty of The US Army Command and General Staff College, Fort Leavenworth, Kansas, USA). Retrieved from handle.dtic.mil/100.2/ADA396542

Jimenez, J. C. (2009). *The significance of values in an organisation.* Retrieved from http://significanceofvalues.com/fostering/index.html

Kerns, C. (2003). *Creating and sustaining an ethical workplace culture. Graziadio Business Review, 6*(3). Retrieved from https://gbr.pepperdine.edu/2010/08/creating-and-sustaining-an-ethical-workplace-culture/

Knickmeyer, E. (2006, June 3). 30 Days to Review Combat Ethics. *Washington Post*, p. A10.

Lloyd, B. (2010). Power, responsibility and wisdom: Exploring the issues at the core of ethical decision making. *Journal of Human Values, 16,* 1–8.

Luthans, F., & Avolio, B. J. (2003). Authentic leadership development. In K. S. Cameron, J. E. Dutton, & R. E. Quinn (Eds.), *Positive organizational scholarship* (pp. 241–258). San Francisco, CA: Berret-Koehler Publishers Inc.

May, D. R., Chan, A., Hodges, T., & Avolio, B. J. (2003). Developing the moral component of authentic leadership. *Organisational Dynamics, 32,* 247–260.

Military (n.d.). *In Wikipedia.* Retrieved from http://en.wikipedia.org/wiki/Military.

Military History (n.d.). In *Wikipedia.* Retrieved from http://en.wikipedia.org/wiki/Military_history

Military Organization. (n.d.). *In Wikipedia.* Retrieved from http://en.wikipedia.org/wiki/Military_organization

Rao, S. R. (2010). *Factors affecting ethical choices.* Retrieved from http://www.citeman.com/12950-factors-affecting-ethical-choices

Sanwal, P. K. (2011). A value system and code of conduct for the armed forces. *USI Digest, XIII* (25), 92–97.

Walumbwa, F. O., Avolio, B. J., Gardner, W. L., Wernsing, T. S., & Peterson, S. J. (2008). Authentic leadership: Development and validation of a theory-based measure. *Journal of Management, 34*, 89–126.

What are Indian Values? (n.d.). *In wiki.answers.* Retrieved from wiki.answers.com/Q/What_is_Indian_Value_System

14

Familial Pathways to Soldier Effectiveness

Archana and Updesh Kumar

The effectiveness of military organization depends upon the physical and mental fitness of its soldiers, who are confronted with arduous operational demands that not only affect their own well-being but also disrupt the well-being of their family members. In order to attain effectiveness, a soldier and his family must be able to adapt to the environmental demands placed on him that varies from time to time and place to place. Thereby, it is pertinent to understand the factors that facilitate soldier's familial well-being and quality of life for enhancing his effectiveness.

Well-being, Quality of Life and Military Families

Well-being is a subjective evaluation of one's current status in the world and an individual's appraisal of his own life captures the essence of well-being (Diener, 1984). It comprises three components, namely emotional, social and psychological (Keyes & Lopez, 2002). Emotional well-being consists of perceptions of avowed happiness and satisfaction with life, along with the balance of positive and negative affects. The coupling of satisfaction and affect serves as a meaningful and measurable conceptualization of emotional well-being. And social well-being covers the dimensions of coherence, integration, actualization, contribution and acceptance. While psychological well-being involves the components of self-acceptance, personal growth, purpose in life, environmental mastery, autonomy and positive relations with others. These researchers also suggest that the complete mental health is conceptualized through combinations of high levels of emotional, social and psychological well-being.

Well-being is closely related to the concept of quality of life. Different perspectives were put forth about the quality of life. For example, economists focused on an individual's socio-economic status; a medical researcher was concerned with how an individual assesses his own physical and psychological well-being and social scientists were interested in looking into the interaction between the person and his environment (Cummins, 2000). Quality of life is a dynamic term suggesting that the same circumstances may not affect different people in the same way, nor may the same individual always react in a consistent manner over time (Carr, Gibson, & Robinson, 2001).

Quality of life for military families largely resides in their ability to adjust to the multiple demands of the military life. Every soldier deserves a quality of life equal to his service (WRAL Raleigh, 2008). Military personnel have a better health and quality of life than the civil population, despite their repeated exposure to considerable hazards and a stressful work environment (Mageroy, Riise, & Johnsen, 2007). Segal (1986) identifies both the military and the family as 'greedy' institutions that place significant demands on an individual in terms of loyalty, time and energy. Hence, the conflict between the two is inevitable. Family issues affect retention more than readiness, including perception of unit morale, confidence in other unit members and overall preparedness for combat. The greatest predictor of a soldier's commitment to the military is the spouse's commitments to the same. In a research carried out by Bourg and Segal (1999), it was found that spouse satisfaction was significantly affected by perceived military interference with family needs. Younger spouses reported the most work–family conflict, reflecting a lack of experience or expectation management. Assignment to a combat unit and the presence of children were also factors that increased reported military–family conflict.

LaGrone (1978) has put forth the concept of 'military family syndrome', suggesting that military families suffer from greater psychosocial difficulties than the general population. The frequent moves of military personnel create stress for their families, and it becomes more challenging when skills and abilities of military families fail to meet the demands of military life. Often military families are acquainted with the nature of stressors they encounter in such life due to their past experiences associated with such life style. But at times when the intervals between stressors are small and coping resources are inadequate, then these experiences constitute an accumulation of stressors that serves as a risk factor in adapting to military life. And during such time prior experiences with stressors, instead of being helpful, may contribute to pile-up due to insufficient time for recovery.

Challenges for Military Families

Adapting to the demands of military lifestyle not only creates stress but also leads to negative outcomes for family members. Prominent among these are geographic mobility, periodic separations from family, long and unpredictable duty hours, pressures for military families to conform to accepted standards of behaviour, sole parenting and deployment (Booth, Segal, & Bell, 2007). Not everyone in the military experiences all of these demands at once. The stressors associated with the challenges of maintaining a military career and a stable household put a strain on all family members. The stressors that military spouses face are unmatched in the civilian world and some of these challenges are discussed in the succeeding text.

Recurrent Relocation

Military is a structured organization, in which military personnel and their families are often required to relocate. Frequent relocation often disrupts family life and affects their supportive relationships in the society. Members of the military cannot refuse to relocate as these families have limited decision-making power (Hosek, Asch, Fair, Martin, & Mattock, 2002). Due to continuous change of stations, military spouses and their children face difficulty in adapting to new locations. Children's adjustment to a move depends on their age. For example, adolescents experience social rejection, which often breaks them down emotionally; particularly girls seem to have more difficulty adjusting than boys, since they place more importance on social relationships (Brown & Orthner, 1990).

Constant relocation also deters military spouses from having a career. Some women seek contentment by engaging themselves in household activities and by taking care of their family, whereas those with career find it difficult to bring a balance between the work–family interfaces. Regardless of employment status, many women whose husbands are absent for extended periods find themselves working twice as hard at home (Zvonkovic, Solomon, Humble, & Manoogian, 2005). However, it is not necessary that every relocation brings stress; some families feel that moving can have a positive experience also. For example, for some children this relocation helps in enhancing their academic performance, if a new environment provides a better educational system and offers more valuable connections with teachers (Cornille, 1993). Researchers suggest that relocation is related to lower psychological well-being and marital happiness (Jensen, Lewis, & Xenakis, 1986).

Separation from Family Members

Frequent separations place additional demands on family members in terms of managing the household and taking on the role of single parent. Even separation for a short duration affects family's life and creates emotional turmoil for many of its members. When one parent is deployed, the other parent is likely to encounter separation strain, loneliness, role overload, role shifts, financial concerns, changes in community support and increased parenting demands (Vormbrock, 1993). In marital relationship, the quality of intimacy and the communication between couples is highly effective and influential in maintaining their relations. Separation accompanied by lack of communication often reduces the intimacy of couples and military spouses consider separation as one of the major sources of dissatisfaction with military life. Importantly, Andres, Moelker and Soeters (2012) reflected that it is the quality of seperation rather than the quantity that matters and beyond the frequency of communications, the degree to which service members and partners engage in active interactions (e.g., keeping each other informed and involved, inquiring how the other feels, expressing affection) helps in maintaining intimate bond. The spouses of military personnel express their failure to handle the separation period as it is negatively related to well-being and often results in the feelings of loneliness, anxiety and depression for some women (Orthner, 2002). This period also affects the children's ability to adjust to the existing circumstances. During separation, children in military families often display minor to serious behaviour problems, including anxiety, sleep disturbances, phobias and increase in physical ailments (Kelley, 1994). Children's responses also reflect their mother's reactions. If the mother's reaction to her spouse's deployment is depression, then the children may mirror her depressive symptoms or behaviours, especially if they manifest as parental inattentiveness and unresponsiveness (Riggs, 1990). The adjustment problem is more prominent in younger children as compared to the infants, as the infants are too young to notice the absence of parent.

During the initial stage of separation, military families experience mixed emotions and feel anxious about their ability to cope with the demands of military life style. After this transitional period most families adjust to separation, adapting to their circumstances and establishing routines and new sources of support, both formal and informal. Many establish mutually supportive relationships with fellow military spouses. For those who work, co-workers are often a source of social support, although some spouses find their expanded responsibilities at home require them to reduce their work hours or cease to work altogether (Hosek, Kavanagh, & Miller, 2006).

Frequent Deployments

Deployments are a pivotal part of military life that creates a unique challenge to the basic structural integrity of the family. Due to the increased demands of being a military family, it becomes crucial for the spouses and their children to cope with deployment of service personnel. Deployment length and the extent to which the duration exceeds families' expectations are two important factors that are found to be negatively associated with the family's ability to cope with the situation and their level of satisfaction with military life (Chandran, Lara-Cinisomo, & Jaycox, 2009). When the husband is deployed, the entire family experiences the absence as a major stressor. Spouses who perceive military life as stressful show reduction in psychological well-being, express feelings of loneliness, role overload, financial difficulties, child-related issues, worry over long-distance relationship maintenance and separation anxiety (Burrell, Adams, Durand, & Castro, 2006).

Several researchers have examined children's responses to parental absence because of military deployment. Many of them have reported adverse effects including sadness and being very emotional, sleeping problems, aggressiveness, irritability, depression and decreased school performance (Booth et al., 2007). Empirical research has shown that separation not only causes maladjustment but also can elicit positive effects such as children acting more maturely and being self-sufficient, cooperative and more responsible at home (Jensen, Martin, & Watanabe, 1996). Research suggests that the effects of deployment indirectly affect child outcomes through parental stress and pathology. The heightened stress of one parent coupled with the absence of the other parent is likely to negatively affect child outcomes (Palmer, 2008).

Parenting Stress

Parenting stress results due to decreased satisfaction with the parenting role and a reduced quality of parent–child interaction (Berry & Jones, 1997). The bond and strength of couple relationship plays a very important role in buffering the effects of parenting stress. The stress within the household is mediated by several factors, including the overall strength of the couple relationship, the perceived amount of co-parenting and the household division of labour (Belsky & Hsieh, 1998). The mothers' social support systems and relationship with husband influence how children cope with and adjust to their father's absence. Andres and Moelker (2011) suggest that maternal well-being predicts children's adjustment difficulties in the course of paternal deployment. Higher levels of mother's parenting stress were found to be associated with higher levels of children's adjustment difficulties during the deployment.

Abrupt Reunion

Reunion can be stressful and can elicit ambivalent emotions and feelings of estrangement. Military families often experience difficulties with reunion of service members into the family system after their return from deployment, since they have already adapted to the absence of military personnel. The difficulty with reunion results in feelings of estrangement among family members who have been functioning without the service member during the separation time frame, and this often results in a great deal of distress (Peebles-Kleiger & Kleiger, 1994). Reunion is assumed to be more difficult for children who experienced higher levels of distress during the separation or who emotionally detached themselves to a higher degree from the absent parent. These children may either reject or become anxious in the presence of parents. Since military parents may leave behind an infant and return to a child who has changed significantly and no longer recognizes them. On the other hand, others may return to infants they have never seen and become frustrated by the strict household regime necessary to accommodate the child (Vormbrock, 1993).

Reunion may also create difficulty for the returning spouses. For example, returning spouses may feel superfluous or excluded from their family if they are impeded in reassuming their previous functions. They may experience jealousy and get frustrated because their wives have less time for them. Also, they feel an intense need to normalize their lives but realize that they are unfamiliar with the new management of the household. These feelings of unfamiliarity may increase tensions between the couple to such an extent that subsequent deployments are welcomed as a source of relief from conflict (ibid.). Although some findings suggest that reunion is a joyous experience felt by both the partners and it is accompanied by greater family cohesiveness and may result in quick readjustment (Kelley, 1994).

Parental Absence

Prolonged parental absence is a major concern for military children. Parenting with a balance of warmth and appropriate control helps in enhancing the well-being and quality of life in military children. Military children are more likely to display a good quality of life when their parents model a positive attitude and show more flexibility in adapting to the challenges of military life (Walsh, 2007).

Thus, it is well understood that the relocation, deployment, separation from family and parenting stress are some of the challenges of military families. And, achieving balance within and between the demands of the family and organization shall help in strengthening the forces.

Strengthening the Forces

Military life is stressful, as it affects the well-being of a soldier in multitude of ways. However, each individual possesses different levels of coping ability as well as different perceptions and definitions of stressful events, based upon their previous attempts at adaptation (McCubbin & McCubbin, 1989). According to the model of stress and coping as proposed by Lazarus and Folkman (1984), how one responds to stress depends on one's appraisal of the stressor, the interpretation of the stressor and the coping behaviours that are then implemented. Following this appraisal process, an individual may choose basically one of two methods of coping with a stressor. One method is known as 'emotion-focused coping' and reflects attempts to manage emotional responses during or after experiencing a stressor. A second method of coping is known as 'problem-focused coping', which is a direct attempt to solve the problem rather than focusing on the emotional aspects of the situation. Generally, problem-focused coping has been related to positive outcomes; however, emotion-focused coping may be better for situations in which the stressors cannot be changed. Military spouses use both these strategies in dealing with their stressful experiences (Carver, 1997). Some of the strategies that help in strengthening the forces through the familial pathways are discussed below.

Consolidating Social Support

Social support contributes to an individual's health independent of his or her level of stress. People involved in close, caring relationships are generally happier and healthier (because of their supportive relationships), and enjoy better health and more personal happiness than those who lack such a network. Therefore, lack of social ties, involvement in conflicting relationships or loss of a significant relationship contributes to loneliness, depression, personal distress and unhappiness (Berscheid, 2003). Having a strong social support helps in dealing with stressors. Social support has both structural and functional properties that are tied to mental and physical health outcomes. Structural support refers to variables such as size of the support network, the sources of support and type of support, whereas functional properties refer to the perception of whether or not support exists, and if it does, the degree of its usefulness (Cohen, 1988).

Adequate social support networks are crucial for military families. Military families need social support to cope with separation stressors. Social support systems, including friends, children, relatives, work colleagues and support groups, have been positively linked to separation adjustment for military families (Wood, Scarville, & Gravino, 1995). Women have specially identified children, employment, close friends and family as their main sources of support when separated from their military husbands (Vormbrock, 1993). Fellow military families serve as one of the most valuable sources for social support; they are the ones who precisely know what a deployment means to a family. The positive military environment fosters social support and helps individuals cope with stressors, thus improving their quality of life (Andres, Moelker, & Soeters, 2012).

Sensitizing Leadership

The military unit directly impacts the quality of life of both soldiers and their families. A unit's leadership is directly responsible for the 'command climate' and the positive unit environment fosters social support to help soldiers and their families cope with stressors and contribute to an improved quality of life. Researchers suggest that higher levels of unit cohesion and quality of leadership are related to service members' well-being and ability to cope with deployment. Also families' ability to cope during separation is positively related to length of marriage, predeployment preparedness and employer support (Jones, 2003). Strong unit cohesion has been positively correlated not only with readiness and individual and group performance, but also with personal well-being (Griffith, 2002). Therefore, high-quality leadership and unit support can enhance the well-being and quality of a soldier.

Increasing Positivity

Emotions come in two basic forms, namely positive and negative affect. Positive affect refers to emotions such as cheerfulness, joy, contentment and happiness and negative affect refers to emotions such as anger, fear, sadness, guilt, contempt and disgust. Positive and negative affects form a basic, underlying structure for people's emotional lives and are significantly related to the measures of personality and well-being (Watson, 2002).

Fredrickson's (2001) broaden-and-build theory of positive emotions describes how positive emotions open up one's thinking and actions to new possibilities and how this expansion can help build physical, psychological and social resources that promote well-being. Positive emotion broadens the outlook, offsets negative emotions, enhances resilience and improves one's emotional well-being that are important in dealing with nearly all life challenges. Positive emotions are more psychological in nature and depend on the appraisal and meaning of events in people's lives rather than just physical stimulation of the body. Both positive and negative emotions are incompatible with each other in the sense that it is hard to imagine experiencing both at the same time. Combinations of emotional feelings are certainly possible, but the simultaneous experience of both intense positive and intense negative emotions seems unlikely. Equally, Fredrickson and Losada (2005) hypothesized that the ratio of positive-to-negative emotions and behaviours that people experience during a given time period might be an index of the flourishing-languishing dimension. Flourishing is a state of optimal human functioning that is at the opposite end of the continuum from mental illness. In other words, flourishing is complete mental health. Languishing is a state that divides mental health from mental illness and is characterized by a feeling of emptiness, hollowness or what people used to call melancholy. Languishing individuals have few symptoms of mental illness, but they also have few symptoms of mental health. In other words, there is no serious pathology, but there is little purpose, meaning or zest for life either.

Military spouses are consistently exposed to numerous stressors due to military demands. Fredrickson's broaden and build theory of positive emotions shows the relationships among stress, positivity and depressive symptoms in military spouses during deployment. According to them, positive emotions down regulate the negative effects of stress. This interaction between stress and positivity revealed that higher levels of positivity protect military spouses from developing depressive symptoms at both low and high levels of stress (Fredrickson, 2001).

Building Family Resilience

Resilience is the capacity to adapt successfully in the presence of risk and adversity. It is a process that involves interaction between an individual, his past experiences and current life context. It is important for the military community with regard to keeping personnel fit for duty and promoting their health and well-being (Wiens & Boss, 2006). Similarly, it is important to build family resilience.

Family resilience has been defined as the characteristics, dimensions and properties of families which help them to be resilient to disruption in the face of change and adaptive in the face of crisis situations (McCubbin & McCubbin, 1989). Resilient families have the capacity to analyze and contextualize obstacles and possess a 'can do' spirit that supports initiative-taking and perseverance. By focusing on their strengths and potential for effectiveness, resilient groups nurture confidence in their member's abilities to overcome the odds (Walsh, 2002). When families approach challenges with a sense of optimism, confidence or hope and a positive emotional atmosphere, the evidence suggests that families are likely to do better (ibid.). Families are more likely to be resilient when they have clear allocations of roles, but are also able to adjust those allocations when challenging circumstances require it. Resilient families are also more likely to communicate and manage behaviour and relationships effectively. Specifically, they are able to share information, solve problems together and manage behaviour with appropriate use of warmth and limit setting (Black & Lobo, 2008). Thereby, building resilient families itself promotes the individual resilience in soldiers.

Striking Work–Family Balance

In order to adapt to the military lifestyle, it is essential to have a balance between the work and family life. Work and family domains are inter-dependent and complementary to each other. The imbalance exists when demands in one life domain (family/work) limit one's ability to complete required duties in other domains (family/work). Therefore, it becomes important to focus on how family and work domain can enhance one another. This mutual enhancement emphasizes on work–family facilitation which is based on the notion that synergy exists between the work and family roles. Work–family facilitation is defined as the extent to which experiences in one role improve the quality of life in another role (Frone, 2003). That is how involvement in one role positively influences another role. Facilitation is associated with various positive outcomes such as commitment, physical and mental health and contributes to the understanding of work–family dynamics above and beyond the conflict. Work-to-family and family-to-work facilitation involves psychological spill-over, a transitory phenom-enon, which includes the ways in which family (work) life affects an individual's energy level, attention span and mood, which in turn are brought into the work (family) setting by the individual (Crouter, 1984). Both directions of work–family facilitation are significantly and positively related to global outcomes such as mental health and life satisfaction. Work-to-family facilitation is characterized by 'one's involvement in work that provides skills and behaviours, which posi-tively influence the family, while family-to-work facilitation refers to one's involvement in family that results in positive moods, support and sense of accomplishment which help an individual to cope better, work harder and feel more confident for one's role at work (Wayne, Musisca, & Fleeson, 2004).

Conclusion

Understanding the concerns of military families has become a necessity. Military spouses are consistently exposed to numerous stressors and coping with these stressors requires skills. Therefore, there is a need to develop strategies to maximize the development of skills that are a good match to these stressors in order to promote the well-being of soldier and his family. It is equally essential for the military organization to improvise tools and improve access to support systems in the course of military-induced separations in order to keep feelings of isolation to a minimum. Identifying factors that strengthen resilience in military spouses, children and service members will serve as one of the most effective tools and techniques for promoting and sustaining well-being. Also, following the principles of positive psychology in military organization will help in promoting successful adaptation and achieving optimal functioning during times of stress, thereby enhancing not only the well-being of military families but also soldier's retention, readiness and effectiveness in the Armed Forces.

References

Andres, M. D., & Moelker, R. (2011). There and back again: How parental experiences affect children's adjustments in the course of military deployments. *Armed Forces and Society, 37,* 418–447.

Andres, M., Moelker, R., & Soeters, J. (2012). A longitudinal study of partners of deployed personnel from the Netherlands' Armed Forces. *Military Psychology, 24,* 270–288.

Belsky, J., & Hsieh, K. H. (1998). Patterns of marital change during the early childhood years: Parent personality, co-parenting, and division-of-labor correlates. *Journal of Family Psychology, 12,* 511–528.

Berry, J. O., & Jones, W. H. (1997). The Parental Stress Scale: Initial psychometric evidence. *Journal of Social and Personal Relationships, 12,* 463–472.

Berscheid, E. (2003). The human's greatest strength: Other humans. In L. G. Aspinwall & U. M. Staudinger (Eds.), *A psychology of human strengths: Fundamental questions and future directions for a positive psychology* (pp. 37–48). Washington, D.C.: American Psychological Association.

Black, K., & Lobo, M. (2008). A conceptual review of family resilience factors. *Journal of Family Nursing, 14,* 33–55.

Booth, B., Segal, M. W., & Bell, D. (2007). *What we know about army families.* Report prepared for the family and morale, Welfare & Recreation Command by Caliber. Fairfax, VA: ICF International.

Bourg, C., & Segal, M. W. (1999). The impact of family supportive policies and practices on organizational commitment to the army. *Armed Forces & Society, 25,* 633–652.

Brown, A. C., & Orhner, D. K. (1990). Relocation and personal well-being among adolescents. *Journal of Elderly Adolescence, 10,* 366–381.

Burrell, L. M., Adams, G. A., Durand, D. B., & Castro, C. A. (2006). The impact of military lifestyle demands on well-being, army, and family outcomes. *Armed Forces & Society, 33,* 43–58.

Carr, A. J., Gibson, B., & Robinson, P. G. (2001). Measuring quality of life: Is quality determined by expectations or experience? *British Medical Journal, 322,* 1240–1243.

Carver, C. S. (1997). You want to measure coping but your protocol's too long: Consider the brief cope. *International Journal of Behavioral Medicine, 4,* 92–100.

Chandran, A., Lara-Cinisomo, S., & Jaycox, L. (2009). Children on the home front: The experience of children from military families. *Pediatrics, 125,* 12–23.

Cohen, S. (1988). Psychosocial models of the role of social support in the etiology of physical disease. *Health Psychology, 7,* 260–297.

Cornille, T. A. (1993). Support systems and the relocation process for children and families. *Marriage & Family Review, 19,* 281–298.

Crouter, A. C. (1984). Participative work as an influence on human development. *Journal of Applied Developmental Psychology, 5,* 71–90.

Cummins, R. A. (2000). Objective and subjective quality of life: An interactive model. *Social Indicators Research, 52,* 55–72.

Diener, E. (1984). Subjective well-being. *Psychological Bulletin, 95,* 542–575.

Fredrickson, B. L. (2001). The role of positive emotions in positive psychology: The broaden-and-built theory of positive emotions. *American Psychologist, 56,* 218–226.

Fredrickson, B. L., & Losada, M. F. (2005). Positive affect and the complex dynamic of human flourishing. *American Psychologist, 60,* 678–686.

Frone, M. R. (2003). Work-family balance. In J. C. Quick & L. E. Tetrick (Eds.), *Handbook of occupational health psychology* (pp. 143–162). Washington, DC: American Psychological Association.

Griffith, J. (2002). Multilevel analysis of cohesion's relation to stress, well-being, identification, disintegration and perceived combat readiness. *Military Psychology, 14,* 217–239.

Hosek, J., Asch, B., Fair, C., Martin, C., & Mattock, M. (2002). *Married to the military: The employment and earnings of military wives compared with civilian wives.* Santa Monica, CA: RAND.

Hosek, J., Kavanagh, J., & Miller, L. (2006). *How deployments affect service members.* Santa Monica, CA: RAND.

Jensen, P. S., Lewis, R. L., & Xenakis, S. N. (1986). The military family in review: Context, risk and prevention. *Journal of the American Academy of Child Psychiatry, 25,* 225–234.

Jensen, P. S., Martin, D., & Watanabe, H. (1996). Children's responses to parental separation during Operation Desert Storm. *Journal of the American Academy of Child and Adolescent Psychiatry, 35,* 433–441.

Jones, S. M. (2003). *Improving accountability for effective command climate: A strategic imperative.* Carlisle Barracks, PA: US Army War College.

Kelley, M. (1994). The effects of military-induced separation on family factors and child behavior. *American Journal of Orthopsychiatry, 64,* 103–111.

Keyes, C. L. M., & Lopez, S. J. (2002). Toward a science of mental health: Positive directions in diagnosis and treatment. In C. R. Snyder & S. J. Lopez (Eds.), *The handbook of positive psychology* (pp. 45–59). New York, NY: Oxford University Press.

LaGrone, D. M. (1978). The military family syndrome. *American Journal of Psychiatry, 13,* 1040–1043.

Lazarus, R. S., & Folkman, S. (1984). *Stress, appraisal and coping.* New York, NY: Springer Publishing Company.

Mageroy, N., Riise, T., & Johnsen, B. H. (2007). Health related quality of life in the Royal Norwegian Navy: Does officer rank matter? *Military Medicine, 172,* 835–842.

McCubbin, M. A. & McCubbin, H. I. (1989). Theoretical orientations to family stress and coping. In C. R. Figley (Ed.), *Treating stress in families* (pp. 3–43). New York, NY: Brunner-Mazel.

Orthner, D. (2002). *Deployment and separation adjustment among army civilian spouses* (SAF IV Survey Report). Chapel Hill, NC: University of North Carolina.

Peebles-Kleiger, M. J., & Kleiger, J. H. (1994). Re-integration stress for desert storm families: Wartime deployments and family trauma. *Journal of Traumatic Stress, 7,* 173–194.

Riggs, B. (1990). Routine work-related absence: The effects on families. *Marriage and Family Review, 15,* 147–160.

Segal, M. W. (1986). The military and family as greedy institutions. *Armed Forces & Society, 13,* 09–38.

Vormbrock, J. (1993). Attachment theory as applied to wartime and job-related marital separation. *Child Development, 114,* 122–144.

Walsh, F. (2002). A family resilience framework: Innovative practice applications. *Family Relations, 51,* 130–137.

———. (2007). Traumatic loss and major disasters: Strengthening family and community resilience. *Family Process, 46,* 207–227.

Watson, D. (2002). Positive affectivity: The disposition to experience pleasurable emotional states. In C. R. Snyder & S. J. Lopez (Eds.), *Handbook of positive psychology* (pp. 106–119). New York, NY: Oxford University Press.

Wayne, J. H., Musisca, N., & Fleeson, W. (2004). Considering the role of personality in the work-family experience: Relationship of big-five to work-family conflict and facilitation. *Journal of Vocational Behaviour, 64,* 108–130.

Wiens, T. W., & Boss, P. (2006). Maintaining family resiliency before, during, and after military separation. In C. A. Castro, A. D. Adler & C. A. Britt (Eds.), *Military Life: The psychology of serving in peace and combat* (pp. 13–38). Westport, CT: Praeger Security International.

Wood, S., Scarville, J., & Gravino, K. (1995). Waiting wives: Separation and reunion among Army wives. *Armed Forces & Society, 21,* 217–236.

WRAL Raleigh. (2008). *Army secretary: Soldier moved from eight barracks after review.* Retrieved from http://www.wral.com/news/state/story/2847618/

Zvonkovic, A. M., Solomon, C. R., Humble, A. M., & Manoogian, M. (2005). Family work and relationships: Lessons from families of men whose jobs require travel. *Family Relations, 54,* 411–422.

15

Countering Terrorism: Interrogating Communication Oversight

James Okolie-Osemene

The emergence of Boko Haram terrorism marked a watershed in Nigeria's political, social and economic milieu. There is no doubt about the truism that it took the Federal Government of Nigeria (FGN) many years to grasp the recruitment, membership and factors that sustain operational efficiency of the Islamist sect. The problematic aspect of the Boko Haram monster is the inability of the FGN to tame the sources of Boko Haram's funding and failure of security forces to contain the sources of arms despite their heavy deployment, aerial bombardment and discovery of some Boko Haram camps and bomb factories in Northeastern enclave. Borno State is located in Northeastern part of Nigeria and shares boundary with Cameroon, Lake Chad and Niger Republic. Various scholars of trans-border studies and practitioners in the security sector hold the view that porous borders in most communities contribute to the complexities of counterterrorism. This problem is as a result of unmanned borders and topography of the Northeast. Nigeria's Armed Forces consist of the Army, Navy and Air Force who are trained to protect the nation's territorial integrity. More so, upsurge in violence in recent months makes Borno State the hotbed of Boko Haram terrorism which is at the detriment of rural people and those in the state capital, Maiduguri. It is obvious that Boko Haram members live with the people, attend the same mosques, visit same markets and even enter taxis/buses with residents of Borno, they are neither spirits or faceless as politicians want us to believe.

Terrorist financing is linked with money laundering (Golwa, 2010), terrorism remains a threat that haunts the continent of Africa (Luz, 2013), whereas most countries consistently embark on security sector reforms to meet up with the challenges posed by this global monster that threatens stability of states, especially now that Somali militia groups operate in Kenya's remote and arid North-eastern Province, an area that borders southern Somalia—a former stronghold of the extremist group Al-Shabaab (Gathigah, 2013). It is noteworthy that Al-Shabaab (which means youth in Arabic) was born in Somalia in 2004, as a result of the defeat of the Islamic Courts Union (ICU) during the war in Somalia (2006–2009). The group's responsibility of the 21 September 2013 attack that left more than 70 dead and 170 injured in a mall in Nairobi, Kenya, showed its strength and source of worry to leaders within Africa and other parts of the world (Luz, 2013). Just like Al-Shabaab attacks, foreigners also suffer the activities of the Boko Haram group in Nigeria with grave human insecurity situation. Also, recent incidents showed that just like Kenya which is caught up in the crosshairs of global terrorism mainly due to its association with Western countries and presence of their installations in the country (Kagwanja, 2012), Nigeria and other African nations have also become targets and destinations for those groups that use terror as weapon to punish perceived groups with rival ideologies.

The name 'Boko Haram' is no longer new in many Nigerian homes, media and the entire political landscape of the nation. Many authors have written a lot about the Islamist group Boko Haram. They include the following: Danjibo (2009), Aleyomi (2012), Bamidele (2012), Marc-Antoine (2012), Pham (2012), Amaraegbu (2013), Adeyeye (2013), Barkindo (2013), Newman (2013), Fayeye (2013), Chikwem (2013), Sodipo (2013), amongst others. The list, however, shows that not much scholarly work has been written to accommodate the activities of the Multi National Joint Task Force (MNJTF) which presently battles with Boko Haram in their Borno strongholds. The trouble with Boko Haram sect is gross intransigence and unnecessary violation of fundamental human rights.

The Islamist sect Jama'atu Ahlus-Sunnah Lidda'Awati Wal Jihad is globally known as *Boko Haram* which remains unacceptable to the group. It is worthy of note that the word 'boko' is often mistaken as book. According to Muhammad, something that involves any form of deception (as cited in Newman, 2013). In his work on the etymology of Hausa Boko, Newman (ibid., p.11) opines that 'boko' is a native Hausa word, originally meaning sham, fraud, inauthenticity, education of sham/unimportance). And that it has nothing to do with 'book'. In essence, Boko Haram is a violent non-state armed group which does not have a specific grievance. It rather wants a radical structural change at the detriment of Nigerian state and intergroup relations. It is obvious that Boko Haram means different things to different people. To some, it is anti-Western and anti-government group; some Nigerians see the group as mirror effect of decades of deprivation, economic marginalization or exclusion, whereas to others it remains a group that upholds the principles of Islam (Okolie-Osemene, 2013).

In his study on *Six Lessons of Suicide Bombers*, Brym (2010) listed only nine countries where most suicide/terrorist attacks were concentrated: Lebanon, Sri Lanka, Israel, Turkey, India (Kashmir), Russia (Chechnya), Afghanistan, Iraq and Pakistan. The author did not include Nigeria at that time, given that the intensity of Boko Haram terrorism and level of the group's radicalization was low until after Nigeria's independence anniversary of 2010 and 2011 general elections when the group became more radicalized and sophisticated both in terms of weaponry, attacks and membership. It should therefore be noted that Northern Nigeria of today, especially Borno State, is not different from all the nine countries listed above, especially when talk is about negative peace, human security and stability due to activities of terrorists who operate as armed gangs, assassins, kidnappers and suicide bombers. There is hardly any day that passes without reports of one incident or another in the area. According to Nwozor (2013), Boko Haram has carved a niche for itself as a ruthless terrorist organization. Pham (2012, p.1) says about the intensity of Boko Haram imbroglio, "Since late 2010, the organization has been responsible for a brutal campaign of attacks targeting public officials and institutions and, increasingly, ordinary men, women, and children, wreaking havoc across northern Nigeria."

Terrorism makes the entire North-eastern Nigeria one of the most dangerous places in the Global South. Boko Haram also carries out activities that relate to what Dzurgba (2006) refers to as 'local terrorists' through their kidnapping profession, maiming, assassination of innocent people in Borno State and neighbouring states. Although some people argue that terrorism is a global phenomenon, given the potency of well organized multinational networks spanning beyond Africa, others have the view that terrorism in Africa is Africa made.

Nigeria is at the forefront of counterterrorism action in West Africa because of her diplomatic and political influence in the sub-region (Imohe, 2010). It is not disputable that Nigeria has outstripped other countries in Africa such as Morocco, Libya, Egypt and Somalia, Kenya in the terrorism casualty index due to daily attacks targeted at security forces, residential areas and religious institutions, especially in Northeast where Borno State is located.

Defining Terrorism

Terrorism is as old as humanity and not a new phenomenon but the intensity of terror has increased over the years (Ashara, 2013; Oshanugor, 2004; Shaw, 1997). It is an elaborate phenomenon which has global dimension and attracts responses from various quarters. In her work on 'Disciplining Terror', Stampnitzky (2013) links the historical antecedent of terrorism to political violence, a transformation that eventually led to the war on terror. Ojakorotu (2011, p. 97) notes that, "terrorism has been one of the most malignant features of domestic and international politics for centuries".

Article 1 of European Convention on Suppression of Terrorism, adopted by Council of Europe on 27 January 1977, describes terrorism as follows: a serious offence involving an attack against the life, physical integrity or liberty of internationally protected persons, including diplomatic agents; an offence involving kidnapping, the taking of a hostage or serious unlawful detention; an offence involving the use of a bomb, grenade, rocket, automatic firearm or letter or parcel bomb if this use endangers persons, amongst others. Oshanugor (2004, pp. 4–5) states that

> while acts that convey terrorist impressions are often viewed from different perspectives as those considered terrorists by one group of people or government of a state may be regarded as heroes or freedom fighters by others, some members of society who are bored and/or sadistic may terrorise others to express their frustrations; vent their rage, or engage in symbolic acts of protest against society.

For instance, many governments use assassinations, massacres and other forms of cruelty to sustain power or gain territory (Ashara, 2013, p. 3). Notable examples of such violence were recorded during the Nigerian-Biafran between 1967 and 1970, Libyan crisis of 2011 that later led to the death of Muammar Gaddafi on 20 October 2011, the 1994 Rwandan genocide, the 2011 Ivorian crisis and political instability which culminated in the arrest of then President Laurent Gbagbo, as well as the war in Sudan which gave birth to the Darfur Peace Agreement signed in July 2011 between the Government of Sudan and Liberation and Justice Movement with the aim of establishing compensation for victims.

The forgoing case studies make people to refer terrorism as indiscriminate killing of people for political goal; and containing terror demands that policy makers understand the psychic of the terrorists and find ways of dealing with them (Kwesi, 2013). Various scholars point to the fact that the ambiguity in giving a specific definition to term terrorism led to the cliché that 'one man's terrorist is another man's freedom fighter (Ashara, 2013; Oshanugor, 2004). This is corroborated by the truism that the term 'terrorist' is viewed according to the side of the fence one is sitting and defined according to the methods used in achieving the objectives (Momoh, 1994). In essence, the socio-political positions of actors involved and strategies adopted are critical to explaining terrorism. What scholars and practitioners agree is that terrorism is associated with mass violence and fatalities.

The Jonathan Netanyahu Institute in Israel defines terrorism as the systematic murder, maiming and menacing of the innocent in order to instil fear for political ends; it involves the use of violence to intimidate innocent population for political ends (Momah, 1994). Imohe (2010) describes terrorism as transnational phenomenon which increases threat to human life, state stability and international security. Some notable acts of terror include night/day assassinations at residential areas or public places (including motor parks, markets, airports), suicide bombing, hostage-taking, kidnapping, sabotage of economic infrastructures such as oil pipelines, attack on religious institutions during worship among others.

Boko Haram terrorism raised question on whether the imbroglio is a religious uprising or a political contest for power (Marc-Antoine, 2012). Although Boko Haram advocates for the reformation of Islam and the implementation of Islamic law as the only valuable option for social justice and prosperity (Barkindo, 2013), Capell and Sahliyeh (2007, p. 267) stated that in an effort to understand modern terrorism's increased lethality, we propose that scholars need to look further than religion as a motive and take into account modern terrorists' willingness to use 'suicide terror' as their primary modus operandi. In addition to the role of religion, the tactic of suicide terrorism, which has become popular with many groups active today, accounts for terrorism's new lethality.

This makes Brym (2010) argue that those actively involved in suicide attacks are not crazy and not motivated principally by religious zeal.

However, terrorism differs from country to country depending on the ideologies of the groups involved and political structure of the nation affected. Notably, terrorists consistently devise new ways of carrying out their nefarious activities which they sometimes justify on the basis of perceived grievances or marginalization. Oshanugor (2004) sees the period 1995–1998 as Nigeria's years of terror due to sporadic bomb explosions, unexplained deaths and disappearances, the allegations and counter allegations targeted at Abacha's military regime and also groups that wanted the end of the regime. It should be pointed out that those threats which were more political than economic or social are in no way comparable with the intensity of Boko Haram terrorism which has recorded fatalities more than most terror incidents in other countries.

This chapter defines terrorism as an act that inflicts sorrow on victims with the aim of disintegration. One notable attribute of terrorism is that it undermines social order leading and creates atmosphere of instability due to destructive manifestations that accompany it. Terrorists are individuals or groups whose actions are not only injurious to public interest but inimical to social order and intergroup relations. There is hardly any terrorist act that is not associated with violence and sorrow. The costs of such actions have debilitating social, psychological and physical implications. In terms of Boko Haram context, terrorists are groups who hardly go to government but send warnings to the public, government officials and media, notifying them of imminent attack and be mindful of their utterances. They launch attacks at strategic places, and capitalize on media propaganda before or after attacks. Terrorists just like armed gangs make governments embark on security sector reforms as well as strengthening of early warning system across the nation.

The Boko Haram Threat

According to Nwozor (2013), distinguishing feature of insecurity in Nigeria is the ubiquity of violence unleashed by ethno-religious conflagration, hostage-taking and kidnap-for-ransom and terrorism. But, Sodipo (2013) maintains that Northern Nigeria has been the locus of an upsurge in youth radicalization and virulent militant Islamist groups in Nigeria since 2009. The 2012 Global Terrorism Index shows that Nigeria's ranking on the Global Terrorism Index rose from 16th out of 158 countries in 2008 to 6th (tied with Somalia) by the end of 2011 (ibid.). It should be pointed out that the value placed on peace and human security, as well as the need to curtail Boko Haram's threat to national security, made the government embark on counterterrorism against Boko Haram sect in Borno state through the MNJTF. According to Nwolise (2009), security is important to individuals, groups, nations and the world because without security, all things else are meaningless; it ensures the preservation of life, liberties and states; it is a prerequisite for any meaningful development, sustainable order, peace and social harmony; it ensures the smooth running of strategic installations such as electricity, military, aviation and shipping. In addition, Nwolise (2009) also highlights some consequences of the absence of national security; thus, lives are lost unnecessarily, property and infrastructure are vandalized, looted, destroyed or stolen; citizens take to self-help; funds are spent unnecessarily on personal security; ethnic groups raise ethnic militias to defend group interests; law enforcement agents attract bad labels of professional ineptitude, incompetence and ineffectiveness; people lose confidence in the ability of government to protect them.

From the above assertions, it could be adduced that security is very paramount for social order, cohesion and actualization of effective development plans. For instance, many months of terrorist attacks in Borno took the states many years backwards due to the economic costs of bomb blasts and attendant atmosphere of fear/hot peace.

Theoretically, Animasawun and Luqman (2013) avers that the government responds to the activities of Boko Haram with orthodox approach to terrorism theory premised on the legitimacy/illegitimacy dualism which constructs non-state violence as terror while state violence as legitimate. Terrorism is, therefore, illegitimate and unacceptable to the state. This made President Jonathan to declare on 1 October 2013 that the Federal Government will spare no cost and that no idea will be ignored. He also assured the nation that Boko Haram would not succeed because Nigerians must never lose sight of their freedom and values.

Containing The Terror

For containing the terror, Crime Control Model as identified by Brewer et al. (1996) can be adopted. It emphasizes upon assertive patrol activity, a prominent street-level police presence, abstractive street contact caused by an extensive use of stop and search powers and a somewhat attitude to civil rights. The Control Balance–Deviance Theory is premised on the amount of control people are subjected to, and exercise that influences their type of deviant behaviour (Tittle, 1995). For instance, Animasawun and Luqman (2013) vividly explains that stop and search operations, door-to-door security searches of weapons and sometimes killing of suspected terrorists, issuance of ultimatum, placement of price tags for notorious members of the sect are some of the approaches adopted by the Joint Task Force (JTF) which have had favourable psychological effects on the people in the northern enclave. In the past, the Federal Government had fixed different price tags for relevant information from residents of northern region that could facilitate the arrest of Boko Haram leaders declared wanted. In 2012 for instance, the Federal Government declared a bounty each of N25 million on the head of the deputy leader of Boko Haram Momodu Bama and N10 million on Zakariyya Yau. Consecutively, Momodu Bama (also called Abusaa), Zakariyya and Zakariyya's father, Abatcha Flatari were killed on 14 August 2013 by security forces in Mubi.

Nigeria's Scholar and Nobel Laureate, Prof. Wole Soyinka, believes that Boko Haram members have limited knowledge of Islam and need to be retrained on the content of Islam (as cited in Ajetunmobi, 2013). This is premised on the assertion that Islam is a religion of peace. Even Nigeria's President Jonathan on 15 October 2013, during Eid-el-Kabir message, stated that the Boko Haram imbroglio does not have links with religion and ethnicity, and called on insurgents to embrace peace.

Multinational Joint Task Force (MNJTF)

Joint security operation is one of the strategies adopted by governments in most parts of the world to contain all forms of internal aggressions, especially when terrorism or insurgency is concerned. Such security operation strengthens the state's instrument of violence, because most times non-state armed groups attempt to usurp states' monopoly of violence by launching simultaneous attacks that often overwhelm security forces. For instance, the entry of the Kenyan Defence Forces (KDFs) into Somalia marked the beginning of military expedition to prevent the spread of terror attacks and influence of al-Shabaab Islamists in Somalia which resulted in weakening of the group's nerve (Kagwanja, 2012). The FGN also formed JTF saddled with the responsibility of containing Boko Haram insurrections, with notable confrontational approach in 2003 in Kanama and Geidam as well as other expeditions in Gwoza hills in Borno State which led to the death of 27 Boko Haram members (Animasawun & Luqman, 2013). Such confrontation contributed to the arrest of former Boko Haram leader, Mohammed Yusuf, who died in police custody in 2009 which scholars and political analysts described as extra-judicial killing. According to the Ministry of Defence, the Operation Restore Order was established to cut down on the activities and intensity of the Boko Haram Sect in the North East zone where Borno State is located. The MNJTF is made up of security forces Armed Forces from Chad, Niger and Nigeria, with Nigeria playing a prominent role in the multinational stability operation.

The historical antecedent of MNJTF is traceable to its metamorphosis from a military-led JTF, code-named Operation Restore Order comprising personnel from the Nigerian Armed Forces, Nigeria Police Force (NPF), the Department of State Security (DSS), Nigerian Customs Service (NCS), Nigeria Immigration Service (NIS) and the Defence Intelligence Agency (DIA), that was earlier deployed on 12 June 2011 for operations in Borno and Yobe States. The mandate of MNJTF to cover counterterrorism operations was extended from arms proliferation to management of trans-border crimes in April 2012 (The National Human Rights Commission, 2013). Over 8,000 troops are involved in the counterterrorism in the area. The emergence of MNJTF was occasioned by a series of coordinated bomb blasts, armed attacks and abduction of ordinary citizens and political elites by terrorists which culminated in a total breakdown of law and order in Borno State.

Some political analysts and scholars hold the view that any security arrangement that involves JTF has some shortcomings. According to Nwozor (2013, p. 23)

> The setting up of Joint Task Forces is a de facto declaration of a state of emergency. Joint Task Forces have different rules of engagement, which are outside the normal operational boundaries of regular security agencies. Their mandate is a quasi-empowerment to engage in war. The code names often associated with specific Joint Task Force operations underscore its direction and strategy. Generally, the strategies of Joint Task Forces are anchored in their ad hoc composition and non-allegiance to any specific security agency. Their driving philosophy is shaped by the mindset that only superior force can tackle insecurity.

The mandate of MNJTF is premised on the need to restore public order, prevent internal aggression and protect Nigeria's territorial integrity. It should be noted, however, that only the police have constitutional responsibility to forestall breakdown of law and order in the polity, but the escalation of Boko Haram imbroglio demanded military involvement to contain the spate of radicalization which subjected residents to humanitarian crisis and downplayed human security in Northeast.

Military and Human Rights Imbroglio

The involvement of military in managing or confronting radicalized groups, provides opportunity for large-scale human rights violations, just like in Rivers State where the JTF further aggravated the security situation in high-density neighbourhood at Marine Base and Port Harcourt city due to the lack of previous training in policing or human rights as the team became a catalyst for violations (Ugwu, 2008). The situation is not different from that of present-day Borno state where there have been conflicting reports on the military's involvement in human rights violations. For instance, there is argument that violation of human rights is always in response to security situation, because security forces are usually the targets of attacks and not adequately compensated (Animasawun & Luqman, 2013). Such volatile situation exposes lives of security forces and civilians to danger because it is not always easy to identify terrorists in civilian populated areas when they are not armed. Some residents still believe that though MNJTF are capable of countering the activities of terrorists, majority of civilians still suffer the consequences of the seemingly hot peace, with increased Boko Haram attacks mostly at nights.

It was also gathered that Nigerien soldiers in the Multinational JTF killed an unspecified number of Boko Haram members on some island communities near Lake Chad on Sunday, 6 October 2013, whereas an offensive launched against the Boko Haram fundamentalist Islamist group by Cameroonian soldiers on Tuesday, 8 October 2013, led to the killing of over 180 members around a Nigerian border with Cameroon. The wounded insurgents were reported to have been arrested, whereas the soldiers recovered over 200 rifles, 70 machine guns and heavy military hardware from the terrorists (Soriwei, 2013). The Director of Defence Information, Brigadier General Chris Olukolade, stated that

> Nigeria's neighbouring countries, either through the instrumentality of the Multi National Joint Task Force or through their security instruments, are involved in operations to complement what Nigeria is doing against terrorism; that the MNJTF and Nigerian security agencies are combing everywhere for any strange movement in our territory. (as cited in Soriwei, 2013)

Stakeholders involved in counterterrorism are as follows: Defence Ministry, Chief of Army Staff, Security forces from Chad, Niger, Nigeria and their Defence Ministers, The Director of Defence Information, Director Army Public Relations, Unit Commanders, JTF Spokesman, Lt. Colonel Sagir Musa, Borno State Government, Traditional Rulers and Civilian JTF formed by Vigilante Youths. They work to enhance the effectiveness of Baga-based MNJTF and Maiduguri-based JTF.

After Boko Haram utilized Thuraya phones to coordinate assassination of some Imams and civilians in schools between May and June 2013, the MNJTF coordinated offensive that enabled them to wrest back control of the remote northeast from Boko Haram, thereby destroying their bases and arrested hundreds of suspected terrorists (Opara, 2013). It is unfortunate that MNJTF's information and communication management by ordering the disconnection of mobile phone networks greatly impeded early response by civilians during emergencies and situations in which they needed to report suspicious movements. MNJTF operations are anchored on Defence policy objectives which aim to protect Nigeria's interest under the ambit of the constitution, including the following: protection of Nigeria's sovereignty against internal and external threats, provision of strategic advice and information to government, promotion of security consciousness of Nigerians among others. The above objectives motivated the deployment of armoured personnel carriers, armoured fighting vehicles, to contain the sophistication of Boko Haram.

Meisels (2008) identified various issues that are also relevant to any discourse on Boko Haram: expression of sympathy, justification for Islamists and reduction in civil liberties in exchange for greater security. From the above issues raised by Meisels (2008) in what could be described as security and liberty tension in a period of terror, the MNJTF ordered the disconnection of mobile telephone services, followed by stop and search operations in Borno State. The need to curb terror network in Borno State occasioned the disconnection of mobile phone services to strengthen MNJTF's intelligence against the Boko Haram monster in all parts of the state. Some Nigerians have the perception that individuals sympathetic to Boko Haram have either infiltrated the Nigerian army or are also political office-holders, given the group's coordinated attacks, intelligence and sophistication in terms of weaponry. It is noteworthy that one act that has not been practiced by Boko Haram terrorists is hijacking (of vehicles or planes) which is a recurrent incident in other countries.

Unlike the police who are inadequate in terms of equipment to suppress activities of terrorists, the military seem to have enough equipment to contain the terror monster. But, the war against terror is beyond the acquisition of only sophisticated weapons, and this makes information discrimination and communication critical through early warning and response systems to curtail threats to human security. This is against the backdrop of the truism that terrorists are not spirits, they live in communities and also visit public places like normal citizens.

Challenges to and Limitations of JTF

The four states that witnessed Boko Haram riots from 25 July to 30 July include Bauchi, Kano, Yobe and Borno, but Borno saw the most extensive rioting mainly because it was the base of the movement and its leader (Adesoji, 2010). The situation took a more dangerous dimension in 2013 when the group embarked on massive killings in public places and highways by mounting roadblocks and camouflage as security forces with the aim of killing commuters. For instance, over ten persons were killed by Boko Haram terrorists when they launched an attack along the Damaturu–Maiduguri highway, set ablaze several vehicles, including those conveying food items to Borno State. From media reports, the incident took place kilometres from Benisheik Town where terrorists dressed in military fatigue and riding in armoured tanks killed at least 140 travellers on 17 September 2013 (Audu, 2013).

One wonders how this group succeeds in camouflaging within Borno State without residents having knowledge of their operational base to forestall the consistent breakdown of law and order by the sect. Animasawun and Luqman (2013) opine that the perception of security forces' proclivity for oppressing the masses makes it difficult for the people to dish out valuable information that may aid in military operations. The tip-off that led to killing of Zakariyya Yau's father and Abatcha Flatari who was also said to be the spiritual mentor of the group on 14 August 2013 by security forces in Mubi shows that the group enjoys support from some residents and people at grassroots. The military reported in August 2013 that Boko Harm leader, Abubakar Shekau, may have died at the forests of Sanbisa in Borno State where he was said to have been shot by security forces during a raid on his hide out. Contrary to the claims by men of the MNJTF, the President of Nigeria, Goodluck Jonathan, on 29 September 2013 enthused that he did not have the knowledge of Shekau's alleged death and that journalists should know better than him. Even though some people argued that the President was right that journalists quote as if they are close to them, some Nigerians described such statement as not only synonymous with hopelessness and confusion, but also discouraging and thus portrayed the government as not capable of devising a way to track down Boko Haram leaders that have been on the wanted list for years due to poor intelligence. Some people even went as far as comparing the President with President Obama of United States who has been instrumental to United State's success in devising a plan to track down Osama Bin Laden who was haunted for a decade since the September 2001 terrorist attack in the USA, with the assertion that Nigeria's President could do the same against Boko Haram to avoid conflicting reports. From the President's statement during the televised media chat, it appears that there is a disconnect between the men of MNJTF in Borno State, Military Headquarters and the Presidency. Nigerians expect the security forces to not only carry the Commander-in-Chief of the Armed Forces along but also dish out security reports that would facilitate their operations in Borno communities.

Boko Haram's sophistication manifested in attacks at various police stations and military barracks in Borno State. It is worthy of mention that apart from abduction of expatriates, one of the notable foreign dimensions of the Boko Haram radicalization was the coordinated attack at the United Nations House in Nigeria's Federal Capital Territory Abuja on 26 August 2011 which led to the death of over 33 civilians, 11 UN personnel and many people injured. The attack heightened interest and debates on the proscription of the group as a global terrorist organization.

The challenges of intra-MNJTF operations and prospects of success in the counterterrorism are noteworthy. Most security operations face different challenges and the ability of the team or task force involved to overcome the challenges would definitely determine the success of the operation embarked upon. Terrorism in Borno state is aggravated by porous borders, intelligence failure and the existence of bomb factories in remote communities in Northeast which complicate the counterterrorism efforts. A situation in which there is no early response to signs of security threats, terrorists are likely to take advantage of the development to recruit more members, acquire more weapons, construct safe havens and even attack financial institutions.

It is worthy of mention that Boko Haram's terrorism in Borno is further consolidated by the successes recorded in attacking banks and escaping with unspecified amounts of money. The call by Niger State Governor and Chairman of Northern Governors Forum, Babangida Aliyu, on Boko Haram terrorists to embrace government's amnesty offer fell on deaf ears as the group neglected ceasefire and possible disarmament. The sect consistently planned and executed more attacks with increased casualties among civilians and security forces despite various checkpoints and joint security patrols in Borno State.

Recent observations and reports show that Boko Haram has developed international connections which strengthen its capacity to challenge security forces and threaten Nigeria's internal security (Marc-Antoine, 2012). Notably, this was widely reported by media organizations that the leader of the sect, Abubakar Shekau, while hiding in unknown location within Borno State, cried out for help from the group's 'Brethren in the Middle East'. According to him "our Brethren please come to us and join this war." This was the period that MNJTF operatives stormed most communities in Borno State and launched offensive against members of the sect in their various camps between June and September 2013.

The United States met with FGN to express its willingness to assist Nigeria by offering modern technological assistance to fortify security at the borders with the aim of curbing the activities of terrorists in the country. The leader of the US delegation, and the Counter Terrorism Deputy Secretary, Ms. Anne Witkowsky, said the aim of the visit was to avail Nigeria to modern technology of securing borders by showing the Personal Identification Secure Comparison and Evaluation System (PISCES) and the Demonstration of the PISCES equipment. It is a border control system with 10 fingerprints, which ensures that if the name and other means of identification were falsified, the 10 finger-prints cannot (Iroegbu & Akinwale, 2013). The murderous character of Boko Haram terrorists remains the threat to the operations of the MNJTF who battle daily to end the breakdown of law.

So far, it has not been easy for MNJTF to forestall all attacks through tip-offs to the extent that security forces arrive at scenes of attacks after the acts of terror have been executed by the terrorists. This is aggravated by the fact that it is not part of MNJTF's operational strategy to be present at residential areas at night or in the markets in the day time.

The Twelve Rules

Terrorism is linked to myriads of intrastate political, social and economic issues. Marc-Antoine (2012) argues that the terrorist evolution of Boko Haram was mainly caused by the brutality of the state repression, but the fact that Boko Haram condones violence to achieve ideological ends, including acts of terrorism (Sodipo, 2013), makes it crucial for Nigerian state to consider some of the 'Twelve Rules for Preventing and Countering Terrorism' as outlined by Schmid (2012). Rules are as follows: prevent radical individuals and groups from becoming terrorist extremists by confronting them with a mix of 'carrot and stick' tactics; deny terrorists access to arms, explosives, false identification documents, safe communication, safe travel and sanctuaries; disrupt and incapacitate their preparations and operations through infiltration, communication intercept, espionage and by limiting their criminal and other fund-raising capabilities; keep in mind that terrorists seek publicity and exploit the media and the Internet to propagate their cause, glorify their attacks, win recruits, solicit donations, gather intelligence, disseminate terrorist know-how and communicate with their target audiences; and try to devise communication strategies to counter them in each of these areas. The carrot and stick tactics manifests in setting up of roadblocks, restriction of movement, parking of vehicles and motorcycles around places of worship to curtail bomb blasts among others. Apart from disconnection of mobile telephone networks after declaration of state of emergency in May 2013, the above rules no doubt have not been given much attention by stakeholders including the MNJTF, especially the issues of arms proliferations, funding and operations of foot soldiers and bombers. There are allegations that the MNJTF forces no longer observe rules of engagement due to some reports on human rights violations in Borno State which affect residents.

Conclusion

Terrorists always pursue ideologies that challenge state policies. They recruit members through indoctrination or by threat to safety of relatives. It poses a grave threat to the forces which are supposed to plan and lead the counterterrorism operations. Interrogation of the operational and communication oversight of MNJTF, formulated for taming the Boko Haram monster in Borno State and other states in North-East, offers several concerns in this direction. It calls for re-examination of MNJTF's operations to comprehend BH's modus operandi both in terms of arms acquisition and deployment. Further, eradication of terrorism demands efforts that must go beyond military deployment and manoeuvres, and focus more on restoring people's confidence in them through enlightenment on the economic, security and political costs of terrorism in affected states. This would motivate more people to give MNJTF relevant information on the sect's modus operandi in Borno State. To make this realistic, community-based approach to human security needs to be mainstreamed into MNJTF's cooperation with civilian JTF in Borno State. Such an ability of security forces to identify those responsible for burying arms in cemeteries, and terrorists who hide in uncompleted buildings to carry out nefarious activities, could contain the Boko Haram sect. Equally, the proven model of countering terrorism by community-based approach can be applied across globe to contain the terror threat wherever applicable.

References

Adesoji, A. (2010). The Boko Haram uprising and Islamic revivalism in Nigeria. *Africa Spectrum, 45*, 95–108.

Adeyeye, A. I. (2013). Identity conflict, terror and the Nigerian state: Between fragility and failure. *Journal of Sustainable Development in Africa, 15*, 116–129. Retrieved from http://www.jsd-africa.com/Jsda/Vol15No4-Summer2013B/PDF/ Identity%20 Conflict%20Terror%20and%20the%20Nigerian%20State.Adebowale %20Idowu%20 Adeyeye.pdf

Ajetunmobi, A. (2013, April 19). Demobilise B'Haram by Qur'anic arbitration, not amnesty. *The Punch*. Retrieved from http://www.punchng.com/opinion/demobilise-bharam-by-quranic-arbitration-not-amnesty/

Aleyomi, M. B. (2012). Ethno-religious crisis as a threat to the stability of Nigeria's federalism. *Journal of Sustainable Development in Africa, 14*, 127–140.

Amaraegbu, D. A. (2013). Failure of human intelligence, Boko Haram and terrorism in Nigeria. *Journal of Sustainable Development in Africa, 15*(4), 66–85. Retrieved from http://www.jsd-africa.com/Jsda/Vol15No4-Summer2013B/PDF/ Failure %20 of%20 Human%20 Intelligence.Declan%20Amaraegbu.pdf

Animasawun, G., & Luqman, S. (2013). Causal analysis of radical Islamism in northern Nigeria's fourth republic. *African Security Review, 22*, 216–231.

Ashara, D. U. (2013). *The trauma of Nigeria's terror ordeals: Trends, effects and panacea*. Kano: Center for Crisis Prevention and Peace Advocacy.

Audu, O. (2013, September 30). Suspected Boko Haram members behead 10 travellers along Borno-Yobe highway. *Premium Times*.

Bamidele, O. (2012). Boko Haram catastrophic terrorism: An albatross to national peace, security and sustainable development in Nigeria. *Journal of Sustainable Development in Africa, 14*, 32–44.

Barkindo, A. (2013). Join the caravan: The ideology of political authority in Islam from Ibn Taymiyya to Boko Haram in North-eastern Nigeria. *Perspectives on Terrorism, 7*, 30–43.

Brewer, J. D., Guelke, A., Hume, I., Maxon-Browne, E., & Wilford, R. (1996). *The police, public order and the state* (2nd ed.). New York, NY: St. Martin's Press.

Brym, R. J. (2010). Six lessons of suicide bombers. In M. Hughes & J. Ryan (Eds.), *6 e Readings in Sociology*. USA: The McGraw-Hill.

Capell, M. B., & Sahliyeh, E. (2007). Suicide terrorism: Is religion the critical factor? *Security Journal, 20*, 267–283.

Chikwem, F. C. (2013). Boko Haram and security threat in Nigeria: A new twist of political game in town. *Review of Public Administration and Management, 1*, 156–172. Retrieved from http://www.arabianjbmr.com/pdfs/public/3.pdf

Danjibo, N. D. (2009). Islamic fundamentalism and sectarian violence: The 'Maitatsine' and 'Boko Haram' crises in Northern Nigeria. *IFRA-Nigeria E- Papers, 1*, 12.

Dzurgba, A. (2006). *Prevention and management of conflict*. Ibadan: Loud Books Publishers.

Fayeye, J. O. (2013). The Boko Haram phenomenon and the other side of the coin in the Nigerian experience. *African Journal of Peace and Security, 1*, 37–45.

Gathigah, M. (2013, October 8). Kenya, Somalis caught between terrorism and a border dispute. *All Africa News*. Retrieved from http://allafrica.com/stories/201310080869. html

Golwa, J. H. P. (2010). ECOWAS and management of security in Africa. In C. O. Eze, A. C. Anigbo, & C. Q. Dokubo (Eds.), *Nigeria's security interest in Africa*. Lagos: Nigerian Institute of International Affairs.

Imohe, E. E. (2010). Extra-regional security challenges for Nigeria. In C. O. Eze, A. C. Anigbo, & C. Q. Dokubo (Eds). *Nigeria's security interest in Africa*. Lagos: Nigerian Institute of International Affairs.

Iroegbu, S., & Akinwale, A. (2013, April 30). Terrorism: US to assist Nigeria fortify border security. *This Day*. Retrieved from http://www.thisdaylive.com/articles/terrorism-us-to-assist-nigeria-fortify-boarder-security/146373

Kagwanja, C. (2012). Religious pockets of conflict – dealing with resurgence of radical Islamism and terrorism in Kenya. *Horn of Africa Bulletin*, July–August, 1–5.

Kwesi. (2013, October 2). Conscious Vyb. *Sahara Reporters Radio*.

Luz, N. (2013, October 2). Terrorism in Africa: A threat that haunts the continent. *Awake Africa*. Retrieved from http://awakeafrica.org/archives/3582#sthash.Jh9bpar6.65EaEYCD.dpbs

Marc-Antoine, P. D. M. (2012). Boko Haram et le terrorisme Islamiste au Nigeria: insurrection religieuse, contestation politique ou protestation sociale? *Questions de Recherche / Research Questions, 1*, 33. Retrieved from http://www.ceri-sciences-po.org/publica/question/qdr40.pdf

Meisels, T. (2008). *The trouble with terror*. Cambridge, UK: Cambridge University Press.

Momah, S. (1994). *Global disorders and the new world order*. Lagos: Vista Books.

Newman, P. (2013). The etymology of Hausa boko. *Mega-Chad Research Network/Réseau Méga-Tchad, 1*, 13. Retrieved from http://lah.soas.ac.uk/projects/mega chad/publications/Newman-2013-\Etymology-of-Hausa-boko.pdf

Nwolise, O. B. C. (2009). Peace and security. In I. O. Albert (Ed.), *Praxis of political concepts and cliches in Nigeria's fourth republic: Essays in honour of Dr. Mu'azu Babangida Aliyu*. Ibadan: Bookcraft.

Nwozor, A. (2013). A reconsideration of force theory in Nigeria's security architecture. *Conflict Trends, 1*, 18–25. Retrieved from http://www.accord.org.za/images/downloads/ct/ACCORD-Conflict-Trends-2013-1.pdf

Ojakorotu, V. (2011). The paradox of terrorism, armed conflict and natural resources: An analysis of cabinda in Angola. *Perspectives on Terrorism, 5*, 96–109.

Okolie-Osemene, J. (2013). Biased reporting exacerbates Nigerian conflict. *The Peace Journalist, 2*, 15–16. Retrieved from http://www.park.edu/center-for-peace-journalism/_ documents/2013-10-The-Peace-Journalist.pdf

Opara, S. (2013, June 20). B'Haram: JTF bans thuraya phones in Borno. *The Punch.* Retrieved from http://www.punchng.com/business/business-economy/bharam-jtf-bans-thuraya-phones-in-borno/

Oshanugor, F. S. (2004). *Terrorism: The Nigerian experience 1995-1998.* Lagos: Advent Communications Ltd.

Pham, P. J. (2012). Boko Haram's evolving threat. *Security Brief, 20*, 1–8.

Schmid, A. P. (2012). Twelve rules for preventing and countering terrorism. *Perspectives on Terrorism, 6*, 77.

Shaw, M. N. (1997). *International law.* Cambridge, UK: Cambridge University Press.

Sodipo, O. M. (2013). Mitigating radicalism in Northern Nigeria. *Africa Security Brief, 26*, 1–8. Retrieved from http://africacenter.org/wp-content/uploads/2013/08/ASB-26-Aug-2013.pdf

Soriwei, F. (2013, October 10). Cameroonian troops kill 180 B'Haram members – JTF. *The Punch*, 10. Retrieved from http://www.punchng.com/news/180-bharam-insurgents-killed-by-camerounian-soldiers-military/

Stampnitzky, L. (2013). *Disciplining terror: How experts invented terrorism.* Cambridge, UK: Cambridge University Press.

The National Human Rights Commission. (2013). The Baga incident and the situation in North-East Nigeria. Retrieved from http://premiumtimesng.com/dev/wp-content/files/2013/07/NHRC-Baga-Report.pdf

Tittle, C.R. (1995). *Control balance: Toward a general theory of deviance.* Boulder, CO: Westview Press.

Ugwu, D. (2008). Niger delta: Violence, human rights and development – Another look at the August 2007 violence. In C. Eze, B. Usiofo-Osakwe, B. I. Akosile, L. Umar, & N. Akpan-Ita (Eds.), *Women in peace building.* Lagos: The Book Company Limited.

16

Winning Hearts and Building Peace

Amparo Pamela H. Fabe

The Armed Forces of Philippines (AFPs) has been facing a three-pronged insurgency war—one against the Moro Islamic Liberation Front (MILF), the second against the communist New People's Army (NPA) and the third front against the Abu Sayyaf terror group. Under the AFP's counter-insurgency plan of clear-hold-and-develop, the additional Army battalions are needed to stick or hold an area which has been cleared to ensure the security of the people. Approximately, 40 per cent of the Philippine Army's combat battalions are being used as manoeuvre force. The remaining force is tasked to guard the local residents and vital infrastructure installations in a particular conflict area.

The 12,000-member MILF has been waging a decade-old insurgency to set up a Muslim state in southern Philippines. Christian settlers now outnumber the Muslim inhabitants. The Philippine government started formal negotiations with the MILF in 1997 to end peacefully more than three decades of Muslim separatist rebellion that has killed at least 120,000 people and stunted development in resource-rich Mindanao. The talks collapsed in 2000 when the Philippine military launched strong offensives and captured several of the MILF's jungle bases, including Camp Abubakar in Maguindanao Province. Malaysia rescued the peace process in 2001 and has since been consistently brokering the talks.

The NPA, the armed wing of the Communist Party of the Philippines, has been waging a guerrilla campaign in the countryside for four decades. Military estimates the NPA strength at more than 4,000 men scattered in more than 60 guerrilla fronts throughout the country. The military has been successful in reducing the number of the NPA to just about 4,000 from a peak of more than 20,000 in the 1980s. Communist rebels are active in 69 of 80 provinces across the country. The membership of the NPA, the armed wing of the Communist Party of the Philippines (CPPs), was reduced further to a historic low of 4,111 from 4,702 in 2009. This figure is way below the peak number of more than 25,000 during the 1970s. Of the 4,111 communist rebels, 52 per cent, or 591, have surrendered voluntarily and have availed of the government's integration and livelihood programmes. Communist-affected barangays have also decreased from 1,077 to 1,017 at the end of 2010 (Armed Forces of the Philippines Report, 2010).

The timely strategy, combined with the successful security operations, community-based civil-military operations and the implementation of an active information drive in rural barangays and urban town centres yielded consequent positive outcomes. Three communist guerilla fronts have been dismantled, from 51 in 2009 to 48 by the end of 2010, as a result of intensive military operations. In 2010, the arrested NPA personalities were five regional committee leaders and three front secretaries in Northern Luzon, Davao, Panay and Negros Island (Armed Forces of the Philippines Report, 2010). The AFP claimed that NPA killed an average of one civilian per week in 374 violent incidents in 2012.

On terror threats, the Armed Forces of the Philippines operations in Basilan and Sulu have reduced the Abu Sayyaf ranks from 391 terrorists and 340 firearms in 2009 to 340 elements and 296 firearms by the end of 2010. The continuing and successful counter-terrorism operations resulted in the neutralization of 51 terrorists including key leaders such as Albader Parad and Abdulgafur Jumdail who were killed in two separate encounters with government forces in Sulu. The Armed Forces stated that its counter-terrorism efforts have resulted in a significant reduction in the frequency of terrorist attacks nationwide from 54 in 2009 to 29 in 2010 (Armed Forces of the Philippines Report, 2010).

Approximately between 60 per cent and 75 per cent of the total battalion units of the Armed Forces of the Philippines have been deployed to Mindanao. Military operations continue in interior barrios such as Magpot and Arakan in North Cotabato, in the Davao City districts of Marilog, Baguio, Calinan and Paquibato, in Kataotao, Bukidnon and in Talaingod, Kapalong, Asuncion, Laac and Sto. Tomas in Davao del Norte. The combined Special Forces of the 73rd Infantry Battalion (IB), 56th Infantry Battalion, 37th Infantry Battalion, 64th Infantry Battalion, Scout Rangers and Special Forces were installed in new areas.

Soldiers with 'Hands Behind the Back'

The persistent counter-insurgency efforts waged by the Philippine government forces against private armed groups, Islamic militants and indigenous people's armed groups have yielded atrocities that are regularly committed against Filipino soldiers and civilians. In separate instances, Islamic militants have resorted to beheadings and genital mutilation of Filipino soldiers after killing them during combat operations. For example, an independent fact-finding team on the Basilan incident, comprising of government and MILF representatives, has identified 15 Muslim preachers in the list of 127 suspects in the 10 July 2007 mutilation and beheading of 10 Philippine Marines in Albarka town in Basilan. The Muslim preachers consisted of an ustad, an *aleem*, an imam and a *hadji*. An ustad is a teacher in a madrasa (Islamic school) while an aleem is a muslim scholar. An imam is a person who leads the prayers in mosques, whereas a haji is someone who has performed Islamic rituals in Mecca. The killing of 14 soldiers and the beheading of the 10 Philippine Marines renewed calls for the observance of respect for human rights in combat situations (Philippine fighting forces, 2007).

Marines Commandant Maj. Gen. Nelson Allaga has instructed the Marines involved in the pursuit operations not to retaliate against the perpetrators by beheading them, saying that beheading is "disrespect to human dignity." "I told them we will not do such barbaric act because we are professional soldiers so we should not commit atrocities as they do. We should not disregard human dignity. We pride ourselves as warriors but we respect human dignity" (Philippine fighting forces, 2007).

In a separate incident, the Abu Sayyaf rebels beheaded five out of seven Philippine Marines who were killed in fierce fighting in a Sulu jungle. Those decapitated included a 2nd lieutenant, a sergeant and three privates first class. Commodore Armando Guzman, Naval Forces Western Mindanao commander, described this incident as a barbaric act (Bodies of mutilated, 2011).

Colonel Daniel Lucero, Commander of the Army's 103rd Infantry Brigade based in Lanao who spent time in Basilan before, said it was not unusual for the Abu Sayyaf to mutilate fallen soldiers. "They behead soldiers to dehumanize our troops," Lucero said, citing at least two additional incidents during his Basilan stint. The soldiers who were killed received Gold Cross medals, the third highest recognition for a soldier and were posthumously promoted to the next highest rank (Bodies of mutilated, 2011).

Moreover, Colonel Lucero stated that many soldiers have lost colleagues in the conflict and many are of the opinion that the armed groups they are fighting show little or no respect for International Humanitarian Law (IHL). Some soldiers feel their rebel opponents encourage human rights violations (HRV) as a way to instill fear among the Army troops. Colonel Lucero said that there were reported instances wherein Filipino soldiers were either beheaded or mutilated by the Muslim armed rebels (personal communication, 6 September 2011).

Former President Gloria Macapagal-Arroyo had condemned the beheading of seven Filipino hostages by the Abu Sayyaf group and ordered government troops to focus on neutralizing the bandits. Arroyo had said that the beheading of the seven captives in Sulu had once again demonstrated the ruthlessness of Abu Sayyaf', However, she reiterated that this has only strengthened their resolve to neutralize them ('Philippine president condemns', 2007).

A kidnapping incident carried out by the Abu Sayyaf group which included the Burnham couple at the upscale Dos Palmas resort on Palawan island in May 2001 resulted in the death of a Filipino nurse and the beheading of two Filipino security guards from the resort. In a separate attack, the Abu Sayaff seized 34 civilians in the southern village of Balobo. The rebels beheaded 10 civilian hostages, said Philippine Army spokesman Maj. Alberto Gepilano ('Basilan kidnapping', 2001). In another incident, the MILF gunmen in Lanao del Norte occupied the towns of Kolambugan, Maigo and Kauswagan. One commuter bus was ambushed at an MILF rebel checkpoint near Kolambugan town and at least 14 passengers were mercilessly gunned down, witnesses said ('Guerrillas attack towns', 2008).

The Philippine Army strongly condemned the NPA rebels for the brutal attack on four unarmed and off-duty soldiers in Sitio Cinco, Barangay Mapula, Paquibato District, Davao City. Major Harold Cabunoc, former Philippine Army spokesperson, stressed that the attack was a clear violation of the IHL and Republic Act No. 9851 or "An Act Defining and Penalizing Crimes Against IHL, Genocide and other Crimes Against Humanity" ('Philippine army condemns NPA', 2012).

The peace negotiations between the Philippine government and the separatist MILF suffered another big blow after the massacre of 19 soldiers in the hands of heavily armed Moro rebels in Al-Barka,

Basilan, in Southern Philippines (News Analysis: Basilan massacre, 2011). In another atrocity committed by Muslim armed groups on civilians comprising the 15 members of the Mangadadatu clan and 30 media workers, a witness recounted that the female corpses were beyond recognition. Forensic investigators pointed out the possibility of post-mortem rape as evidenced by the female corpses bearing laceration of the hymen (Morelos, 2010).

According to Khadaffy Mangudadatu, an assemblyman of the Autonomous Region in Muslim Mindanao who testified for the prosecution at the resumption of the Maguindanao massacre trial, the female corpses were "beyond recognition because of the number of wounds inflicted on them". The Philippine National Police (PNP) Crime Laboratory revealed that five female victims of the Maguindanao massacre were positive of semen sample which authorities branded as presumptive evidence of possible rape (Five female victims, 2009).

Based on Republic Act 9851, these acts that were perpetrated by rebel armed groups on Filipino soldiers and civilians are considered war crimes and punishable under Chapter III, Section 4(b) of R.A. 9851 are: (a) violence to life and person, in particular willful killings, mutilation, cruel treatment and torture; (b) committing outrages upon personal dignity, in particular, humiliating and degrading treatment; (c) Taking of hostages and (d) the passing of sentences and the carrying out of executions without previous judgement pronounced by a regularly constituted court, affording all judicial guarantees which are generally recognized as indispensable.

Winning Hearts and Minds

The AFP has started implementing its new anti-insurgency strategy aimed at winning the hearts and minds of the people rather than directly confronting the enemy to achieve peace and stability and finally put an end to the 40-year-old communist insurgency. The AFP Internal Peace and Security Plan (IPSP), also known as "Bayanihan", is being implemented until the end of the term of President Benigno Aquino III in 2016. The IPSP was anchored on the President's national security strategy which emerges from the realization of lasting peace and stability, development and social progress, through a multi-stakeholder approach focused on the protection of the citizens' rights and civil liberties. The plan puts equal emphasis on the combat and non-combat dimension of the campaign such that efforts are not only focused on combat operations, but will likewise give importance to the peaceful settlement of conflict. Thus, the main parameters of success shall place a high importance on the number of friends won as contrasted to the number of enemies killed. The Armed Forces leadership is highly optimistic that IPSP Bayanihan ends what IPSP Bantay Laya started.

   The IPSP Bayanihan is characterized by an increasing involvement of stakeholders both from the government side and the civil society comprising a 'Whole of Nation' approach. The IPSP promotes a "People-Centred Approach," featuring a strict adherence to the IHL, human rights and the rule of law. By the thorough implementation of this plan, the AFP articulates its specific desire and commitment to the peaceful and just settlement of conflicts, and its firm belief that lasting peace and security is a shared vision and undertaking among the players in the security sector, civil society stakeholders and the entire citizenry (Philippine Army, 2011).

   The Armed Forces of the Philippines with its IPSP Bayanihan aims at zero tolerance for HRV and upholds respect for the IHL. It paved the way for the establishment of Human Rights Offices within the AFP and the PNP, and the introduction of several humanitarian interventions provided by government agencies to communities, especially women and children, in conflict-torn areas.

Upholding Human Rights

The Government of the Republic of the Philippines has passed several laws penalizing the most heinous offenses against human rights, such as torture (as penalized by Republic Act No. 9745), enforced disappearance (as penalized by R.A. No. 10353), war crimes, genocide and crimes against humanity (as penalized by R.A. No. 9851) and a law compensating victims of HRV during the Marcos Regime (R.A. No. 10368). Furthermore, during the Universal Periodic Review conducted by the United Nations' Human Rights Commission last May 2012, the observers highlighted a decrease in reported HRV. The government took significant steps towards prosecuting the alleged violators of human rights with the formation of the Inter-Agency Committee (IAC) through Administrative Order No. 35 (s. 2012) comprising the Department of Justice, the PNP, Armed Forces of the Philippines and a representative of the Office of the Presidential Adviser on the Peace Process.

Philippines ratified Protocol I additional to the Geneva Conventions that protects victims of armed conflict. Filipino victims of international conflict—whether they are civilians or wounded, sick, shipwrecked or detained military personnel—will now have more protection. Philippines has signed and ratified more treaties relating to IHL more than any other country in South-East Asia. The country acceded to the Geneva Conventions in 1952, signed both Additional Protocol I and Additional Protocol II in 1977 and ratified Protocol II in 1986. In 2010, it passed the Republic Act 9851 (or Philippine Act on Crimes against IHL, Genocide and Other Crimes against Humanity), incorporating many of the obligations of IHL into domestic law.

The Additional Protocol I imposes constraints for humanitarian reasons, on the way in which military operations may be conducted in internal armed conflicts. Its ratification will result in greater protection for Filipino military personnel deployed abroad in peace-keeping or other military operations undertaken in connection with an internal armed conflict. Armed Forces medical units and medical transportation enjoy enhanced protection.

Another important development in the field of human rights protection is the ratification of the Philippine Senate of the Rome Statute of the International Criminal Court (ICC) in August 2011. The ICC is the first permanent international court set up to prosecute individuals for genocide, crimes against humanity, war crimes and the crime of aggression. The Statute entered into force for the Philippines on 1 November 2011, bringing the total number of States that have joined the Rome Statute system to 117.

The Armed Forces Human Rights Office

In December 2010, the AFP has established the Human Rights Office which serves as the main platform for addressing all human rights and IHL issues involving the Armed Forces. The AFP has its *Human Rights and International Humanitarian Law (IHL) Handbook* as an important component of its efforts to institutionalize human rights concepts. The handbook provides a handy guide for soldiers during the conduct of operations.

According to Colonel Rhoderick Parayno, AFP Chief Human Rights Officer, every Filipino soldier is supposed to know: the need to protect civilians, the need to distinguish combatants from non-combatants and the humane treatment of prisoners. Each ICRC human rights training presents an overview of the Universal Declaration of Human Rights, and the laws related to respect for human rights encompassing R.A. 9851, R.A. 9745 and R.A. 10353 (personal communication, 4 December, 2013).

Colonel Domingo Tutaan Jr., the previous AFP Human Rights Officer, stated that his office is responsible for ensuring that every soldier knows IHL and applies it in the field. Colonel Tutaan draws a parallel on the IHL being applicable during armed conflict and human rights being applicable during both peace and armed conflict. He emphasizes that a Filipino soldier is a guardian of human rights. He admitted that with the long period of martial law, something was tainted… something was needed to restore relations with the public." The colonel clearly sees IHL promotion not just as a moral imperative but it is considered strategically smart. He affirms that "with IHL, this work may help bring an end to the conflict" he claims that it is not defeating the enemy, it is winning the peace. Further, he stated that there is a Human Rights Officer in charge of IHL, who is usually the second in command, in every unit and battalion. The main job description of the AFP Human Rights Officer is to monitor implementation and compliance with IHL, and to report violations. The HRV by soldiers can then be prosecuted under the new RA 9851 law (personal communication, 1 May 2012).

The Philippine military continues to affirm human rights more comprehensively and firmly as a means to improve national security. They have seriously adopted timely policies incorporating human rights in the conduct of security and tactical operations. Through the continuous training of military commanders, soldiers and auxiliary staff, the AFP hopes to address human rights more strategically both in terms of policy and practice. The AFP Human Rights Officers believe that the promotion of human rights finally supports the analysis of conflict and exclusion, and helps win the drive against insurgency and terrorism. Through the use of various technical handbooks, the

Philippine military is equipped with innovative tools to support human rights analysis and assessment, and help promote culturally sensitive approaches in the battlefield (R. Parayno, personal communication, 4 December 2013).

Partnering with Civil Society Organizations

To facilitate the promotion of human rights and IHL within the military organization, the AFP has partnered with the Philippine Commission on Human Rights (CHRs) which was established as "an independent office" by the 1987 Constitution of the Republic of the Philippines. The Philippine CHR exercises the following mandate: (a) monitor the Philippine Government's compliance with international treaty obligations on human rights; (b) provide appropriate legal measures for the protection of human rights of all persons within the Philippines; (c) investigate, on its own or on complaint by any party, all forms of HRV involving civil and political rights and (d) exercise visitorial powers over jails, prisons or detention facilities (Philippine Constitution, 1986).

In addition, the Armed Forces of the Philippines has established linkages with the International Commission of the Red Cross (ICRC) and the European Union Criminal Justice Support Programme for the publication of the handbooks and the training of military officers and personnel on human rights and the IHL. The ICRC has been working in the Philippines for 50 years. Its main advocacy is to visit detainees and assist people in need, many of them displaced because of military conflict. The ICRC reminds all parties to armed conflicts of their obligations under IHL, and the protection of civilians. It acts as a neutral and impartial intermediary to facilitate the handover of people captured and detained in relation with the conflicts to their families. The organization has been working to integrate IHL into national legislation, and promote knowledge of and respect for this body of law among local and national authorities, armed and security forces, university students and civil society.

The AFP also coordinates closely with the Working Group for an ASEAN Human Rights Mechanism, in partnership with the Philippine Representative to the ASEAN Intergovernmental Commission on Human Rights (AICHR). Ambassador Rosario Manalo, the Philippine

Representative to AICHR, emphasized that while ASEAN make economic growth and shared prosperity a high priority, it is important to keep in mind the states' commitment to upholding, respecting, promoting and protecting human rights. The AFP has linked with We Act 1325 or Women Engaged in Action on 1325 (United Nations Security Council Resolution 1325), a national network supporting the National Action Plan of United Nations Security Council Resolution 1325. The AFP has existing links with Amnesty International Philippines, the Ateneo Human Rights Center, BALAY Rehabilitation Center, Families of Victims of Involuntary Disappearance (FIND), Medical Action Group, PhilRights, Task Force Detainees-Philippines (TFDP), Women's Legal Bureau and WEDPRO. These civil society organizations continue to facilitate in monitoring and documenting human rights and IHL violations, providing relief and rehabilitation services to internally displaced persons, and advocating for IHL and legal remedies to HRV.

Training Soldiers on Human Rights

The Filipino soldiers regularly undergo refresher training on human rights as part of its campaign against the communist insurgency. The main focus of military operations is to defeat the enemies and at the same time, help the soldiers concentrate on their training on the observance of human rights. Training officers from the International Committee of the Red Cross tackle questions and areas of concerns by the soldiers such as how to operationalize observance of the IHL at the field level of the AFP and PNP. Soldiers also seek guidance regarding the provision of adequate protection of victims and witnesses. They also undergo training regarding the designation of special courts to try cases involving crimes punishable under RA 9851 or the Philippine IHL law (R. Parayno, personal communication, 4 December 2013).

The AFP-HRO promotes advocacy and training seminars in all AFP area commands and Philippine Navy units in the field. These activities help foster awareness, as well as informing soldiers about the principles of the Comprehensive Agreement on Human Rights and IHL. These seminars help soldiers about the proper process of filing complaints against the enemies who committed violations of this agreement.

Interventions for Protecting Human Rights: Current Status

Apart from the regular AFP courses for all military personnel from the pre-entry, basic, advance, specialization courses up to the career courses for promotion and re-assignment, the AFP through the AFP Human Rights Office (AFP-HRO) conducts a continuing advocacy and information dissemination campaign to individual and formed units, particularly on the principles of IHL, and on the Republic Act 9851, which is the Philippine Act against Genocide, War Crimes, and Crimes Against Humanity. The same IHL training likewise include the rules on arrest (Rule 113 of the Rules of Court of the Philippines), rights of persons arrested, detained, or under custodial investigation (Republic Act 7438), and the Anti-Torture Law (Republic Act 9745).

Aside from the regular training and advocacy campaigns, the AFP-HRO conducts its regular Company Commanders' Symposium and the Battalion/Brigade Commanders Symposium, wherein the commanders in the field are continuously briefed and updated on IHL and human rights issues. Furthermore, the AFP partners with non-governmental organizations/civil society organizations/governmental organizations in the conduct of IHL training, such as the International Committee of the Red Cross (ICRC), Sulong CARHRIHL and the Government of the Philippines Monitoring Committee (GPHMC). All of the Philippine military personnel are continuously inculcated on the principles of IHL, human rights and the rule of law. Every AFP unit/command has a designated Human Rights Officer, who is usually the Deputy Commander or the Executive Officer.

The AFP through the Human Rights Office is in continues engagement with stakeholders, including the CHR, in the spirit of transparency and accountability. This approach is in close consonance with the strategic imperatives of the "Internal Peace and Security Plan-BAYANIHAN". The CHR, in its annual "Ulat sa Bayan" (Report to the People) issued during its anniversary last May 2013, reported that there has been a significant drop or marked reduction in the incidence of extrajudicial killings, and forced disappearances and acts of torture from 2010 to 2013. This may be attributed to the ongoing campaign of the AFP to adhere to IHL and human rights, and to abide to the rule of law (R. Parayno, personal communication, 4 December 2013).

The AFP continues to value its partnerships with the International Committee of the Red Cross (ICRC), ASEAN Human Rights and the Hanns Seidel Foundation to foster the aims of the HR office of the AFP. The ICRC is in continuous partnership with the AFP-HRO. It has jointly conducted with the AFP numerous training programmes, training of non-commissioned officers as "Mobile Training Teams" regarding the observance of IHL to military personnel and CAFGU units assigned in the different detachments/patrol bases of the AFP, publication of training manuals and code of competence under IHL, publication of manuals called "Essentials for Commanders" to ensure compliance with IHL. The Hanns Seidel Foundation, in partnerships with the AFP, PNP, CHR and civil society groups, has, since 2008, conducted "community based dialogue sessions" in the different regions of the Philippines to foster constant coordination with all stakeholders to address and resolve issues on human rights, IHL and the rule of law. At present, this partnership has been conducting "top level policy discussions" to further enhance the resolve to end extra judicial killings, and forced disappearances, torture and other grave violations to the right to life, liberty and security of persons. In fact, an agreement had been reached by the recent issuance of the "La Breza Declaration on Human Rights Cooperation on December 2012, wherein the CHR, PNP, AFP and the Bureau of Jail Management and Penology agreed to speed up the human rights cooperation between and among them. The Defense Institute of International Legal Studies (DIILSs) of the United States of America, together with the JUSMAG, has also partnered with the AFP-HRO in the conduct of IHL Training and maritime law enforcement (R. Parayno, personal communication, 4 December 2013).

The International Labor Organization (ILO), together with the Department of Labor and Employment (DOLE) and the AFP, is now in the process of finalizing the move to conduct a nationwide campaign on the "Joint DOJ, DILG, DND, DOLE, AFP and PNP Guidelines" to observe and respect workers' rights and activities. The UNICEF, together with the Council for the Welfare of Children (CWC) of the Philippine Government, partnered with the AFP in disseminating the recently issued AFP Letter Directive 25 that mandates the protection of children and the prohibition on the occupation of schools and hospitals by armed units of the AFP. This is consonant to the IHL principle that schools and hospitals are protected objects. The AFP continues to partner with various entities to further enhance the culture of human rights, IHL, and the rule of law in the Philippines.

Finally, the AFP welcomes partnerships with NGOs, whether local or international, to pursue adherence and observance of the law. Mechanisms had been set up to address IHL and HR issues, that is why the AFP has adopted the BAYANIHAN concept to ultimately address these concerns and resolve the cases. Mechanisms such as the National Monitoring Mechanism (NMM) are a tripartite system of validating reported violations that is jointly investigated by the CHR, NGOs and the government. This complements the inter-agency committee created under the Administrative Order Number 35 signed by President Benigno Aquino on November 2012. There are mechanisms pertaining to the following: (a) implementation and compliance on the Convention Against Torture, which is known as the National Preventive Mechanism led by the CHR; (b) monitoring, reporting and response system (MRRS) led by the CWC and the UNICEF to prevent grave child rights violations (GCRVs) as enunciated in United Nations resolutions; (c) the Complaints Monitoring Working Group (CMWG) of the GPHMC on the agreement entered into by the Philippine Government and the CPP/NPA/NDF known as the Comprehensive Agreement on the Respect of Human Rights and International Humanitarian Law (CARHRIHL).

Positive Outcomes of the Interventions

The constant training on 'respect for human rights' has yielded immediate positive outcomes for the Philippine military. For example, the Armed Forces of Philippines (AFPs) have reported that no branch of the military service has been involved in any instance of HRV during the first quarter of 2012. Col. Arnulfo Burgos, Jr., AFP spokesperson, attributes this positive development to the sincere efforts of the military command to educate and train its soldiers on the protection of human rights and adherence to international humanitarian and rule of law. AFP-HRO has been tasked to verify HRV reports involving military personnel that were referred to the CHR (personal communication, 1 June 2011).

The AFP-HRO has a Board of Inquiry which was also replicated down to the battalion levels to conduct investigations that would validate the reports of the CHR. The AFP-HRO has vowed to maintain its close coordination with the CHR to reconcile the numbers of alleged

330 Amparo Pamela H. Fabe

HRVs recorded by both parties. Any soldier who will be involved in any misconduct or malpractice will be dealt with through the military justice system. The AFP is tasked to apply disciplinary actions such as dismissal from military service and revocation of privileges. Once an involved soldier goes outside military bounds, the case will come from the higher civilian court which will undergo due process of law.

In addition, the AFP-HRO is very active in teaching the active-duty AFP personnel who are assigned in garrisons and in the field about local laws such as the Republic Act No. 9851 or the Philippine Act on Crimes Against IHL, Genocide, and Other Crimes Against Humanity; R.A. 9745 or the Anti-Torture Act of 2009; R.A. 7438 or the Act defining the rights of the arrested, detained or persons under custody as well as the duties of the arresting, detaining and investigating officers; and Rule 113 under the Revised Rules of Criminal Procedure.

Conclusion

Upholding human rights by the forces can provide a winning edge in seeking cooperation of the people and building peace in conflict areas. It is seen that the respect for human rights and the observance of the IHL by the Armed Forces of the Philippines in all of its operations in the battlefield represents a significant paradigm shift that started in 2010. This shift took place despite the consistent atrocities being committed on Filipino soldiers at the hands of communist rebels and Islamic militants. The implementation of an Internal Security and Peace Plan from 2011 to 2016 paved the way for more engagement with respect for human rights in peace and conflict zones. This positive trend towards safeguarding human rights in combat zones led to the establishment of the AFP-HRO in 2010 and the appointment of one Human Rights officer in each battalion and unit. The reduction in the number of reported HRV by members of the Philippine military in 2012 reflects the seriousness with which the soldiers have imbibed the respect for human rights. Perceptibly, the attitudinal and behavioural change in the forces is encouraging for winning the hearts and minds of the people and for building peace and security in a nation marred by conflicts.

References

Armed Forces of the Philippines. (2010). *Armed Forces of the Philippines Report —2010.* Quezon City: Philippines

Basilan kidnapping. (2001, August 5). *The Commercial Appeal*, p. 12.

Bodies of mutilated marines recovered in Sulu. (2011, July 30). Retrieved from http://www.philstar.com/headlines/711145/bodies-mutilated-marines-recovered-sulu

Commission on Human Rights. (2010). *Primer on Philippine commission on human rights.* Manila, Philippine: Commission on Human Rights.

Fact-finding team tags Abu Sayyaf in beheading of marines. (2007, August 8). *Businessworld*, p. 23.

Guerrillas attack towns in southern Philippines; The Moro Islamic Liberation Front kills more than a dozen people, military says. (2008, August 18). *Los Angeles Times*, p. 7.

Morelos, M. (2010, December 9). Corpses beyond recognition after Maguindanao Massacre. *McClatchy—Tribune Business News.*

News Analysis: Basilan massacre of Philippine troops big blow to peace talks with Moro rebels. (2011, October 1). *Xinhua News Agency*, p. 23.

Philippine army chief says troops to take refresher training on human rights. (2010, July 5). *BBC Monitoring Asia Pacific*, p.12.

Philippine Army. (2011). *Armed forces of the Philippines internal peace and security plan.* Manila, Philippine: Research Department of the Philippine Army.

Philippine fighting forces "thousands" to flee as army hunt Marines killers. (2007, August). *BBC Monitoring Asia Pacific.*

Philippine president condemns Abu Sayyaf's beheading of hostages. (2007, April 20). *BBC Monitoring Asia Pacific*, p.17.

Philippine President strongly condemns beheading of five soldiers. (2011, July 16). *BBC Monitoring Asia Pacific*, p. 26.

Philippine troops rescue 13 captives from guerillas. (2001, August 2). *The Associated Press*, p. 6.

Philippines: AFP records zero HRV for first 4 months of 2012. (2012, May 16). *Asia News Monitor*, p. 15.

———: Arroyo said should rethink peace initiative after Moro ambush. (2007, August). *BBC Monitoring Asia Pacific*, p. 14.

———: ICRC welcomes PHL ratification of protocol 1 protecting victims of armed conflicts. (2012, March 3). *Asia News Monitor*, p. 9.

———: Philippine army condemns NPA attack on defenseless soldiers in Davao City. (2012, November 12). *Asia News Monitor*, p. 19.

PNP: Five female victims of Maguindanao massacre tested positive for semen sample. (2009, December 8). *Asia News Monitor*, p. 7.

The Philippine Constitution. (1986). Manila: The Philippine Congress and the Philippine Senate.

17

Beyond Century: The Future of Military Psychology

Nidhi Maheshwari and Vineeth V. Kumar

Future conflicts, especially in Asian countries such as India, will enthuse military psychologists to uptake arduous challenges in service of the soldier. In the upcoming era of 'smart bombs' and 'killer robots', it is not only the psychology of the man behind the machine but that of machine itself, which has to be deciphered and managed. Both the real (man) and the artificial intelligence (machine) need to be in sync for appropriate mission success. Hence, irrespective of the nature of missions, manned or unmanned, psychology has an enormous role to play both as an art and science of future wars on Indian or other soils. Matthews (2014) argues in his book, *Head Strong: How Psychology is Revolutionizing War,* that psychology will be the deciding science in future wars. Nations that embrace psychology and turn to it to improve military selection, training, decisionmaking, resilience, leadership and cultural understanding will succeed compared with nations that focus only on building bigger and more lethal weapons.

Futuristic War Demands and Supplies

Network-centric warfare, information warfare, non-platform centric warfare, cyber warfare, nuclear-bio-chemical warfare, high-end electronic warfare (EW), etc., are some of the catchall terms to signify future conflicts. While Western nations have begun slipping into such high-end warfare, eastern countries are heading towards it. Use of Electro-Magnetic Pulses (EMPs), Robot soldiers such as British Lethal Autonomous Weapons Systems (LAWSs), Weapons of Mass Destruction (WMD), Kinetic energy weapons, Laser weapons and security systems, Particle Beam weapons, Sonic weapons, Nuclear-Biological-Chemical

(NBC) weapons, AI-based Cyborgs, advanced surveillance and stealth weapons include the dynamic threats that the century ahead is going to propose in front of the soldier and his psyche. At some instances, he has to use the same while at other instances he has to guard himself against them. Such generation of warfare shall increase ambiguity, uncertainty, repulsion, stress and mental workload for the future warriors at a lightning speed. Along with the use of such weapons, advances in the arena of human enhancement through regenerative medicine, cybernetic implants, human brain programming and critical organ transplants, etc., might pose greater moral, legal and operational threats than ever before. The soldier has to work as a system or console with an independent mind, yet rooted to a national and noble cause.

Interestingly, as the enemy will have no defined battle space, battle time and fighting strength, the soldier too will have an ill-defined working ground. As an exemplar, the drone operators, who target their covert enemies while sitting in their basement, and then, soon step out of it to their bed rooms or dining rooms at dinner with family and children; such a dynamic working ground is a reality. Equally, job of the commander even gets tougher as he will have to widen his expanse of operations and manage mixed teams of inter-disciplinary and virtual nature. This will build a new generation of operational stress for the Forces. Clearly, deployment in such a scenario shall project the following dynamic challenges for the soldier:

1. Fighting alone at several nodes,
2. Meek replacement, repair and replenishment possibility,
3. Unawareness about the zone of safety,
4. Requirement of super-skilled maneuvering,
5. Wariness of passive detection systems,
6. Strong gesture and neural control,
7. Guarding against monotony and complacency,
8. Working ethically in unethical environments,
9. Instantaneous and independent decision making,
10. Operating with/in virtual groups.

Close to the above-mentioned challenges, it should be kept in focus that alongside technology the enemy too will be marching ahead in warfare tactics pace by pace with the Forces. And so, the only edge in defeating the enemy can be procured by maintaining psychological robustness coupled with the technological grip. Laurence and Matthews (2012) believe that sophisticated weapon systems, platforms and technology offer critical advantages only if they can be mastered. Thus, physiological functioning, information processing, cognition, decision making and so forth remain key ingredients to victory.

Futuristic Errands for Military Psychology

Again selection, training, adaptation, performance and sustenance will remain the focal areas for military psychology in future warfare scenario. And, military psychologists will be more immersed than ever before in all the concerned areas. Especially for developing countries such as India, wherein the socio-economic gap alongside cultural fanaticism is widening with technological advances, military psychology has to bear much load to train and sustain the soldier. Traces of fundamentalist ISIS being operational in India, frequent cross-border activities in plains and at high altitude, suicide bombings, breach of maritime boundaries, misrepresentation of Indian Territory in the world map, etc., are indicative of different types of threats for the Forces. India and her neighbours are just a playground for the disgruntled lot who is trained on latest technology elsewhere against the State. Psychologists have not only the onus to select the right soldier for the right job but also to train and sustain him in such varied conditions. They have to go interdisciplinary while developing theoretical and methodological apparatus for their research works and applying them from laboratory to land. Neuroergonomics, cultural engineering, cognitive engineering, forensic-based intelligence, neural physics, etc., can be some of the emergent hybrids to the knowledge base and applications of military psychology.

Selection

Selecting men for appropriate military jobs of the coming times require skilled professionals who can think beyond the obvious job summaries and estimate the extent of enhancement in the soldier after requisite training. Assessing men for covert operations, capture or kill missions such as the recent Operation Neptune Spear of Navy SEALs (Sea, Air and Land) in Pakistan, conducting Military Information and Support Operations (MISOs), designing cultural engineering programmes, winning the hearts and minds of the alien group, performing crisis negotiations, nabbing and interrogating suicide bombers, mapping human terrains, conducting cognitive engineering, evacuating hostages, hacking the enemy's mind, directing relief and rescue missions as in Japan, Nepal or Uttarakhand, etc., will be some of the regular job summaries for military psychologists in the approaching times. Both the cognitive and non-cognitive attributes would be of prominence to assess the future warriors. Following are some of them:

1. 3-D/ 4-D environmental scanning and tolerance
2. Individual survivability under extreme conditions
3. Cultural sensitivity
4. Stronger ego strength
5. Sense of justice/ Moral decision making
6. Risk-orientation
7. Adaptability to new gadgets/ systems
8. Alertness
9. Ingenuity
10. Cognitive flexibility
11. Emotional regulation
12. Tactical creativity
13. Communication/ negotiation skills

Likewise, innovative selection processes have to be developed which not only covertly assesses the candidate's aptitude on the said skills to serve the forces but also his commitment and integrity to do the same despite prospective obstacles.

Training

Training need analysis reveals that education and instruction should be consistent with the future operational context. It has to be specific for the specific rather than application of the general to the specific. War games, Internet simulations, fuzzy games, serious games can mark a definite impact onto the preparedness of the soldier for future warfare. However, mode of training has to be swift with the advances in weapon systems and technology. Simulation training has to be more real time and should ignite psycho-physiological changes of the same intensity and frequency as that during a real life mission. Training 'soldier as a system' with mind-enabled tools should be a priority because well-equipped soldier with all gadgetry might have to engage personally on one-to-one basis with the enemy. The soldier needs to function independently and covertly on sophisticated weapon systems and technologies, with swift decision-making capability and pure intent-based actions towards the desired goals as defined by the organization. Thus, psychology would be of help to designing such automated systems and training soldiers to such systems including force multipliers such as EW equipments, Unmanned Aerial Vehicles (UAVs), aerial warships and satellite systems, etc. Melton (1957) anticipated that future systems will involve greater automation, and, programming of operations and maintenance suggest that military psychology will become involved in system design and system functioning problems.

More so with the soldier's increasing role in humanitarian operations, psychologists need to train the soldier to shed off his masculine orientation ('*think soldier think male*' stereotype) and move towards more androgynous orientation. Also, soldier has to rub his shoulders more with personnel of non-military organizations, social groups and media in upcoming Military Operations Other Than War (MOOTW). Hence, he needs to be trained for better communication, being compassionate while handling victims and survivors, considerate to the needs of refugees, females and children, etc., well in advance, so that the desired behaviour becomes spontaneous during peace making/peace-keeping and other humanitarian efforts. Also, psychologists as trainers would serve the purpose in training future men for the following:

1. Advanced patrolling and ambush techniques
2. Advanced search and destroy tactics
3. Simulation-based psycho-physical stress resilience
4. Handling and feeding media
5. Eliciting intelligence in the hinterland
6. Cultural shaping and sensitivity

7. 6s sensitivity training
8. Humanitarian and ethical laws
9. Remote viewing
10. Task-mastery and proficiency over gadgets
11. Effective group communication
12. Lone survival
13. Directive style of leadership
14. Social engineering
15. Cognitive hacking
16. Psychological operations
17. Neural programming

For countries in the Asian subcontinent, training institutes need to upgrade their syllabus in light of the upcoming challenges and training needs as mentioned above. Also, *neuro-psychologists* should be employed as trainers on a regular basis to the institutes.

Adaptation

Adapting to the upcoming challenges would again be a test for the soldier. Art lies both in the quantity and quality of adaptation. The soldier needs to be more resilient and robust as fear of unknown would be prominent during this era. He has to not only survive but also excel in trying situations of future warfare. Often, his close team or unit which is otherwise a significant stress-buster may not be at his arm's length to pacify him in times of crises. Moreover, adaptation becomes customary while functioning in mixed teams in an alien land with scarce social resources. Efforts by psychologists need to be made to provide the soldier enough alternatives to his privy so that he hardly gives up in the face of future adversities. Conversely, building cohesive teams should be another focal area, especially in expeditionary forces. Leadership has to be strengthened by making it more transparent and diversified. Inoculation of leaders to the psychological operations of the enemy is an area of considerable research. Countering propaganda and rumour intervention will be applicable more than any other time as non-state actors would aim more at the collateral damage rather than the State forces. Thereby, psychologists have to fully suffice to the demands of the organization in adapting soldiers to their environment. Following can be some of the group and individual adaptation strategies to train upon the soldiers of any country for present and future:

1. Cognitive retraining
2. Learned positive reappraisal of situations
3. Attitude modulation and perception management
4. Situational awareness
5. Relaxation training through goggles, helmets, etc.
6. Rhythm healing
7. Yogic breathing and style of life

The above-mentioned adaptation strategies will be extra productive in future situations as soldiers would be tasked to function in small groups or on an individual basis without much social support.

Importantly in countries such as India, it is high time that *operational psychologist* be placed in respective units of the soldiers where he rubs his shoulders with the soldier and accompanies the soldier during his adaptation to the vagaries of the operational environment. Additionally, *psychological support to military families* must gear up as they bear the secondary stress and trauma of deployment besides the soldier. Alert and agile support system at community level with a keen supervision of the command and control will be a responsibility of military psychologists to aid the families face future adversities appropriately.

Performance

Performance of the soldier has to be extraordinary at all times in every situation as exception is a norm for him. With the possibility of cognitive hacking, remote neural monitoring, brain-washing and advanced brain–computer interface (BCI), he has to think beyond the enemy lines while moving on his mission with commitment. Planning, decision making, cognitive flexibility, situation awareness, etc., shall be some of the imperatives to better functioning because often his plans will be excellent but action to plans will be amiss due to dynamic targets. Performance will be based on preparedness which can be achieved by mentally rehearsing the actions, taking assistance of VR programmes and practicing cognitive restructuring in the face of obligatory events. Equally, combat performance has to aim at minimum destruction with maximum psychological lethality. Target would shift from the enemy's body to his mind through use of enhanced non-lethal weapons wherein psychologists have a significant role to offer.

Importantly enough, use of primitive warfare strategies and charismatic 'hero' warrior image cannot be ruled out in the future as the jamming and raiding of techno-based devices and gadgets is a possibility. Thus, psychologists have not only to shape the future performance of a soldier on one hand but also to sharpen the conventional performance of a soldier without gadgets on the other hand.

Sustenance

In light of Google patenting the robots with personalities, future warriors would face a new generation of challenge while sustaining in battlefield. Psychologists have much to deliver. Like present times, sustenance would be crucial to any combat mission in the future also. Motivation and morale can often be marred by the ambiguity, uncertainty, boredom and isolation associated with the mission. Often, they might find it difficult to fight with their adversary in an unknown land, clear the area off the terrorists with hostile civilians around them, hit robots or unmanned vehicles who might come in swarms without any emotion of being dead or disclosing the strength of the enemy forces, fight an insider defying humanitarian laws or witness brutal and unprecedented death of the buddy along with the intruder attempting to hack one's mind. Under such circumstances, psychology can come in a big way to the rescue of a disheartened soldier. Following are some of the sustenance strategies that the psychologists can train upon the future warriors of both individualistic and collectivistic societies:

1. Mindfulness-based stress reduction
2. Guided meditation/imagery
3. Revisiting one's precious belongings or cherished moments
4. Spiritual discourses
5. Personally valued songs or prayers through various apps on one's cellphone or ipod

Also, sinuous reintegration with families after deployment, re-employment after active service, coupled with attracting youth to the services and minimizing prejudices and stereotypes in military would be some other upcoming arenas for military psychology to get deployed in the Armed Forces.

Conclusion

Military psychology bears the onus for preserving the mental and moral health of the Armed Forces. Building tech-savvy warriors as well as savvy warriors sans technology remains the agenda for military psychology to delve upon. The discipline has proved its worth for nearly a century and is going to make headway beyond century also as an independent and customary subject for the Forces to conquer.

References

Laurence, J. H., & Matthews, M. D. (Eds.), *The Oxford handbook of military psychology* (pp. 92–113). New York, NY: Oxford University Press.

Matthews, M. D. (2014). *Head strong: How psychology is revolutionizing war.* New York, NY: Oxford University Press.

Melton, A. W. (1957). Military psychology in the United States of America. *American Psychologist, 12,* 740–746.

About the Editors and Contributors

Editors

Nidhi Maheshwari, Ph.D., is a Scientist at the Strategic Behaviour Division, Defence Institute of Psychological Research (DIPR), Defence Research and Development Organisation (DRDO), Ministry of Defence, Government of India, Delhi. She specializes in the area of Strategic Behaviour Analysis, particularly Assessment and Management of Psycho-bio-social Markers of Combat Stress Behaviours, Combat Motivation and Morale, Rumour and Propaganda Management (Psychological Operations), as well as Special Forces Profiling. She has obtained her doctorate degree from University of Rajasthan, Jaipur. As a psychologist, Dr Maheshwari has been employed at various Services Selection Board (SSB) for the assessment and selection of officers of Indian Army, Navy and Air Force. As the principal investigator, she has pursued several research projects for Indian Armed Forces. Also, several need-based assessment tools and manuals have been developed and delivered by her. Besides her contribution to various books and journals, she has authored a field guide on assessment and management of combat stress behaviours for Indian soldiers. She has conducted various training programmes on strategic issues in numerous fields and forward locations of Northern and Eastern Command of the Indian Armed Forces. She has also been a guest speaker cum faculty to various seminars/courses at institutions such as National Security Guard (NSG), Army War College and Military Intelligence Training School and Depot, apart from various leading universities of India. She has been a recipient of the research fellowship namely NET-JRF by University Grants Commission in 2001. She is also, the Young Scientist Awardee (2006) and the Technology Group Awardee (2011). In 2013, her book on military psychology entitled *Sainya Manovigyan* was awarded the DRDO Rajbhasha Award. She is a member of the American Psychological Association (APA), International Association of Applied Psychology (IAAP), the Indian Science Congress Association (ISCA), the Centre for Land Warfare Studies (CLAWS), the Administrative Staff College of India (ASCI), the Industrial Psychiatry Association of India (IPAI) and the National Academy of Psychology (NAOP).

Vineeth V. Kumar, Ph.D., is Associate Professor at the School of Management, BML Munjal University, Gurgaon, Haryana. He has over 13 years of experience in teaching, training and research in behavioural sciences. He has been awarded doctorate in psychology from University of Rajasthan. His subjects of interest include psychometric testing and assessment, applied psychology, organizational behaviour, positive psychology and military psychology. He has published several research papers in peer-reviewed national and international journals. His pursuits have been in the arena of positive psychology-based interventions for nurturing holistic well-being. Dr Kumar has been a faculty with Amity University; SRM University, Tamil Nadu and the Indian Institute of Health Management and Research (IIHMR), Jaipur. Over the years, he has been actively undertaking management development and training programmes for management professionals, defence scientists and service personnel. He has organized several international conferences in the arena of positive psychology-based interventions and has been the editor of the *Journal of Human and Work Management* and associate editor of the *Amity Journal of Applied Psychology*. He is a member of the APA and the Indian Science Congress Association (ISCA).

Contributors

Suresh A. is Scientist 'D' at the DIPR (DRDO), Delhi, India.

Dinika Anand is a Researcher at the DIPR (DRDO), Delhi, India.

Archana is Scientist 'D' at the DIPR (DRDO), Delhi, India.

Emerald M. Archer is an Assistant Professor and Chair at the Department of Politics and History, Woodbury University, CA, USA.

Soumi Awasthy is Scientist 'F' and Head, Intelligence and Aptitude Division, DIPR (DRDO), Delhi, India.

Paul T. Bartone, is Senior Research Fellow at the Center for Technology & National Security Policy (CTNSP), Institute for National Strategic Studies, National Defense University, USA.

Jarle Eid is Dean, Faculty of Psychology at the Department of Psychosocial Science, University of Bergen, Norway.

Amparo Pamela H. Fabe is Senior Fellow at the Philippine Institute for Peace, Violence and Terrorism Research (PIPVTR), Manila, Philippines.

Sneha Goswami is Researcher at the DIPR (DRDO), Delhi, India.

Arunima Gupta is Scientist G and Additional Director of the DIPR (DRDO), Delhi. She is also Head, Personality Division, DIPR.

Sigurd W. Hystad is Associate Professor of Psychology at the Department of Psychosocial Science, University of Bergen, Norway.

Swati Johar is Scientist 'C' at the DIPR (DRDO), Delhi, India.

Gurpreet Kaur is Scientist 'D' at the DIPR (DRDO), Delhi, India.

Updesh Kumar is Scientist 'F' and Head, Mental Health and Follow-up Division, DIPR (DRDO), Delhi, India.

Eyal Lewin is Assistant Professor at the Department of Middle Eastern Studies and Political Science, Ariel University, Israel.

T. Madhusudhan is Serving Colonel and Senior Advisor at the Department of Psychiatry, Command Hospital (Southern Command), Pune, India.

Michael D. Matthews is Professor of Engineering Psychology at the Department of Behavioral Sciences and Leadership, United States Military Academy (USMA), West Point, USA.

Joseph Miller is ex-US Marine and Iraq War veteran. He is a tactics writer, veteran's advocate and memoirist. Currently, he is a history graduate student at the University of Maine, USA.

James Okolie-Osemene is Research Fellow at the French Institute for Research in Africa (IFRA), Nairobi, Kenya.

Vidushi Pathak is Researcher at the DIPR (DRDO), Delhi, India.

Anju Rani is Researcher at the DIPR (DRDO), Delhi, India.

Ashutosh Ratnam is a serving Major and resident Doctor at the Department of Psychiatry, Armed Forces Medical College (AFMC), Pune.

Sucheta Sarkar is Researcher at the DIPR (DRDO), Delhi, India.

Sujata Satapathy is Clinical Psychologist and Assistant Professor at the Department of Psychiatry, All India Institute of Medical Sciences (AIIMS), Delhi, India.

Ron Schleifer is Senior Lecturer at the School of Mass Communication, Ariel University, Israel.

Pankaj Kumar Sharma is a serving Major and Resident at the Department of Psychiatry, Armed Forces Medical College (AFMC), Pune, India.

N. P. Singh is Scientist G and Head, Strategic Behavior Division, DIPR (DRDO), Delhi, India.

Index